NEUROPSYCHOLOGY OF CHILDHOOD EPILEPSY

ADVANCES IN BEHAVIORAL BIOLOGY

Recent Volumes in This Series

Volume 37 KINDLING 4
Edited by Juhn A. Wada

Volume 38A BASIC, CLINICAL, AND THERAPEUTIC ASPECTS OF ALZHEIMER'S AND PARKINSON'S DISEASES, Volume 1
Edited by Toshiharu Nagatsu, Abraham Fisher, and Mitsuo Yoshida

Volume 38B BASIC, CLINICAL, AND THERAPEUTIC ASPECTS OF ALZHEIMER'S AND PARKINSON'S DISEASES, Volume 2
Edited by Toshiharu Nagatsu, Abraham Fisher, and Mitsuo Yoshida

Volume 39 THE BASAL GANGLIA III
Edited by Giorgio Bernardi, Malcolm B. Carpenter, Gaetano Di Chiara, Micaela Morelli, and Paolo Stanzione

Volume 40 TREATMENT OF DEMENTIAS: A New Generation of Progress
Edited by Edwin M. Meyer, James W. Simpkins, Jyunji Yamamoto, and Fulton T. Crews

Volume 41 THE BASAL GANGLIA IV: New Ideas and Data on Structure and Function
Edited by Gérard Percheron, John S. McKenzie, and Jean Féger

Volume 42 CALLOSAL AGENESIS: A Natural Split Brain?
Edited by Maryse Lassonde and Malcolm A. Jeeves

Volume 43 NEUROTRANSMITTERS IN THE HUMAN BRAIN
Edited by David J. Tracey, George Paxinos, and Jonathan Stone

Volume 44 ALZHEIMER'S AND PARKINSON'S DISEASES: Recent Developments
Edited by Israel Hanin, Mitsuo Yoshia, and Abraham Fisher

Volume 45 EPILEPSY AND THE CORPUS CALLOSUM 2
Edited by Alexander G. Reeves and David W. Roberts

Volume 46 BIOLOGY AND PHYSIOLOGY OF THE BLOOD–BRAIN BARRIER: Transport, Cellular Interactions, and Brain Pathologies
Edited by Pierre-Olivier Couraud and Daniel Scherman

Volume 47 THE BASAL GANGLIA IV
Edited by Chihoto Ohye, Minoru Kimura, and John S. McKenzie

Volume 48 KINDLING 5
Edited by Michael E. Corcoran and Solomon Moshé

Volume 49 PROGRESS IN ALZHEIMER'S AND PARKINSON'S DISEASES
Edited by Abraham Fisher, Israel Hanin, and Mitsuo Yoshida

Volume 50 NEUROPSYCHOLOGY OF CHILDHOOD EPILEPSY
Edited by Isabelle Jambaqué, Maryse Lassonde, and Olivier Dulac

NEUROPSYCHOLOGY OF CHILDHOOD EPILEPSY

Edited by

Isabelle Jambaqué
Hôpital Saint-Vincent de Paul
Paris, France

Maryse Lassonde
Universite de Montreal
Montreal, Quebec, Canada

and

Olivier Dulac
Hôpital Saint-Vincent de Paul
Paris, France

Kluwer Academic/Plenum Publishers
New York, Boston, Dordrecht, London, Moscow

Library of Congress Cataloging-in-Publication Data

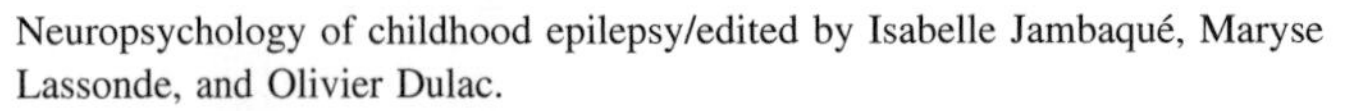

Neuropsychology of childhood epilepsy/edited by Isabelle Jambaqué, Maryse Lassonde, and Olivier Dulac.
p. ; cm. — (Advances in behavioral biology; 50)
Includes bibliographical references and index.
ISBN 0-306-46522-1
1. Epilepsy in children—Congresses. 2. Clinical neuropsychology—Congresses. 3. Pediatric neurology—Congresses. 4. Developmental neurobiology—Congresses. I. Jambaqué, Isabelle. II. Lassonde, Maryse. III. Dulac, Olivier, 1946– IV. Series.

RJ496.E6 N48 2001
618.92′853—dc21

2001023703

Proceedings of the Neuropsychology of Childhood Epilepsy, held October 13–16, 1996, in Asnieres sur Oise, France

ISBN 0-306-46522-1

233 Spring Street, New York, New York 10013

http://www.wkap.nl/

10 9 8 7 6 5 4 3 2 1

A C.I.P. record for this book is available from the Library of Congress

Printed in the United States of America

CONTRIBUTORS

Albert P. Aldenkamp, Faculty of Social and Behavioral Sciences, University of Amsterdam, and Department of Behavioral Science and Psychological Services, Epilepsy Centre, Kempenhaeghe, P.O. Box 6I, NL-5590 A.B. Heeze, The Netherlands

Sophie Bayard, Groupe de Recherche en Neuropsychologie Expérimentale, Département de Psychologie, Université de Montréal, C.P. 6128, Succ. Centre-Ville, Montréal, Qué., H3C 3J7, Canada

Yezekiel Ben-Ari, INMED Institute National de la Santé et de la Recherche Médicale, Unité 29, Avenue de Luminy, B.P. 13, 13273 Marseille Cedex 09, France

Rebecca L. Billingsley, Department of Psychology, University of Toronto at Mississauga, 3359 Mississauga Road North, Mississauga, Ontario, L5L 1C6, Canada, and Department of Psychology, The Hospital for Sick Children, 555 University Avenue, Toronto, Ontario, M5G 1X8, Canada

Daniela Brizzolara, Dipartimento di Medicina della Procreazione e dell'Età Evolutiva, University of Pisa, IRCSS Stella Maris, Via dei Giacinti 2, 56018 Calambrone (Pisa), Italy

Paola Brovedani, Dipartimento di Medicina della Procreazione e dell'Età Evolutiva, University of Pisa, IRCSS Stella Maris, Via dei Giacinti 2, 56018 Calambrone (Pisa), Italy

Christine Bulteau, Service de Neuropédiatrie, Hôpital Saint-Vincent de Paul, 82 Avenue Denfert Rochereau, 75674 Paris Cedex 14, France, and INSERM U169, Hôpital Paul Brousse, 16 Avenue Paul Vaillant Couturier, 94800 Villejuif, France

Claudia Casalini, Dipartimento di Medicina della Procreazione e dell'Età Evolutiva, University of Pisa, IRCSS Stella Maris, Via dei Giacinti 2, 56018 Calambrone (Pisa), Italy

Catherine Cassé-Perrot, Centre Saint-Paul, 300 Boulevard Sainte Marguerite, 13009 Marseille, France

Esper A. Cavalheiro, Neurologia Experimental, UNIFESP-EPM, Rua Botucatu, 862, 04023-900 São Paulo, SP, Brazil

Catherine Chiron, INSERM U29, Hôpital Saint-Vincent de Paul, 82 Avenue Denfert Rochereau, 75674 Paris Cedex 14, France, and Service Hospitalier Fréderic Joliot, Département de Recherche Médicale, Centre à l'Énergie Atomique, Orsay, France

Francine Cyr, Département de Psychologie, Université de Montréal, C.P. 6128, Succ. Centre-Ville, Montréal, Qué., H3C 3J7, Canada

Elaine De Guise, Groupe de Recherche en Neuropsychologie Expérimentale, Département de Psychologie, Université de Montréal, C.P. 6128, Succ. Centre-Ville, Montréal, Qué., H3C 3J7, Canada

Georges Dellatolas, INSERM U169, Hôpital Paul Brousse, 16 Avenue Paul Vaillant Couturier, 94800 Villejuif, France

Thierry Deonna, CHUV, Neuropediatric Unit, Rue du Bugnon 46, 1101 Lausanne, Switzerland

Anne de Saint-Martin, INSERM U398, Clinique Neurologique, Hôpitaux Universitaires de Strasbourg, 67091 Strasbourg, France, and Service de Pédiatrie 1, Hôpital de Hautepierre, Hôpitaux Universitaires de Strasbourg, 67098 Strasbourg, France

Charlotte Dravet, Centre Saint-Paul, 300 Boulevard Sainte Marguerite, 13009 Marseille, France

Olivier Dulac, Service de Neuropédiatrie, Hôpital Saint-Vincent de Paul, 82 Avenue Denfert Rochereau, 75674 Paris Cedex 14, France

Connie C. Duncan, Section on Clinical and Experimental Neuropsychology, LBC, National Institute of Mental Health, 15 North Drive, MSC 2668, Bethesda, MD 20892-2668, USA, and Department of Psychiatry, Uniformed Services of the Health Sciences, Bethesda, MD 20514-4799, USA

Jean-Luc Gaïarsa, INMED Institute National de la Santé et de la Recherche Médicale, Unité 29, Avenue de Luminy, B.P. 13, 13273 Marseille Cedex 09, France

Antoinette Gelot, Unité de Neuropathologie, Hôpital Saint-Vincent de Paul, 82 Avenue Denfert Rochereau, 75674 Paris Cedex 14, France

Guy Geoffroy, Service de Neurologie, Hôpital Sainte-Justine, 3175, Ch. Côte Sainte-Catherine, Montréal, Qué., H3T 1C5, Canada

Rienzo Guerrini, Dipartimento di Medicina della Procreazione e dell'Età Evolutiva, University of Pisa, IRCSS Stella Maris, Via dei Giacinti 2, 56018 Calambrone (Pisa), Italy

Jean-Paul Guillemot, Groupe de Recherche en Neuropsychologie Expérimentale, Département de Psychologie, Université de Montréal, C.P. 6128, Succ. Centre-Ville, Montréal, Qué., H3C 3J7, Canada, and Département de Kinanthropologie, Université du Québec à Montréal, C.P. 8888, Succ. Centre-Ville, Montréal, Qué., H3C 3P8, Canada

Christoph Helmstaedter, University Clinic of Epileptology Bonn, Sigmund Freud Strasse 25, D-53105 Bonn, Germany

Ann Hempel, Minnesota Epilepsy Group, P.A., 310 Smith Avenue, N., Suite 300, St. Paul, MN 55102-2383, USA

Maria Teresa Hernandez, Groupe de Recherche en Neuropsychologie Expérimentale, Département de Psychologie, Université de Montréal, C.P. 6128, Succ. Centre-Ville, Montréal, Qué., H3C 3J7, Canada

Lucie Hertz-Pannier, Service Hospitalier Fréderic Joliot, Department of Medical Research, CEA, Orsay, France, and Pediatric Radiology Department, Hôpital Necker-Enfants Malades, 149 Rue De Sèvres, 75143 Paris Cedex 15, France

Edouard Hirsch, INSERM U398, Clinique Neurologique, Hôpitaux Universitaires de Strasbourg, 67091 Strasbourg, France

Isabelle Jambaqué, Service de Neuropédiatrie, Hôpital Saint-Vincent de Paul, 82 Avenue Denfert Rochereau, 75674 Paris Cedex 14, France

Anna Kaminska, Service de Neuropédiatrie, Hôpital Saint-Vincent de Paul, 82 Avenue Denfert Rochereau, 75674 Paris Cedex 14, France

Virginie Kieffer-Renaux, Service de Neuropédiatrie, Hôpital Saint-Vincent de Paul, 82 Avenue Denfert Rochereau, 75674 Paris Cedex 14, France

Maryse Lassonde, Groupe de Recherche en Neuropsychologie Expérimentale, Département de Psychologie, Université de Montréal, C.P. 6128, Succ. Centre-Ville, Montréal, Qué., H3C 3J7, Canada

João Pereira Leite, Departamento de Neurologica, Psiquiatria et Psicologia, Faculdade de Medicine de Ribeirão Preto-USP, Av Bandeirantes 3900, Campus Universitário, 14049-900 Ribeirão Preto, SP, Brazil

Michael Lendt, University Clinic of Epileptology Bonn, Sigmund Freud Strasse 25, D-53105 Bonn, Germany

Franco Lepore, Groupe de Recherche en Neuropsychologie Expérimentale, Département de Psychologie, Université de Montréal, C.P. 6128, Succ. Centre-Ville, Montréal, Qué., H3C 3J7, Canada

Miriam Levav, Section on Clinical and Experimental Neuropsychology, LBC, National Institute of Mental Health, 15 North Drive, MSC 2668, Bethesda, MD 20892-2668, USA

Anne Lortie, Service de Neurologie, Hôpital Sainte-Justine, 3175, Ch. Côte Sainte-Catherine, Montréal, Qué., H3T 1C5, Canada

Rita Massa, INSERM U398, Clinique Neurologique, Hôpitaux Universitaires de Strasbourg, 67091 Strasbourg, France, and Instituto di Neurologia, Università degli Studi di Cagliari, Cagliari, Italy

Claude Mercier, Service de Neurologie, Hôpital Sainte-Justine, 3175, Ch. Côte Sainte-Catherine, Montréal, Qué., H3T 1C5, Canada

Marie-Noëlle Metz-Lutz, INSERM U398, Clinique Neurologique, Hôpitaux Universitaires de Strasbourg, 67091 Strasbourg, France

Mohammad A. Mikati, Adult and Pediatric Epilepsy Program, Department of Pediatrics, American University of Beirut, P.O. Box 113-6044/B52, Beirut, Lebanon

Allan F. Mirsky, Section on Clinical and Experimental Neuropsychology, LBC, National Institute of Mental Health, 15 North Drive, MSC 2668, Bethesda, MD 20892-2668, USA

Laurent Mottron, Clinique Spécialisée de l'Autisme, Hôpital Rivières des Prairies, 7070 Boulevard Perras, Montréal, Qué., H1E 1A4, Canada

Amal Rameh, Adult and Pediatric Epilepsy Program, Department of Pediatrics, American University of Beirut, P.O. Box 113-6044/B52, Beirut, Lebanon

Olivier Revol, Département de Pédopsychiatrie, Hôpital Neurologique et Neuro-Chirurgical Pierre-Wertheimer, 69394 Lyon Cedex 03, France

Gail L. Risse, Minnesota Epilepsy Group, P.A., 310 Smith Avenue, N., Suite 300, St. Paul, MN 55102-2383, USA

Eliane Roulet Perez, CHUV, Neuropediatric Unit, Rue du Bugnon 46, 1011 Lausanne, Switzerland

Hannelore C. Sauerwein, Groupe de Recherche en Neuropsychologie Expérimentale, Département de Psychologie, Université de Montréal, C.P. 6128, Succ. Centre-Ville, Montréal, Qué., H3C 3J7, Canada

Mary Lou Smith, Department of Psychology, University of Toronto at Mississauga, 3359 Mississauga Road North, Mississauga, Ontario, L5L 1C6, Canada, and Department of Psychology, The Hospital for Sick Children, 555 University Avenue, Toronto, Ontario, M5G 1X8, Canada

Anne Van Houte, Catholic University of Louvain, Pediatric Neurology Service, Saint Luc University Clinics, 10 Hippocrate Avenue, 1200 Brussels, Belgium

Markus Wolf, Universitäts Kinderklinik, Neuropädiatrie, 72070 Tübingen, Germany

CONTENTS

PART I: NEURAL BASIS OF COGNITIVE DEVELOPMENT AND EPILEPSY

PART II: NEUROPSYCHOLOGY OF VARIOUS TYPES OF EPILEPSY

A. Focal Epilepsies

B. Generalized Epileptic Syndromes

C. Transitory Cognitive Impairments

D. Epileptogenic Encephalopaties

PREFACE

This book is devoted to the neuropsychological description of childhood epilepsy, a neurological condition that constitutes one of the most prevalent forms of chronic and disabling childhood illnesses. Indeed, one child out of 20 experiences one or more seizures before the age of 5, and one in a hundred develops epilepsy as a chronic disorder. Approximately half of these children with epilepsy display academic difficulties and/or behavioral disorders. Moreoever, it is now believed that a sizable proportion of children with learning disability suffer from undiagnosed epilepsy.

While a great number of textbooks have been devoted to various medical aspects of childhood epilepsy (diagnosis, genetics, etiology, drug and surgical treatment, etc.), there have been no comprehensive accounts of the cognitive consequences of this condition. Advance of medical knowledge has shown that childhood epilepsy should not be considered as a single disorder but encompasses a whole range of different conditions that exhibit specific clinical EEG and outcome characteristics. It is not becoming apparent that these various clinical entities have different cognitive expression that yet need to be specified. The purpose of this book is to provide a complete up-to-date analysis of this multi-faceted pathology.

The first section of Part I may be regarded as a necessary introduction to the neuropsychology of childhood epilepsy. In this section, the electro-clinical patterns of various types of epilepsy are summarized together with a brief description of neuroradiological findings pertaining to childhood epilepsy. The pervasive consequences of early epilepsy on neural processes are then described. Finally, the last section describes brain maturation processes covering the age range during which a child is susceptible to developing epilepsy. Based upon animal models, essential findings regarding the ontogenesis of neural pathways, synapses and neurotransmitters are presented within the context of their potential susceptibility to early brain damage.

The first section of Part II is devoted to the characterization of the neuropsychological profile that accompanies focal epilepsies, as defined by the site of the epileptic process. The cognitive characteristics of temporal, frontal and parieto-occipital lobe epilepsies are thus described in this first section. In generalized epilepsies, age at onset, type of seizures and occurrence of severe convulsions that define the various epileptic syndromes are the main factors that determine the extent and nature of cognitive impairments. The second section addresses this issue. There is growing evidence that electrical events occurring between seizures can also contribute to a temporary disorganization of cognitive abilities. The third section of Part II covers both the behavioral and cognitive impairments linked to these transient events that often go unnoticed. When these electrical events become continuous, a condition termed epileptogenic encephalopathy, severe and permanent cognitive dysfunctions may occur even in the absence of any seizures. Various neuropsychological syndromes, akin to those observed in the adult patient (aphasia, agnosia and apraxia), develop in the course of these epileptogenic encephalopathies and their expression is related to age at onset and the brain are that is most affected by this epileptic process. The last section of Part II describes these concomitant neuropsychological syndromes.

Part III deals with the medical, surgical and socio-educational management of childhood epilepsy. A chapter is devoted to the study of the effects of medical as it is now generally known that cognitive functions may be altered by drug treatments. Similarly, since neurosurgery is increasingly being used as part of the arsenal treatment in epilepsy, the neuropsychological consequences of the various procecures are described in Part III. Social and educational consequences of childhood epilepsy constitute an important issue that also needs to be addressed. Absenteeism, learning difficulties, cognitive dysfunctions and/or behavioral disorders may all have an impact on a child's self-esteem and upon his or her quality of life. The final part of this book therefore presents an account of the psychological consequences of epilepsy.

In editing this book, we intended to address ourselves to a multidisciplinary readership. Various professional and caretakers dealing with different aspects of childhood epilepsy, be it for diagnosis, rehabilitation and/or educational programs, may benefit from this book. Experimental and cognitive neuropsycholgists should also be provided with valuable information regarding cerebral reorganization following early brain damage. Finally, this book constitutes an important complement to the medical knowledge of epilepsy and may therefore be of great use to medical specialists involved in the care of children with epilepsy.

As with other works of this type, a number of people and organizations were involved in its realization. The book was partly based on an international workshop held at the Abbaye de Royaumont, France, in October 1996. The symposium was made possible by funding from the International League Against Epilepsy, Naturalia et Biologia, Sanofi and Glaxo-Wellcome Laboratories, the Fonds FCAR from Québec, and the Epilepsy Unit of the Hospital Saint-Vincent de Paul in Paris. We wish to thank Mrs. Marie-Claude Herry for her invaluable help in the organization of the symposium, Dr. Hannelore Sauerwein and especially Ms. Sophie Bayard for their revisions of some parts of this book.

Isabelle Jambaqué
Maryse Lassonde
Olivier Dulac

1

MECHANISMS, CLASSIFICATION AND MANAGEMENT OF SEIZURES AND EPILEPSIES

Olivier Dulac

Service de Neuropédiatrie
Hôpital Saint Vincent de Paul
82 Avenue Denfert Rochereau
75674 Paris Cedex 14
France

INTRODUCTION

Clinical manifestations linked to chronic recurrence of paroxysmal discharges within neuronal networks of the brain define epilepsy. Causes comprise various combinations of brain damage, genetic predisposition and maturation phenomena characteristic of the child's brain. The wide range of clinical expression includes acute events, the epileptic seizures, and/or progressive deterioration of motor, sensory or cognitive functions. Various patterns are observed, consisting of recurrent seizures as in adults, acute episodes of prolonged or recurrent seizures called status epilepticus, and intermediary conditions in which distinct seizures are combined with progressive deterioration of brain functions, but the latter are not linked to the seizures themselves but to a combination of seizures and EEG so-called "interictal" paroxysmal activity. The latter condition, called epileptogenic encephalopathy is very specific of pediatric epilepsy. Epilepsy may last from days to decades, and consequences range from simply an odd experience to complete loss of cognitive abilities that may persist throughout life although the epilepsy has died out.

1. MECHANISMS

Paroxysmal discharges within a given neuronal network may be either excitatory (excitatory paroxysmal discharge, EPD) or inhibitory (inhibitory paroxysmal discharge, IPD), IPD being triggered by EPD. The electroencephalographic (EEG) correlate consists of a spike for EPD and a slow wave for IPD. The variable combinations of EPD and IPD may produce four types of patterns:

Neuropsychology of Childhood Epilepsy, edited by Jambaqué et al.
Kluwer Academic / Plenum Publishers, New York, 2001.

- rhythmic repetitive EPD produces a tonic event. The EEG pattern consists of low amplitude rapid rhythmic activity
- progressively increasing inhibition of this tonic event by IPD produces a tonic-phasic event. The EEG pattern consists of rapid rhythmic activity of increasing amplitude intermingled with increasing amplitude slow wave activity
- EPD followed by IPD produces a phasic, therefore brief event. The EEG pattern consists of a spike wave
- repetition of a phasic event may be more or less rhythmic and lasting. The EEG pattern consists of more or less rhythmic spike and slow wave activity.

This distinction is of major importance because it determines both the elementary clinical expression and the outcome: the phasic activity modifies the properties of synapses with stabilization of the affected pathway and therefore the hyperexcitable neuronal network, and prevents it from fine pruning which is critical for the development of cognitive functions. Inhibition, on the contrary, prevents stabilization of the pathway, but it also prevents normal activity of the brain: it therefore interrupts normal maturation although it does not contribute to reinforce the epilepsy.

The elementary clinical expression of these various events depends of the area of the brain that is involved. Tonic or tonic-clonic events produce both positive and negative expression of the corresponding motor, sensory or cognitive function: for instance, a discharge within the motor strip produces both uncontrollable motor activity, and inhibition of voluntary motor activity in one given muscle group. Phasic events in motor or sensory areas respectively produce myoclonus, phosphenes or acouphenes. In other areas, isolated phasic events produce brief inhibition of the corresponding cognitive function that is usually overlooked and may only be expressed by brief inhibition of tapping for instance. Repeat phasic events affecting large areas of the brain for long periods produce major inhibition of the corresponding motor, sensory and cognitive functions, eventually combined with myoclonus.

The extent of the discharge varies. Tonic-clonic and phasic events may involve the whole brain or begin in one or several restricted area(s) or structures of the brain and, following anatomical pathways, gain wider areas of the cortex and/or subcortical areas. Prolonged phasic events involving large areas of the cortex may disinhibit subcortical areas in which they are also likely to trigger paroxysmal, usually tonic activity. Thus, localized epileptogenic activity may extend to wide areas of the brain and produce generalized clinical expression through two distinct mechanisms. One consists of repeat secondary generalization of seizures as in adulthood. The other combines deterioration of cognitive functions due to continuous phasic activity in large parts of the cortex and generalized usually tonic seizures of subcortical origin. It may produce myoclonic activity when it affects the motor strip. These "epileptogenic encephalopathies" exclusively affect the developing brain, thus infants and young children.

The area or structure that holds the epileptogenic focus depends on its nature. It may be a lesion within the cortex, rarely in subcortical structures. It may not only involve a structure but a whole pathway, i.e., a cortico-thalamic loop in the genetically determined idiopathic generalized epilepsy, thus mainly affecting the motor strip or premotor area, and subcortical structures.[13] The type of epileptogenic structure determines particular rhythms of spike-wave activity: the cortico-thalamic loop generates a 3 Hz rhythm because of a thalamic pacemaker whereas secondary generalization from a cortical focus produces less rhythmic activities of lower frequency (0.5–2.5 Hz).

The developing brain has, more than the adult brain, the ability to organize a neuronal network that may be functional from the motor or cognitive point of view, or epileptogenic. This ability mostly concerns the critical period of development of the function to which it is dedicated. After the end of the epilepsy, the affected structure may have lost its ability to acquire the normal function to which it was destined once the critical period of development is over. In fact, it seems that epileptogenicity is increased during the critical period because synchrony and hyperexcitability are necessary for the development of the given function.[22] Epileptogenicity decreases at the end of the critical period and this contributes to the disappearance of the epilepsy.

The sequence of maturation of the different areas of the brain contributes to determine the various patterns observed in clinical epileptology. Subcortical areas become mature before the first cortical areas. The right hemisphere becomes mature before the left one, as far as visual and speech areas are concerned.[10] Maturation within the cortex begins in the lower part of the motor strip, followed by the auditory and visual, and then the parieto-occipital areas. The frontal lobe becomes mature progressively by the end of the first year of life.[10,11]

The age at which epilepsy begins depends on the etiology and its location within the brain, both determining the stage of development at which it is likely to express itself. For instance, a dysplastic structure becomes epileptogenic earlier than an ischemic lesion and an ischemic lesion in the occipital cortex, earlier than in the frontal convexity.[27]

2. CLASSIFICATION AND MANAGEMENT

2.1. Epileptic Conditions

Three types of epileptic conditions are on record:

- Epilepsy consists of repeat paroxysmal events called seizures. Between them, the neurological, motor, sensory or cognitive condition is usually normal. Any permanent defect may be due to preexisting brain lesion, to postictal deficit of frequent or severe seizures such as in Sturge-Weber disease,[1] or to a progressive brain lesion produced by the cause of the seizure disorder, like inborn errors of metabolism and Rasmussen's progressive focal encephalitis.[2]
- Status epilepticus is a prolonged epileptic condition during which motor, autonomous or cognitive disorders, including consciousness are affected for periods lasting hours or days. It may consist of repeat seizures without recovery of consciousness between seizures, or of a single long lasting seizure. Hemiconvulsion-Hemiplegia (HH) syndrome is the best example of the latter.[21]
- Epileptogenic encephalopathy is an intermediary condition in which cognitive functions are affected in conjunction with continuous spike and slow wave activity whereas seizures are either rare or clearly not the only cause of the cognitive deterioration. The age determines the pattern of spike and slow wave activity: before the age of 2, lack of myelin prevents bilateral synchrony of paroxysmal activity, thus the pattern is that of asynchronous spikes and slow waves called "hypsarrhythmia".[18] Later, completion of myelination increases significantly the axonal conduction velocity and permits bilateral synchrony, producing the pattern of generalized slow spike waves.

In addition, any focal brain damage generates slow wave activity whereas fast activity is characteristic of cortical dysplasia, and the genetically determined 3 Hz spikewave bursts may also modify the pattern of spike and slow wave activity.

2.2. Etiology

Three types of etiology contribute to the occurrence of epilepsy:

- brain lesions that may be focal, multifocal or diffuse, and may be malformative or clastic.[16]
- genetic predisposition which rarely exhibits a simple Mendelian mode of inheritance. It usually is polyfactorial, as in idiopathic generalized epilepsy.
- maturational phenomena which increase excitability in the area of the brain that is experiencing motor, sensory or cognitive acquisition, during the critical period of development which is specific for each given area.

In a given patient, etiology may be single or multiple. For instance, some types of epilepsy are generated by a single dominantly or recessively inherited gene. Others combine either genetic predisposition or brain lesion with maturational phenomena such as, respectively myoclonic-astatic epilepsy or infantile spasms.

In general, idiopathic cases are those for which clinical and EEG characteristics permit to claim the lack of any brain lesion, symptomatic cases are those for which a brain lesion can be demonstrated, and cryptogenic cases are those that do not meet one of these two categories, thus the cause is "hidden".[12]

The various types of epilepsy are characterized by the combination of a specific age of onset, various types of epileptic seizures, and interictal clinical condition and EEG. These conditions that result from non-random combinations and are called "epilepsy syndromes" have been classified by the International League Against Epilepsy.[12,31] Based on classical concepts, this classification distinguishes partial and generalized epilepsies. However, for a significant number of patients, particularly those for which the onset is in the youngest age group and for cryptogenic and symptomatic cases, this distinction is not relevant because they combine features of both focal and generalized epilepsy.

2.3. Idiopathic Epilepsies[16]

2.3.1. Benign Neonatal and Infantile Convulsions. Benign neonatal idiopathic and familial[31] convulsions and benign familial and non familial infantile[33,36] convulsions produce clusters of focal or generalized seizures that, up to now, have been found to have no impact on motor, sensory or cognitive functions. Familial cases are autosomal, dominantly inherited.

2.3.2. Idiopathic Generalized Epilepsies (IGE). This group was the first identified as an epileptic syndrome. It shares generalized seizures and spike waves, and usually the outcome is favorable. Infantile absence epilepsy consists of frequent typical absences as a single seizure type, beginning between 4 and 8 years, mainly in girls with previously normal development and no evidence of brain damage. Seizures are rapidly controlled by treatment and have no or mild impact on cognitive functions, and later occurrence of GTC is a rare event. EEG shows long runs of 3 Hz spike-wave ictal discharges and brief bursts of generalized spike waves.

Juvenile absence epilepsy exhibits less frequent absences, briefer ictal discharges, and later occurrence of GTC is frequent. Whether childhood and juvenile absence epilepsies are neurobiologically different syndromes or part of a single spectrum remains open to question.[6] There may be a specific group of female patients with later GTCs and chronic course with eyelid myoclonus and absences.[3] In addition, the nosological place of epilepsy with myoclonic absences[31] and absences with perioral jerks[28] is difficult to determine. In half the cases of myoclonic absences it mainly involves mentally retarded patients and is highly resistant to conventional AEDs.[34]

Benign myoclonic epilepsy of infancy begins between 6 months and 3 years of age. Massive myoclonus is the only type of seizures, affecting the upper limbs and rarely causing the patient to fall, even if s/he has gained the ability to walk. These jerks occur when awake but also when falling asleep, and this wakes the child up, causing poor quality of sleep. Myoclonic seizures may be precipitated by touch or noise,[31] rarely by photic stimulation. They are combined with generalized spike waves on EEG. Development is usually normal, but a number of patients have speech delay, particularly those with early onset of the seizures. Quality of seizure control is supposed to determine future development, and indeed the earlier administration of valproate has improved outcome, but some patients with poor outcome may have had earlier onset because of pre-existing brain dysfunction, overlooked before the first seizures because of the young age of onset.

Epilepsy with generalized tonic-clonic seizures (CTCS) is less frequent in childhood than in adolescence and occurrence is evenly distributed between 3 and 11 years of age. It often begins with febrile convulsions (FC), has low seizure frequency and an excellent prognosis for seizures. Half the patients exhibit generalized spikes waves.

Treatment of IGE is based on the administration of valproate (VPA), although ethosuximide (ESM) is eventually preferred for infantile absence epilepsy. In cases resistant to VPA and ESM, alone or in combination, lamotrigine (LTG) is often helpful, alone or in combination with VPA.

The prognosis for seizure control in IGE varies in the different syndromes, the best prognosis being for infantile absence epilepsy. For psychosocial integration and schooling it is also variable, even for patients who have complete seizure control. Desguerre *et al.*[15] in an epidemiological study found that patients who had recovered from idiopathic generalized epilepsy with convulsive seizures experienced more schooling difficulties than patients with infantile absence or idiopathic partial epilepsies.

2.3.3. Idiopathic Partial Epilepsies in Childhood. This group shares age of onset, partial seizures, focal interictal spikes and favorable outcome. The diagnosis is based on clinical and EEG grounds, and in order to avoid misdiagnosis, it is important to be very strict with gathering the various diagnostic criteria given by Dalla Bernardina:[31] lack of neurological or cognitive defect, age of onset after 18 months, a single type of seizure in a given patient, no tonic or reflex seizures, and normal interictal clinical condition. The frequency of seizures may be high at onset, but it never increases over time. Status epilepticus may occur as the initial seizure but has no lasting impact. The EEG shows normal background activity, and focal or multifocal spikes that are activated by sleep but remain with similar morphology. There are no polyspikes or periods of suppression of activity following the spikes. Continuous spike waves during slow sleep may develop, particularly with carbamazepine (CBZ) therapy, and this condition may produce atypical absences, drop attacks and cognitive deterioration.[29] Evoked potentials, either somesthetic or visual according to the type of epilepsy, are occasionally giant, but their latency and morphology are unchanged.

Benign partial epilepsy with centrotemporal spikes (BECT) is the most frequently observed, accounting for 10–20% of the cases of childhood epilepsy. After a first seizure there is an 85% risk of recurrence but over two thirds of the patients have few seizures and do not need any treatment.[24] The psychosocial benefits of appropriate early diagnosis have been clearly shown.

The diagnosis of benign partial epilepsy with occipital paroxysms (EOP) may be difficult because the nosological limits are unclear. The seizures are characterized by lateral deviation of the eyes, vomiting and unilateral convulsions, not visual signs, and EEG shows occipital spikes blocked by opening the eyes.

Idiopathic partial epilepsy with affective seizures (benign psychomotor epilepsy) is even less frequent than EOP, and may be difficult to identify. Seizures of a single type consist of an expression of terror with mastication, laughter, crying or salivation that occur during daytime and at night with high frequency at onset.

2.3.4. Idiopathic West Syndrome. Lack of brain damage can only be suspected on indirect basis comprising: lack of pre- or perinatal history, symmetrical spasms and hypsarrhythmia occurring after normal initial development and without major deterioration, and lack of interictal EEG focus, including after intravenous administration of diazepam. Evolution is favorable from both the epilepsy and cognitive points of view.[20]

2.4. Symptomatic or Cryptogenic Partial Epilepsies

Onset may occur at any age, including the neonatal period and depends of both etiology and the topography of the focus: dysplastic lesions tend to produce epilepsy of earlier onset than those due to clastic causes, and occipital epilepsy is likely to begin earlier than epilepsies whose foci are located on the frontal convexity. Symptomatic or cryptogenic partial epilepsy's do not determine epilepsy syndromes *stricto sensu*, but clinical manifestations of both seizures and interictal, motor, sensory or cognitive defects depend of the topography of the focus, the age of occurrence, frequency and severity of fits, and etiology. A non selected series of localization related epilepsy in children showed that patients with occipital epilepsy exhibit visual processing defects whereas those with frontal lobe epilepsy suffer from attention disorder, thus both may have reading difficulties resulting from distinct mechanisms.[8] When drug treatment proves ineffective, surgery may be helpful to relieve the patient from seizures and improve selective functional defects.

2.5. Severe Convulsive Epilepsy in Infancy

2.5.1. Epilepsy with migrating partial seizures in infancy begins between 2 and 6 months of age with increasing seizure frequency. After a few months, the patient exhibits episodes of very frequent partial seizures, occurring nearly continuously with episodes of status epilepticus and severe hypotonia, cerebellar syndrome and major psychomotor delay.

2.5.2. Severe myoclonic epilepsy in infancy (SMEI, Dravet syndrome) begins in the middle of the first year of life, with convulsive seizures that are often triggered by fever and are occasionally unilateral but shift from one side to the other in consecutive seizures. Lack of myoclonus and spike waves during the first two or three years of the disease is misleading, and the condition can easily be confused with either febrile convulsions or partial epilepsy.[37] These patients are at high risk to develop status epilepticus, either clonic especially during febrile episodes in the first years of life or myoclonic at a later stage. After the second year of life, myoclonic jerks, ataxia and mental delay, and occasional photosensitivity appear. The nosological distinction with conditions in which intractable generalized seizures occur during the first year of life in patients who do not develop myoclonus, and whose motor and cognitive outcome is similar.

2.6. Epileptogenic Encephalopathies

2.6.1. Neonatal epileptogenic encephalopathy with suppression bursts begins in the first months of life. Inborn errors of metabolism or extensive malformations such as hemimegalencephaly and Aicardi syndrome may be disclosed. Patients whose seizures persist for over 3 months develop severe dystonia of axial and limb muscles and remain with no acquisitions.

2.6.2. West syndrome (WS) occurs usually between 3 and 12 months of age, although 2% of the cases begin later, up to four years of life.[4] It combines epileptic spasms with hypsarrhythmia.[18] In addition, a number of patients exhibit partial seizures. When a focus can be disclosed, spontaneously or after intravenous administration of diazepam, it may involve occipital or temporal areas. Cognitive deterioration often consists of loss of visual and/or auditory contact, and neuropsychological follow-up shows that agnosia is the predominant cognitive disorder of these patients. The treatment is based on vigabatrin, alone or in combination with steroids.

2.6.3. Myoclonic epilepsy ("myoclonic status") in non-progressive encephalopathies begins in the first months of life and consists of severe axial hypotonia and polymorphous abnormal movements, combing dystonia and myoclonic jerks of the face and distal muscles. Anoxia, various kinds of brain malformations and chromosomal abnormalities including Angelman syndrome are involved. The background EEG activity is slow, and polygraphy shows bilateral myoclonus to be correlated with brief bursts of diffuse slow waves with superimposed spikes, continuous in the waking state and persisting during sleep. During evolution, patients exhibit long lasting episodes of myoclonic status with worsening of tone, which can strongly mimic a progressive encephalopathy.

2.6.4. Lennox-Gastaut syndrome (LGS) consists of tonic seizures, atypical absences and slow spike waves beginning between 2 and 8 years of age. The EEG during atypical absences is often difficult to distinguish from the so-called interictal activity. The spike activity predominates in the frontal areas. Cognitive deterioration at onset is characterized by attention disorder and hyperkinesia, followed during the course of the disorder by slowness, stereotypes, loss of judgment and of the ability to anticipate and control behavior.

2.6.5. Myoclonic astatic epilepsy (MAE) begins between 2 and 5 years of life. Generalized tonic clonic seizures are the only type of seizure for the first months of the disease. Then myoclonic astatic seizures occur with increasing frequency, causing the patients to fall several times a day, and s/he becomes hyperkinetic and ataxic.[17] Some patients go into status epilepticus, with massive jerks and atypical absences, but recover completely after a few months.[19] Others suffer erratic myoclonus with drowsiness and generalized tonic seizures that last for months, with considerable deterioration of cognitive functions.[26]

2.6.6. In continuous spike waves during slow sleep (CSWS) which may be either symptomatic or cryptogenic, the seizure type varies, and may lack altogether. The spike activity may predominate in the rolandic area and be correlated with negative myoclonus[23] or mouth apraxia[14] which disappear when CSWS remits. In other instances, the spike activity predominates in temporal or frontal areas. Aphasia characterizes the Landau-Kleffner syndrome in which spike activity predominates in the temporal areas, whereas a severe frontal syndrome has been reported in cases with frontal predominance of spiking activity.[32] In both instances, cognitive functions take months to recover once CSWS has remitted.

During the course of the epilepsy, the type of syndrome may vary from focal to generalized or the reverse. Particularly, West, Lennox-Gastaut and CSWS are likely to follow or precede focal epilepsy since in fact they merely represent secondary generalization from single or multiple foci. It is the generalization that impacts the most the cognitive functions.

In conclusion, 70 to 80% of infants and children with epilepsy have the characteristics of an identifiable epilepsy syndrome, the smallest proportion being in infancy. The main difficulty, in addition to infantile epilepsy, is to identify the type of epilepsy that causes the child to fall: Lennox-Gastaut, CSWS, myoclonic-astatic epilepsy and late onset West syndrome are the most difficult to distinguish.

3. PATHOPHYSIOLOGY OF COGNITIVE TROUBLES DETERMINED BY EPILEPSY

Any pathophysiological speculation regarding cognitive alterations must take into account the correlation between the age of onset, the topography of predominating spike activity, and the type of cognitive deterioration observed. Indeed, the functions that are altered in a given syndrome are those that are experiencing rapid development when epilepsy occurs.

The period during which the syndrome may develop is limited, and there is a tendency to spontaneous recovery, or to switching to another type of epilepsy after a certain age. Spasms and hypsarrhythmia usually disappear in early childhood, CSWS before puberty. Regarding LGS, although tonic seizures persist into adulthood, atypical absences and slow spike waves tend also to diminish in the second decade. The period of risk for the occurrence of each given syndrome corresponds therefore to what may be considered as a "critical period of development", in the same way and with similar basis as for the development of cognitive functions. Furthermore, the function affected by a given epilepsy depends of both the area of the brain that is specifically involved and the age, thus the maturation stage shown by cerebral blood flow investigations showing that maturation is earlier in the posterior than in the anterior areas.[9]

Similarly, the usual age of disappearance of a given syndrome is also determined by the age at which the "critical period of development" of the affected cognitive function finishes. If the epilepsy has persisted during this whole period, spontaneous cessation of seizures occurs at the cost of loss of the ability to recover the corresponding function, i.e., gnosia, language and judgment. In addition, abnormal development of elementary functions may result in abnormal development of more elaborate functions, i.e., auditory agnosia precludes the later development of language, visual agnosia prevents normal communication with the environment to develop.

In the first years of life, the brain is undergoing rapid maturation. It is not the miniature of an adult brain. This early postnatal period is indeed a "critical period" for epilepsy, for several reasons related to developmental characteristics (for a review, see 35). During the early postnatal period in animals, which could correspond to the end of pregnancy or maybe the second year of life in humans, the cortex is more excitable than later in life, and a focal discharge is more likely to become generalized, in an asynchronic way, to the whole cortex. Indeed, NMDA receptors are in excess, in relation with the development of the neuronal network, as shown in visual pathways in animals during the critical period of development. In addition, the GABA B inhibitory system is not functional before the second week of life in the rat.[5] Axonal collaterals are redundant and subcortical structures that contribute to prevent generalization later in life, such as the substantia nigra, are ineffective. Myelination of the hemispheric white matter is immature during the first two years of life.

The same neuronal networks are involved in both cognitive functions and epilepsy, and therefore both activities are in competition.[7] Any potentially epileptogenic structure, such as brain dysplasia, is likely to initiate epileptic phenomena from the stage of development when the surrounding cortex starts undergoing rapid maturation, and therefore experiences an increase in excitatory pathways activity necessary for this maturation, and then during the whole "critical period". The occurrence of epilepsy within this developing network is likely to produce much greater disorganization than epilepsy affecting a mature and therefore previously structured network, i.e., that of an adult brain.[25]

The respective roles of spikes versus slow wave activity are difficult to assess. Since spike-waves in the motor area may affect tone with time locked correlation to the slow wave of the spike wave complex,[23] one can easily anticipate the negative impact of similar slow wave activity in areas of the brain involved in cognitive, not motor functions. This phenomenon persists for months, during wakefulness, sleep or both, thus preventing reinforcements that are so necessary in the learning process. Furthermore, both hemispheres are involved, which prevents one hemisphere from taking over a function that the other hemisphere cannot support. Therefore, cessation of WS, LGS or CSWS is not sufficient for cognitive functions to recover, although it is a prerequisite. Cessation of spike and slow wave activity permits learning abilities to recover provided the critical period of development is not over, and from then on, the corresponding cognitive function may resume development.

REFERENCES

1. Adamsbaum C, Pinton F, Rolland Y, Chiron C, Dulac O and Kalifa G (1996): Accelerated myalination in early Sturge-Weber syndrome: MRI-SPECT correlations. Pediatric Radiology 26:759–762.
2. Andermann F (1991): Chronic Encephalitis and Epilepsy. Boston: Butterworth-Heineman.
3. Appleton R, Panyiotopoulos CP, Acimb AB and Beirne M (1993): Eyelid myoclonia with absences: an epilepsy syndrome. Journal of Neurology, Neurosurgery and Psychiatry 56:1312–1316.
4. Bednarek N, Motte J, Plouin P, Soufflet C and Dulac O (1998): Evidence for late onset infantile spasms. Epilepsia 39:55–60.
5. Ben Ari Y, Tseeb V, Raggozzino D, Khazipov R and Gaiarsa JL (1994): g-aminobutyric acid (GABA): a fast excitatory transmitter which may regulate the development of hippocampal neurons in early

postnatal life. In van Pelr I, Corner MA, Uylings HBM, Lopes da Silva FH (eds): "Progress in Brain Research." Amsterdam: Elsevier, pp 261–273.

6. Berkovic SF, Andermann F, Andermann E and Gloor P (1987): Concepts of absence epilepsies: discrete syndromes or biological continuum? Neurology 37:993–1000.
7. Binnie C, Kasteleijn-Nolst Trenité DGA, Smit AM and Wilkins AJ (1987): Interactions of epileptiform EEG discharges and cognition. Epilepsy Research 1:239–245.
8. Bulteau C *et al.* (in preparation).
9. Chiron C, Raynaud C, Mazière B *et al.* (1992): Changes in regional cerebral blood flow during brain maturation in children and adolescents. Journal of Nuclear Medicine 33:696–703.
10. Chiron C, Jambaqué I, Nabbout R, Lounes R, Syrota A and Dulac O (1997): The right brain hemisphere is dominant in human infants. Brain 120:1057–1065.
11. Chugani HT, Phelps ME and Mazziotta JC (1984): Positon emission tomography with 18F-2-fluorodeoxyglucose in infantile spasms. Annals of Neurology 16:376–377.
12. Commission of Classification and Terminology of the International League Against Epilepsy (1989): Proposal for revised classification of epilepsies and epileptic syndromes. Epilepsia 30:389–399.
13. Coulter DA and Lee CJ (1993): Thalamocortical rhythm generation in vitro: extra- and intracellular recordings in mouse thalamocortical slices perfused with low Mg++ medium. Brain Research 631:137–142.
14. Deonna T, Roulet E, Fontan D and Marcoz JP (1993): Speech and oral deficits of epileptic origine in benign partial epilepsy of childhood with rolandic spikes. Neuropediatrics 24:83–87.
15. Desguerre I, Chiron C, Loiseau J, Dartigues JF, Dulac O and Loiseau P (1994): Epidemiology of idiopathic generalized epilepsy. In Malafosse A, Genton P, Hirsch E, Marescaux C, Broglin D, Bernasconi R (eds): "Idiopathic Generalized Epilepsies: Clinical, Experimental and Genetic Aspects" London: John Libbey, pp 19–26.
16. Diebler C and Dulac O (1987): Pediatric Neurology and Neuroradiology. Cerebral and Cranial Diseases. Berlin: Springer-Verlag.
17. Doose H, Gerken H, Morstmann T and Völtzke E (1970): Centrencephalic myoclonic-astatic petit mal. Neuropediatrics 2:59–78.
18. Dulac O, Chugani H and Dalla Bernardina B (1994): Infantile Spasms and West Syndrome. London: Saunders.
19. Dulac O, Plouin P and Chiron C (1990): Forme "bénigne" d'épilepsie myoclonique chez l'enfant. Neurophysiologie Clinique 2:77–84.
20. Dulac O, Plouin P and Jambaque I (1993): Predicting favorable outcome in idiopathic West syndrome. Epilepsia 34:747–756.
21. Gastaut H, Poirier F, Payan H *et al.* (1960): HHE syndrome, hemiconvulsions, hemiplagia, epilepsy. Epilepsia 1:418–447.
22. Goodman CS and Shatz CJ (1993): Developmental mechanisms that generate precise patterns of neuronal activity. Cell 10:77–98.
23. Guerrini R, Dravet C, Genton P, Bureau M, Roger J, Rubboli G and Tassinari CA (1993): Epileptic negative myoclonus. Neurology 43:1078–1083.
24. Hamada Y, Okuno T, Hattori H and Mikawa H (1994): Indication for antiepileptic drug treatment of benign childhood epilepsy with centro-temporal spikes. Brain Development 16:159–161.
25. Holmes GL, Sarkisian M, Ben-Ari Y, Liu Z and Chevassus-Au-Louis N (1999): Consequences of cortical dysplasia during development in rats. *Epilepsia* 40:537–544.
26. Kaminska A, Ickowicz A, Plouin P, Bru MF, Dellatolas G and Dulac O (1999): Nosological delineation of non symptomatic Lennox-Gastaut syndrome and myoclonic-astatic epilepsy. Epilepsy Research 36:15–29.
27. Koo B and Hwang P (1996): Localization of focal cortical lesions influences age of onset of infantile spasms. Epilepsia 37:1068–1071.
28. Panyiotopoulos CP, Ferrie C, Giannakodimos S and Robinson R (1997): Perioral myoclonia with absences: a new syndrome? In Wolf P (ed): "Epileptic Seizures and Syndromes." London: John Libbey, pp 143–153.
29. Perucca E, Gralm L, Avanzini G and Dulac O (1998): Antiepileptic drugs as a cause of worsening of seizures. Epilepsia 39:5–17.
30. Ricci S, Cusmai R, Fusco L and Vigevano F (1995): Reflex myoclonic epilepsy: a new age-dependant idiopathic epileptic syndrome related to startle reaction. Epilepsia 36:342–348.
31. Roger J, Bureau M, Dravet C, Dreifuss FE, Perret A and Wolf P (1992): Epileptic Syndromes in Infancy, Childhood and Adolescence. Second Edition. London, Paris: John Libbey.

32. Roulet Perez E, Davidoff V, Despland PA and Deonna T (1993): Mental and behavioural deterioration of children with epilepsy and CSWS: acquired epileptic frontal syndrome. Developmental Medicine and Child Neurology 35:661–674.
33. Vigevano F, Fusco L, Di Capua M *et al.* (1992): Benign infantile familial convulsions. European Journal of Pediatrics 151:608–612.
34. Wallace SJ (1994): Epilepsy with myoclonic absences. Archives of Disease in Childhood 70:288–290.
35. Wasterlain CG and Shirasaka Y (1994): Seizures, brain damage and brain development. Brain Development 16:279–295.
36. Watanabe K, Negoro T and Aso K (1993): Benign partial epilepsy with secondarily generalized seizures in infancy. Epilepsia 34:635–638.
37. Yakoub M, Dulac O, Jambaqué I, Chiron C and Plouin P (1992): Early diagnosis of severe myoclonic epilepy in infancy. Brain Development 14:299–303.

2

CONVENTIONAL AND FUNCTIONAL BRAIN IMAGING IN CHILDHOOD EPILEPSY

Lucie Hertz-Pannier[1,2] and Catherine Chiron*[,1,3]

[1]Service Hospitalier Fréderic Joliot
Department of Medical Research
CEA, Orsay, France
[2]Pediatric Radiology Department
Hôpital Necker-Enfants Malades
149 Rue De Sèvres
75143 Paris Cedex 15
France
[3]Service de Neuropédiatrie and INSERM Unit 29
Hôpital Saint Vincent de Paul
82 Avenue Denfert Rochereau
75674 Paris Cedex 14, France

INTRODUCTION

The past two decades have seen drastic developments in medical brain imaging, particularly in the areas of structural imaging with MRI (magnetic resonance imaging), functional imaging with PET (positron emission tomography), SPECT (single photon emission computed tomography) and functional MRI (fMRI), and for metabolic studies with MRS (magnetic resonance spectroscopy). Most of these techniques constitute non invasive means of examination which are progressively transferred from research to clinical practice and are beginning to be adapted from adults to children (for more details about the techniques see the chapter Cerebral Maturation and Functional Imaging, this volume).

PET uses positron emitter tracers in order to study the brain glucose metabolism (with ^{18}F-FDG, fluorodesoxyglucose), the distribution of various neuroreceptors (such as benzodiazepine receptors with ^{18}F-Flumazenil, dopamine receptors with ^{11}C-Raclopride), or the regional cerebral blood flow (rCBF) with H_2O^{15} (activation studies).

The SPECT technique uses gamma emitter tracers, to study rCBF either with ^{133}-Xenon at rest or during stimulation tasks, or with ^{99m}Tc-HMPAO or ^{99m}Tc-ECD at

* To whom correspondence should be addressed

Neuropsychology of Childhood Epilepsy, edited by Jambaqué et al.
Kluwer Academic / Plenum Publishers, New York, 2001.

rest between seizures (interictal SPECT) and during seizures (ictal SPECT). It can also be used to map receptors with the same tracers as those used for PET except that they are labeled with ^{133}Iodine.

MRI has recently experienced a revolution in terms of temporal resolution. Indeed, ultra-fast techniques such as the Echo-planar imaging make it possible to acquire one image in less than 100 milliseconds and offer the possibility to monitor physiological events taking place over a few seconds. This process is achieved via perfusion imaging using endogenous contrast media, the functional brain MRI being based on intrinsic contrast that depend on blood oxygenation. MRI spatial resolution has also considerably increased thus improving the anatomical accuracy of the technique to a few hundreds of microns, or even less when using dedicated coils. Finally, new contrasts (such as FLAIR, Magnetization Transfer Imaging, diffusion imaging) have improved both the depiction and characterization of brain lesions.

^{1}H-MRS (proton Magnetic Resonance Spectroscopy) offers the possibility to obtain biological spectra of cerebral tissues *in vivo* by detecting and localizing non invasively metabolites such as NAA (N amino-aspartate, a neuronal marker) or GABA (g-aminobutyric acid) in the brain.

1. BRAIN IMAGING AND CHILDHOOD EPILEPSY

In childhood epilepsy, neuroimaging improvements lead to major advances in four directions: i) diagnosing a cerebral lesion associated with epilepsy, ii) localizing the epileptogenic focus, iii) preoperative mapping of motor and cognitive functions, iv) localizing the neural substrates of the neuropsychological disorders associated with epilepsy.

1.1. Lesion Localization

The detection and characterization of cerebral lesions causing epilepsy dramatically benefit from MRI progresses: gyral abnormalities and heterotopias are better analyzed using 3D T1 and T2 weighted sequences with high resolution, multi-planar reconstructions and surface renderings. The depiction of cortical dysplasias is greatly improved by FLAIR techniques that can show subtle loss of gray matter-white matter delineation. Hippocampal sclerosis is now more often depicted in young patients on high resolution T2 or FLAIR images. Quantitative measurements of hippocampal volumes help in doubtful cases.

1.2. Localization of the Epileptogenic Focus

Localization of the epileptic focus, which in numerous cases does not coincide with the "visible" cerebral lesion, relies mostly on neurophysiological investigations (video electroencephalography, deep electrodes) but has improved using both PET and SPECT, and mostly ictal SPECT. Nuclear imaging has a great sensitivity for the detection of focal abnormalities in epileptic patients, even when MRI fails to disclose any lesion ("cryptogenic epilepsies"). Interictally, glucose metabolism and rCBF are decreased in the epileptogenic focus whereas they increase ictally. MRS can also detect subtle interictal changes in the focus.

1.3. Preoperative Mapping

Preoperative mapping mainly concerns motor, language and memory functions and currently relies on invasive intracranial recordings. Non-invasive techniques were recently developed in this field and provide available preliminary results in children. However, SPECT and H_2O^{15}-PET cannot be used on a large scale in children, because of radiation constraints and technical limitations. Because of its non-invasive character, functional MRI has become the technique of choice and will probably replace the other techniques for activation studies in the near future.

1.4. Localization of Neurospychological Disorders

Functional disorders associated with epilepsy can now be characterized by childhood neuropsychology. Functional imaging may provide complementary data by visualizing the dysfunctioning areas of the brain and may be used to follow the patients longitudinally. Unfortunately, this field of research is still less developed than the previous ones, due to various methodological unsolved issues (such as the establishment of normative data).

These techniques, when combined with clinical, neuropsychological and neurophysiological data, offer the possibility of a complete non invasive work-up in childhood epilepsy, with three main probable consequences:

- an increasing number of epilepsies will be classified as symptomatic (improved lesion detection, and ictal focus localization) and will be amenable to neurosurgical treatment.
- in terms of functional mapping, there will be a decreased need for invasive techniques that will then only be used in selected cases. Non-invasive techniques can be used in a larger number of cases than invasive ones, and therefore may help optimizing the management of a greater number of children.
- these improvements should reinforce the current trend toward earlier surgery in children, and help to alleviate the debilitating consequences of severe epilepsy in early childhood.

2. TECHNICAL, ETHICAL AND METHODOLOGICAL LIMITATIONS IN PEDIATRIC FUNCTIONAL NEUROIMAGING

2.1. Cooperation and Sedation

Using PET and SPECT, sedation is needed under 6 years of age in order to avoid head movement during image acquisition. When a barbiturate is administered before the tracer injection such as in SPECT with ^{133}-Xenon, it produces a decrease in global CBF by about 15%, but the regional CBF distribution is not modified.[11] When using ^{99m}Tc-SPECT or FDG-PET, sedation does not modify rCBF nor metabolism since it is administrated after the injection of the isotope.

In functional MRI studies, the cooperation of the child must be obtained in most cases (activation studies), which makes sedation undesirable. This, however, is at the same time a limiting factor since cooperation cannot be obtained easily in very young or debilitated patients. Only studies of passive sensory stimulation (such as visual stimulations) have been performed in sedated neonates and infants (see chapter on Cerebral Maturation and Functional Imaging), but the effects of sedation on fMRI signal changes are still incompletely understood.

2.2. Ethics and Gold Standards

Research studies in children are limited by ethical considerations, such as the existence or not of a direct individual benefit for the subject himself. Most studies evaluating new investigation procedures such as functional neuroimaging do not meet that criterion and are therefore very difficult to perform in pediatric populations. Preliminary data from adult studies can be used but they do not account for child-specific maturation processes.

In epilepsy, the gold standard techniques for ictal localization and functional mapping are invasive, expensive, and difficult to perform in children (deep electrodes, cortical stimulations, Wada testing). Therefore they are restricted to high-risk selected cases, and the other patients do not benefit from extensive work up.

2.3. Normative Data

Because of the very stringent constraints for the administration of radioactive tracers in children, control populations for PET and SPECT studies cannot be obtained among normal volunteers like in adults. When studying pediatric patients, controls can be selected in another pathological population, provided the latter also directly benefits from the study. When comparison with normal subjects is required, authors usually refer to the few studies published from small populations of children considered normal *a posteriori*.[11] On the other hand, fMRI promises to be a unique method for activation studies in healthy children,[9] because it does not require any exogenous tracer, is performed without sedation and has no known side-effects.

3. LESION DETECTION

MRI has become the first (and most often the only) imaging modality in epilepsy, because recent improvements of MRI technology (very high spatial resolution using surface coils, and new contrasts such as Inversion-Recovery and FLAIR) now make it possible to detect subtle brain abnormalities that are not easily seen on CT scans. Beside improved description of gross anomalies (such as lissencephaly, hemimegalencephaly, schizencephaly, porencephaly, major atrophy, etc.), MRI is particularly useful for the detection of hippocampal sclerosis and of focal malformations of cortical development (MCD) like cortical dysplasia, heterotopia, and polymicrogyria.

Although rarer in children than in adults, hippocampal sclerosis is more and more often detected using optimized techniques in patients with temporal lobe epilepsy and frequently a history of prolonged febrile seizures in early infancy (Fig. 1).[10] Simple visual analysis of hippocampal structures on high resolution heavily T2-weighted or FLAIR images in a coronal plane perpendicular to the temporal lobe axis shows a hypersignal of the hippocampus (which can be restricted to the body or involve also the head, the tail, and sometimes the amygdala and the parahippocampal gyrus). Hippocampal atrophy and loss of normal hippocampal internal structures are better seen on heavily T1-weighted images (such as Inversion-Recovery), and on multiplanar reconstructions

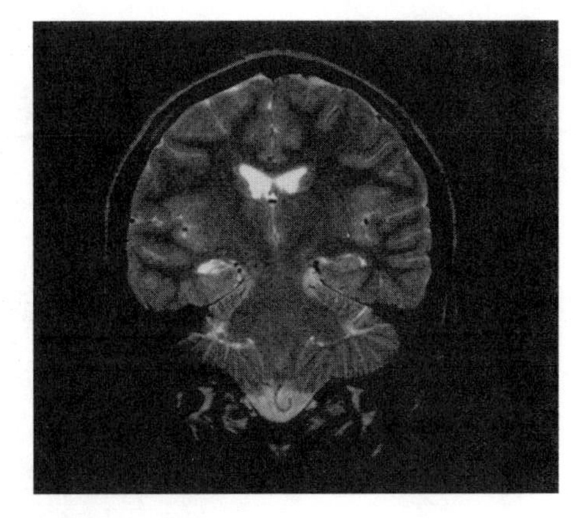

Figure 1. Right hippocampal sclerosis in a 15-year-old boy with partial complex seizures. Hypersignal and volume loss of the right hippocampal formation on this T2 weighted coronal image.

from 3 dimensional data sets that allow precise delineation of hippocampal boundaries.[30] Whereas visual analysis alone has a sensitivity of 80 to 90% in experienced centers, along with an up to 94% specificity, quantitative hippocampal volumetry has been proposed to improve the sensitivity of the detection of hippocampal atrophy (up to 95%, 21), but the method relies on manual drawing of hippocampal boundaries which explains a fairly high inter-observer variability and the large range of normal values. Most centers use left to right volumic ratios to detect unilateral hippocampal atrophy. Bilateral hippocampal sclerosis still remains difficult to diagnose using MRI. Hippocampal atrophy is strongly correlated to CA1, CA3 and CA4 neuronal loss seen on histopathological examinations,[8] and T2 hypersignal is a reflection of hippocampal gliosis. There is a strong correlation between the severity of left hippocampal atrophy on MRI and verbal memory tests in adults.[31,39] Most importantly, it has been established that in adult patients with concordant clinical, EEG and MRI localization, the outcome after temporal lobectomy was favorable in up to 97% of the cases, with no need for invasive intracranial EEG recordings. However, no clear data have yet been collected in children about hippocampal volumetry.

The understanding and unraveling of the malformations of cortical development both largely benefit from MR Imaging. These pathologies are almost always associated with various types of epilepsy depending upon the localization, extension and type of the malformation. While the natural history of some of theses disorders begins to be better understood, thanks to embryological studies and genetic investigations, their description still relies almost entirely on their macroscopic appearance as demonstrated on pathology specimens and on *in vivo* MR imaging, and their classification remains controversial. Lissencephaly (Fig. 2) is characterized by the loss of gyral and sulcal development, more or less diffuse, associated to a thickened cortex comprising only 4 layers. It has a genetic origin, and the patients usually present with refractory epilepsy and severe mental retardation. Gray matter heterotopias consist of clusters of normal neurons in abnormal locations (outside the cerebral cortex), that can be subependymal or subcortical, focal or diffuse laminar,[3] and present as islets of gray matter with no signal abnormalities. Epilepsy is usually associated with diffuse and subcortical heterotopias. Polymicrogyria, which result from an abnormal organization of normal neurons within the cortical ribbon (with the loss of the normal 6 layers) during gestation is encountered in a very large variety of diseases, including infections (e.g., CMV), ischemic insults, malformations (e.g., Aicardi syndrome), and genetic diseases. They present on MRI as focal

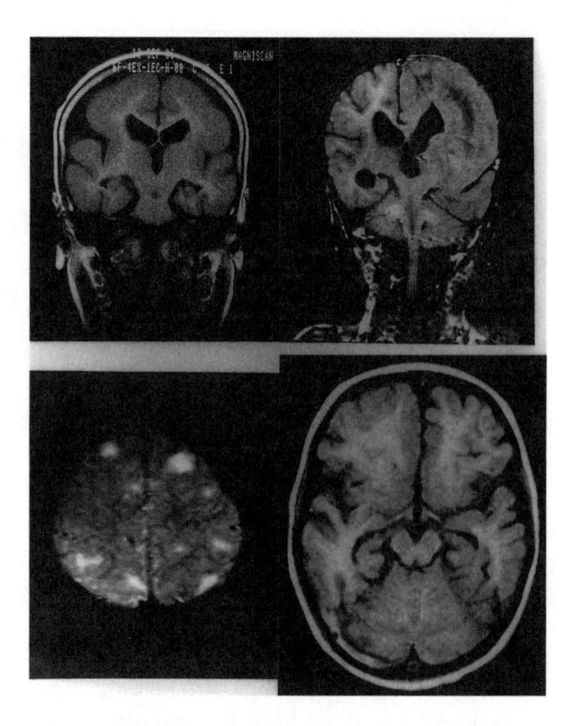

Figure 2. Malformations of cortical development. From top to bottom and from left to right:

- Lissencephaly: Almost absent sulcation and very thick cortex on this coronal T1 weighted image. The signal of both gray and white matter is normal.
- Hemimegalencephaly: Enlargement of the left hemisphere with poor sulcation and thickened cortex associated with ex-tended white matter signal abnormalities. Enlarged lateral ventricle.
- Tuberous sclerosis: Numerous hyperintense cortical and subcortical tubers (T2 weighted image).
- Right frontal cortical dysplasia: Thickened cortex and poor sulcation, with blurred gray-white matter delineation on this T1 weighted image.

or diffuse thickened cortex with a normal signal and small bumps on the surface ("cauliflower"-like), associated with abnormal sulci, and prominent drainage veins. Schizencephaly clefts are usually bordered by polymicrogyric cortex. Cortical dysplasias are a very good example of lesions that have greatly benefited from recent advances in MR studies. Histologically, they are characterized by abnormal organization of cortical layers, with blurred gray-white matter delineation, abnormally large neurons, and giant glial cells. They can be either focal or more diffuse, like in hemimegalencephaly (Fig. 2).[1] At MRI, they present with a poor delineation between the cortical ribbon and the subcortical white matter on high resolution 3D T1 and T2 images (Fig. 2), often associated with T2 high signal of the underlying white matter (abnormal myelination), and thickened cortex. These abnormalities are not sharply defined from the adjacent normal cortex, and they may be difficult to depict in young infants, because of the lack of myelination. They are highly epileptogenic, presenting either as partial or secondarily generalized seizures (like in infantile spasms). They may display hypometabolism or hypoperfusion on PET or SPECT, which can help to draw attention to a particular region at MRI, using surface coils and very high-resolution images. In tuberous sclerosis, MRI appears particularly useful in "formes frustes" and in very young patients, because of the frequent negativity of CT scans in the absence of calcifications. Subependymal nodules can be detected in neonates and even in fetuses,[40] as foci hyperintense on T1 weighted images and hypointense on T2 weighted images. The MRI detection of tubers depends upon myelination stage. In infants, the lesions are usually hyperintense on T1 and hypointense on T2 sequences (unmyelinated white matter). Later on, their contrast is inverted and become hypo or isointense on T1 and hyperintense on T2 images (reduced myelination and gliosis: Fig. 2). The FLAIR sequence may be very helpful by clearly delineating high intensity tubers from surrounding white matter and low signal cortex. Those lesions do not enhance after Gadolinium injection.

4. FOCUS LOCALIZATION

Partial epileptic seizures usually comprise two components which involve different regions of the brain, namely ictal onset and propagation of the epileptic discharge. In refractory cases planned for surgery, removing the ictal onset zone is the absolute condition for favorable outcome. Intracranial EEG recording is the gold standard for localizing the ictal onset zone, because it can describe electric events within milliseconds. But this technique is highly invasive and cannot be easily used in very young children. The challenge of functional neuroimaging is to localize the ictal focus non-invasively.

In adults, the best results are obtained in temporal lobe epilepsy (Table 1). A meat-analysis of published studies shows that ictal SPECT localizes the ictal onset zone in almost all cases (focal hyperperfusion),[16] being superior to interictal FDG-PET (focal hypometabolism) and MRS (focal decrease in NAA) and having now replaced invasive EEG recording in unilateral temporal lobe epilepsy. In bilateral temporal lobe epilepsy and in extratemporal epilepsies, sensitivity is not so good or not yet as established (Table 1), and intracranial EEG remains necessary to localize the ictal onset zone. However, sensitivity can be increased by improving postprocessing analysis, for instance, using a sophisticated statistical analysis (SPM, statistical parametric map) for PET (Table 1) or subtraction images of ictal—interictal and coregistration to MRI for SPECT. Using this procedure, O'Brien *et al.*[35] raised the sensitivity of ictal SPECT from 39% to 88% in bilateral temporal and extratemporal epilepsies.

Table 1. Sensitivity of different techniques for localizing the ictal onset zone in adult epilepsy (meta-analysis)

FDG-PET	TLE	79%
	TLE (spm)	94%
interictal SPECT	TLE	64%
	exTLE	33%
ictal SPECT	TLE	98%
	exTLE	88%
1H-MRS	TLE	93%

FDG-PET = fluoro-deoxyglucose positron emission tomography.
SPECT = single photon emission computed tomography.
1H-MRS = proton magnetic resonance spectroscopy.
TLE = temporal lobe epilepsy.
exTLE = extratemporal epilepsy.
spm = Statistical Parametric Mapping (see text).

In children, the use of these techniques still remains limited, because of particular difficulties related to this population. Most partial epilepsies are extratemporal at this age; seizures are usually shorter than in adults thus decreasing the feasibility of ictal SPECT, and accessibility to both PET and MRS is reduced. Sensitivity of interictal FDG-PET seems to be approximately the same as in adults (about 75%) but the number of cases is small and limited to a few centers. Preliminary reports with MRS showed high sensitivity (more than 90%) in children with temporal lobe epilepsy.[13] The most convincing results relate to ictal SPECT. First reports revealed a hyperperfused focus consistent with the ictal onset zone in more than 95% of the patients[33,36] compared with 25% to 80% using interictal SPECT. More recently, two larger series of respectively 48 and 27 peri-ictal or ictal SPECT emphasized the optimization of the technique by using the subtraction of ictal—interictal images coregistered to MRI (Fig. 3).[35,41] With this procedure, SPECT provided consistent localization of ictal onset zone in respectively 92% and 93% of the cases. Ictal SPECT seems particularly useful for deciding the placement of intracranial electrodes in extratemporal epilepsies in children; it also improves our knowledge of the semiology of partial seizures in pediatrics, particularly in infants.

Other non-invasive techniques are in progress but still limited to research. Neuroreceptor mapping studies in adults have reported interictal abnormalities restricted to the focus, such as decreased benzodiazepine (BZD) receptor density using PET—^{11}C-Flumazenil or increased opioid neuroreceptors using ^{11}C-Carfentanil in temporal lobe epilepsy. BZD mapping seems to be more sensitive than interictal FDG-PET, but loss of binding may not simply be due to neuronal loss in the epileptogenic focus.[27] Original but

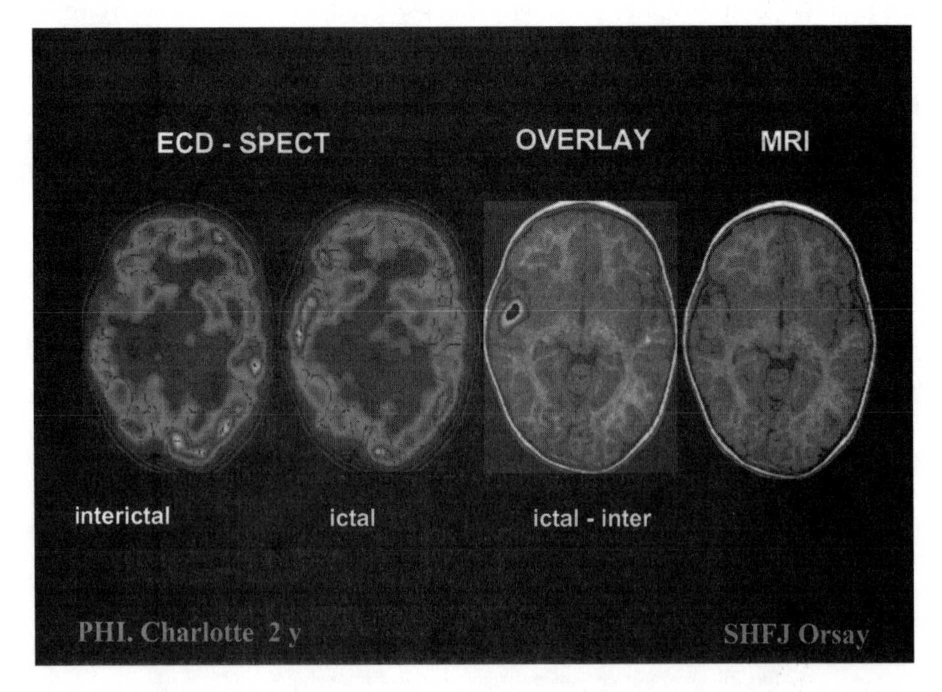

Figure 3. Ictal and interictal SPECT. From left to right, interictal SPECT showing a right temporal hypoperfusion, ictal SPECT showing a right temporal hyperperfusion, subtraction image locating the ictal onset focus, and MRI showing a subtle cortical abnormality in the right temporal pole, in a 2-year-old girl presenting with refractory partial seizures. After right temporal lobectomy limited to the pole, the child is seizure free.

isolated data using ^{11}C-methyl-tryptophane (a marker of serotonine synthesis) in epileptic children with tuberous sclerosis and multifocal lesions showed increased uptake in 5/8 epileptogenic tubers but none in the non-epileptogenic ones.

Using fMRI and BOLD effect, Jackson *et al.* were able to record partial motor seizures in a child with Rasmussen syndrome.[22] Apart from a consistent BOLD signal increase in the contralateral motor cortex, they could assess the temporal spread of the ictal discharge. fMRI is the only neuroimaging technique able to follow the seizure spread as EEG does, because of its high temporal resolution. However, this approach has strong technical and safety limitations such as implementation of EEG in the magnet, discrimination between spikes and artifacts, or synchronization of MR acquisition to the seizure onset. To date, the technique remains not accessible for seizures with head movements.

5. ACTIVATION STUDIES AND PRESURGICAL MAPPING

Non-invasive studies of brain activation can be obtained using PET or fMRI, two techniques that display image perfusion changes related to neuronal activity. This field of research has considerably expanded in the recent years, but most studies have concentrated on healthy adult volunteers, and there are few series published in patients with either technique. This is all the more true in children where activation PET studies are limited to a couple of centers worldwide that have concentrated on plastic reorganization of functions in brain-lesioned patients.[12] The spatial resolution of PET images does not provide sufficiently precise anatomic localization to be used in pre-surgical work-ups.

Functional MRI appears a unique tool for evaluating non-invasively the cortical organization of many functions in children, either sick or healthy. Indeed, fMRI contrast relies on the local and transient increase in blood oxygenation and CBF that occurs during neuronal activation (BOLD, for Blood Oxygenation Level Dependent contrast). This means no exogenous contrast agent must be injected. MRI high spatial resolution allows good anatomical description of activated regions in single subjects by simply overlaying functional results on anatomical MR images acquired in the same session. Because no absolute quantification of hemodynamic changes can be currently obtained using BOLD fMRI, data analysis is made semi-quantitatively by comparing the differential activity between two difference states (called "activation" and "reference" states, the sequence of which constitutes the activation paradigm). So far, all clinical studies have used "block paradigms" in which the patient performs the tasks repeatedly over the stimulation and reference periods (usually 30 to 40 seconds each), and the comparison is made on the sum of all activated and reference periods. To gain statistical significance, cycles of activation and reference blocks are repeated several times in a single trial. Sequential images are acquired at a high speed during the performance of the paradigm (for example, images of the whole brain can be acquired every 2 to 4 seconds over a 4 to 6 mn period). More sophisticated single-event paradigms, which allow monitoring of the brain response during the processing of a single stimulus, have been recently developed in healthy adults[7] and may prove very useful in patients, because they permit to account for inter-stimulus response variability and offer extended possibilities of experimental designs. But their implementation remains difficult in clinical environments due to very large data volumes and non-customized analysis. The most critical aspect of fMRI studies resides in the choice of the activation and reference tasks, since data analysis most often relies on "cognitive subtraction".[37] In this analysis, the resulting activated areas are thought to sustain the cognitive components that are involved in the activated state but NOT in the reference state. For example, the comparison of an auditorily cued semantic decision task and a simple tone discrimination task shows mainly regions involved in semantic processes.[4] On the other hand, the comparison of a less specific language task (such as sentence generation to a given noun) compared to simple "rest" will show a larger functional network that includes participation of receptive and expressive language (phonemic discrimination, phonological encoding, lexical retrieval, semantic analysis, syntax, along with verbal working memory and prearticulatory processing). However, the underlying assumptions of cognitive subtraction may not be fulfilled in all cases because of the non-linearity of many brain processes (interactions may also be studied). In language fMRI studies, most tasks must be performed silently to avoid artifacts due to face movements. A precise control of task performance may be obtained using computer-based paradigms by monitoring responses performed by the patient pressing on joystick buttons. Still, compliance to the tasks in MR environment is a challenging issue in young children. Despite the use of dedicated image registration algorithms, head motion remains another critical limiting factor, especially in uncooperative, young or debilitated patients.

fMRI mapping of the primary motor cortex has been tested in several series of adult and pediatric patients with either tumors or epilepsy foci located in the central region. These studies have shown consistent activation in regions predicted by the electrophysiological data (Penfield's homonculus), when the lesion was relatively remote from the functional areas. However, plastic changes of the cortical organization could also been demonstrated in cases of lesions within the motor cortex in excellent agreement with the results of cortical stimulation in the same patients. These results support the hypothesis of a good localizing power of fMRI.

Concerning language functions, investigators so far have concentrated on the assessment of language hemispheric specialization,[2,4,14] using semantic decision tasks or verbal fluency tasks in limited series of adult patients with epilepsy. Activated areas were located in widespread regions including the inferior frontal gyrus (Broca's area), the dorsolateral prefrontal cortex, the supplementary motor area, the inferior parietal lobule (gyrus supramarginalis, and angular gyrus) and the basal temporal lobe (Fig. 4). Lateralization of language networks was evaluated using laterality indices on activated regions, and proved highly correlated with the results of intracarotid Amytal test (Wada test). Only one similar study has been published in children to date,[19] showing results comparable to adults in young patients aged 8 to 18 years. These preliminary studies suggest that fMRI is a powerful technique for lateralizing language networks and may therefore replace invasive Wada test in the future,[20] although further refinements are needed before the technique enters clinical routine. Memory assessment is also part of Wada testing in adults, and fMRI memory tests are being developed in adult epilepsy patients,[15] but no data have been reported in children to date.

Presurgical cortical mapping of eloquent regions sustaining language functions may be needed when surgery is to be performed in the dominant hemisphere, and usually uses intraoperative cortical stimulation in adults, and subdural grids in children. These techniques, however, carry significant risks and are difficult to perform. fMRI may be used to map non-invasively language areas, and colocalization of fMRI activated regions and significant stimulation sites have been reported within 1 to 2 cm in preliminary reports.[17] However, strict comparison of both techniques remains difficult, since cortical stimulation during naming tasks discloses only limited regions critical to language functions, whereas fMRI does not provide hierarchical information on the numerous activated regions, which may not all be essential to language.[6]

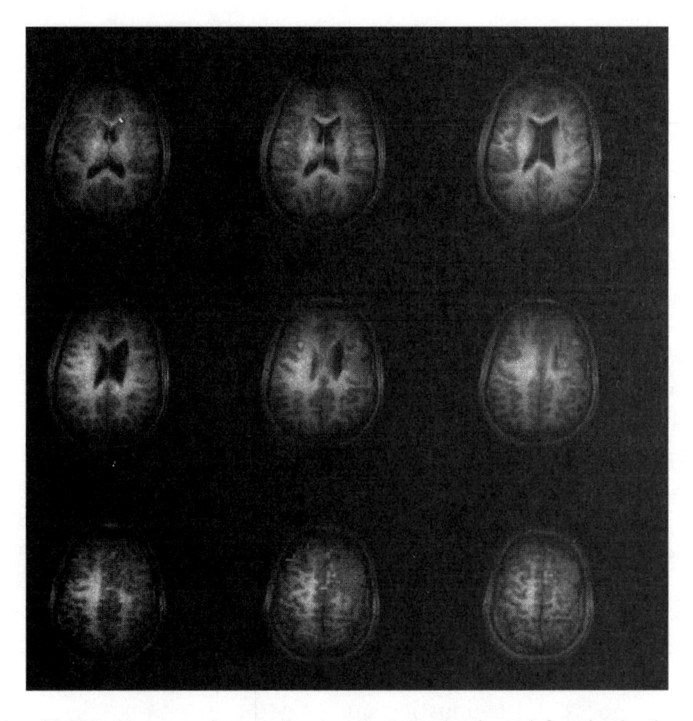

Figure 4. Functional MRI of language in an epilepsy patient. Predominant left activation located in prefrontal cortex and supplementary motor area in a 15-year-old right-handed girl with refractory left frontal epilepsy during a task of silent word generation versus rest. Notice that one activated area is seen within the abnormal cortex.

In conclusion, fMRI is very promising for presurgical mapping, according to the few preliminary studies that have been performed to date. While sophisticated activation paradigms currently developed in normal adult volunteers may provide more specific information on the role of the various activated regions within language functional networks, the development of comparable studies in adult and pediatric patients, combined with extensive pre- and postoperative neuropsychological testing, will undoubtedly lead to a better understanding of high level function disorders that accompany most severe epilepsies, and improve their clinical management.

6. NEUROIMAGING AND NEUROPSYCHOLOGICAL DISORDERS

The correlations between neuropsychological deficits in epilepsy and functional imaging have been unfortunately much less explored than the localization of the ictal onset zone. Functional imaging techniques could, however, help to evaluate the cognitive consequences of the propagation of the discharges. Correspondences between selective cognitive deficits and neuroimaging focal abnormalities have been more studied in stroke adult patients and in childhood developmental disorders than in epilepsy. The best correspondences between cognitive sequelae of epilepsy and functional neuroimaging have been found using interictal images. The few studies developed in childhood epilepsy will be described in the different chapters of this book.

To summarize, using SPECT, a frontal syndrome has been found associated with unilateral frontal hypoperfusion in cryptogenic frontal epilepsy,[23] visual agnosia with occipital hypoperfusion in lesional occipital epilepsy[26] and transcortical aphasia with hypoperfusion of Broca's area in cryptogenic temporo-frontal epilepsy.[25] The extent of hypoperfusion may regress several months after seizure control while neuropsychological skills improve.

In cryptogenic West syndrome, patients develop normally until they exhibit seizures and then experience a deterioration of neurologic functions, which most frequently involves visual agnosia and indifference to interpersonal contact. This corresponds to a relative rCBF decrease in posterior cortex associated with abnormal interictal CBF increase in the whole brain.[24] Later on, the type of mental sequelae corresponds well to the topography of hypoperfused areas; patients with predominant verbal disorders have temporal rCBF defects whereas those with predominant non-verbal disorders have parieto-temporo-occipital rCBF defects and those with autistic sequelae also exhibit frontal defects.[24]

In Continuous Spike Waves during Sleep (CSWS), several PET and SPECT studies have been conducted, all showing focal abnormalities although EEG features were generalized. Most of them found symmetrical hypermetabolism or hyperperfusion in the thalamus associated with similar but asymmetrical abnormalities of higher level in cortical areas corresponding well with the predominant cognitive pattern: usually in temporal lobe in the most frequent form, the Landau-Kleffner syndrome with auditory agnosia,[18] but also in frontal regions in patients with predominant behavioral disorders or in parietal areas in patients with dyspraxic troubles.[32]

Non-invasive activation studies in epileptic patients are still limited to presurgical mapping but should be developed to assess neuropsychological deficits in the future, particularly using fMRI. Another promising approach consists in studying the cognitive functionality of an epileptic lesion as well as the surrounding brain. Using visual and frontal activation tasks with H_2O^{15}-PET in patients with cortical developmental malformations involving respectively the occipital and frontal cortices, Richardson *et al.*[38] showed that malformed cortex may participate in normal cognitive functions but with a widespread atypical organization of the surrounding normal cortex. Finally, mapping endogenous neuromediators with PET could shed light on the neurobiological basis of seizures and also of attention, a psychological component very frequently impaired in epileptic children. Endogenous opioids are released in the left parieto-temporo-occipital cortex during reading-induced seizures.[28] A relationship between performance level and endogenous dopamine release has recently been demonstrated in striatum during a goal-directed motor task using a video game.[29]

REFERENCES

1. Adamsbaum C, Robain O, Cohen P, Delalande O, Fohlen M and Kalifa G (1998): Focal cortical dy plasia and hemimegalencephaly: histological and neuroimaging correlations. Pediatric Radiology 28:583–590.
2. Bahn M, Lin W, Silbergeld D *et al.* (1997): Localization of language cortices by functional MR imaging compared with intracarotid Amobarbital hemispheric sedation. American Journal of Roentgenology 169:575–579.
3. Barkovich A and Kjos B. (1992): Gray matter heterotopias: MR characteristics and correlation with developmental and neurological manifestations. Radiology 182:493–499.
4. Binder J, Rao S, Hammecke T *et al.* (1995): Lateralized human brain language systems demonstrated by task subtraction functional magnetic resonance imaging. Archives of Neurology 52:593–601.

5. Binder J, Swanson S, Hammecke T *et al.* (1996): Determination of language dominance with functional MRI: A comparison with the Wada test. Neurology 46:978–984.
6. Bookheimer S, Zeffiro T, Blaxton T *et al.* (1997): A direct comparison of PET activation and electrocortical stimulation mapping for language localization. Neurology 48:1056–1065.
7. Buckner R, Bandettini P, O'Craven K *et al.* (1996): Detection of cortical activation during averaged single trials of a cognitive task using functional magnetic resonance imaging. Proceedings of the National Academy of Sciences USA 93:14878–14883.
8. Cascino G, Jack C, Parisi J *et al.* (1991): Magnetic Resonance imaging-based volume studies in temporal lobe epilepsy: pathological correlations. Annals of Neurology 30:31–36.
9. Casey BJ, Cohen JD, Jezzard P, Turner R, Noll DC, Trainor RJ, Giedd J, Kaysen D, Hertz-Pannier L and Rapoport JL (1995): Activation of prefrontal cortex in children during a nonspatial working memory task with functional MRI. Neuroimage 2:221–229.
10. Cendes F, Andermann F, Dubeau F *et al.* (1993): Early childhood prolonged febrile convulsions, atrophy and sclerosis of mesial structures, and temporal lobe epilepsy: an MRI volumetric study. Neurology 43:1083–1087.
11. Chiron C, Raynaud C, Maziere B *et al.* (1992): Changes in regional cerebral blood flow during brain maturation in children and adolescents. Journal of Nuclear Medicine 33:696–703.
12. Chugani H, Müller R and Chugani D (1996): Functional brain reorganization in children. Brain Development 18:347–356.
13. Cross JH, Conelly A, Jackson GD, Johnson CL, Neville BGR and Gadian DG (1996): Proton magnetic resonance spectroscopy in children with temporal lobe epilepsy. Annals of Neurology 39:107–113.
14. Desmond J, Sum J, Wagner A *et al.* (1995): Functional MRI measurement of language lateralization in Wada-tested patients. Brain 118:1411–1419.
15. Detre J, Maccotta L, King D *et al.* (1998): Functional MRI lateralization of memory in temporal lobe epilepsy. Neurology 50:926–932.
16. Devous MD, Thisted RA, Morgan GF, Leroy RF and Rowe CC (1998): SPECT brain imaging in epilepsy: a meta-analysis. Journal of Nuclear Medicine 39:285–293.
17. Fitzgerald D, Cosgrove G, Ronner S *et al.* (1997): Location of language in the cortex: a comparison between functional MR Imaging and Electrocortical stimulation. American Journal of Neuroradiology 18:1529–1539.
18. Gaggero R, Caputo M, Fioro P, Pessagno A, Baglietto MG, Muttini P and De Negri M (1995): SPECT and epilepsy with continuous spike waves during slow-wave sleep. Childs Nervous System 11:154–160.
19. Hertz-Pannier L, Gaillard W, Mott S *et al.* (1997): Non invasive assessment of language dominance in children and adolescents with functional MRI: a preliminary study. Neurology 48:1003–1012.
20. Hertz-Pannier L, Chiron C, Van de Moortele P, Dulac O, Brunelle F and Le Bihan D (1998): 3 Tesla fMRI study of language dominance in children with epilepsy: is Wada test always the gold standard? Neuroimage 7:S459 (Abstract).
21. Jack C, Bentley M, Twomey C and Zinsmeister A (1990): MR imaging-based volume measurements of the hippocampal formation and anterior temporal lobe: validation studies. Radiology 176:205–209.
22. Jackson GD, Connelly A, Cross JH, Gordon I and Gadian DG (1994): Functional magnetic resonance imaging of focal seizures. Neurology 44:850–856.
23. Jambaqué I and Dulac O (1989): Syndrome frontal réversible et épilepsie frontale chez un enfant de 8 ans. Archives Françaises de Pédiatrie 46:525–529.
24. Jambaqué I, Chiron C, Dulac O, Raynaud C and Syrota A (1993): Visual inattention in West syndrome: a neuropsychological and neurofunctional imaging study. Epilepsia 34:692–700.
25. Jambaqué I, Chiron C, Kaminska A, Plouin P and Dulac O (1998): Transient motor aphasia and recurrent partial seizures in a child: language recovery upon seizure control. Journal of Child Neurology 13:296–300.
26. Jambaqué I, Mottron L, Ponsot G and Chiron C (1998): Autism and visual agnosia in a child with right occipital lobectomy. Journal of Neurology, Neurosurgery and Psychiatry 65:555–560.
27. Koepp MJ, Richardson MP, Labbe C, Cunningham VJ, Ashburner J, Van Paesschen W, Revesz T, Brooks DJ and Duncan JS (1997): ^{11}C-flumazenil PET, volumetric MRI and quantitative pathology in mesial temporal lobe epilepsy. Neurology 49:764–773.
28. Koepp MJ, Richardson MP, Brooks DJ and Duncan JS (1998a): Focal release of endogeneous opioids during reading-induced seizures. Lancet 19:952–955.
29. Koepp MJ, Gunn RN, Lawrence AD, Cunningham VJ, Dagher A, Jones T, Brooks DJ, Bench CJ and Grasby PM (1998b): Evidence for striatal dopamine release during a video game. Nature 393:266–268.
30. Kuzniecky R and Jackson G (1995): Temporal lobe epilepsy. In Kuzniecky RI, Jackson GD (eds): "Magnetic Resonance in Epilepsy." New York: Raven Press, pp 107–182.

31. Lencz T, McCarthy G, Bronen R *et al.* (1992): Quantitative magnetic resonance imaging in temporal lobe epilepsy: relationship to neuropathology and neuropsychological function. Annals of Neurology 31:629–637.
32. Maquet P, Hirsch E, Metz-Lutz MN, Motte J, Dive D, Marescaux C and Franck G (1995): Regional cerebral glucose metabolism in children with deterioration of one or more cognitive functions and continuous spike and waves discharges during sleep. Brain 118:1497–1520.
33. Menzel C, Steidele S, Grunwald F *et al.* (1996): Evaluation of technetium-99m-ECD in childhood epilepsy. Journal of Nuclear Medicine 37:1106–1112.
34. O'Brien TJ, Zupanc ML, Mullan BP, O'Connor MK, Brinkmann BH, Cicora KM and So EL (1998a): The practical utility of performing peri-ictal SPECT in the evaluation of children with partial epilepsy. Pediatric Neurology 19:15–22.
35. O'Brien TJ, So EL, Mullan BP *et al.* (1998b): Subtraction ictal SPECT co-registered to MRI improves clinical usefulness of SPECT in localizing the surgical seizure focus. Neurology 50:445–454.
36. Packard AB, Roach PJ, Davis RT *et al.* (1996): Ictal and interictal technetium-99m-bicisate brain SPECT in children with refractory epilepsy. Journal of Nuclear Medicine 37:1101–1106.
37. Petersen S, Fox P, Posner M, Mintun M and Raichle M (1988): Positron Emission Tomographic studies of the cortical anatomy of single-word processing. Nature 331:585–589.
38. Richardson MP, Koepp MJ, Brooks DJ, Coull JT, Grasby P, Fish DR and Duncan JS (1998): Cerebral activation in malformations of cortical development. Brain 121:1295–1304.
39. Sass K, Spencer D, Kim J *et al.* (1990): Verbal memory impairment correlates with hippocampal pyramidal cell density. Neurology 40:1694–1697.
40. Sonigo P, Elmaleh A, Fermont L, Delezoide A, Mirlesse V and Brunelle F (1996): Prenatal MRI diagnosis of fetal cerebral tuberous sclerosis. Pediatric Radiology 26:1–4.
41. Véra P, Kaminska A, Cieuta C, Hollo A, Stievenart JL, Gardin I, Ville D, Mangin JF, Plouin P, Dulac O and Chiron C (1999): Optimizing the localization of seizure foci in children using subtraction ictal SPECT co-registered to MRI. Journal of Nuclear Medicine 40:786–792.

3

LONG TERM EFFECTS OF STATUS EPILEPTICUS IN THE DEVELOPING BRAIN

João Pereira Leite[1] and Esper A. Cavalheiro*[,2]

[1]Departamento de Neurologia
Psiquiatria e Psicologia
Faculdade de Medicina de Ribeirão Preto-USP
Av Bandeirantes 3900, Campus Universitário
14049-900 Ribeirão Preto, SP, Brazil
[2]Neurologia Experimental, UNIFESP-EPM
Rua Botucatu, 862, 04023-900 São Paulo, SP
Brazil

INTRODUCTION

Status epilepticus (SE) is a state of continuous or repeated epileptic seizures without recovery of consciousness between attacks.[17,19] Although the minimum duration necessary to establish a diagnosis of SE has varied considerably, most recent works have accepted seizures lasting longer than 30 minutes.[2,41,42] It represents a major neurological emergency that requires aggressive seizure control because numerous clinical series and experimental studies demonstrate that morbidity and mortality are correlated to the duration time before seizures are brought under control.[3,29,39]

SE occurs mainly in young children, particularly in the first years of life,[1,2,12,28,31] a period in which SE and other epileptic events can be more detrimental to further brain development.[43] Neurologic sequelae of SE reported in the literature include intellectual impairment, brain damage-related deficits and long-term development of recurrent seizures.[19,24] The neurologic morbidity and mortality from convulsive SE seem to be related to three distinct factors: 1) adverse systemic metabolic and physiologic effects of repeated generalized tonic-clonic seizures; 2) brain damage caused by the underlying illness or acute insult that may induce SE; 3) direct excitotoxic damage from repetitive electrical discharges of the seizures themselves.[29,23]

There are some controversies regarding the neurologic outcome after an episode of SE and part of the conflicting data is due to the different ways of sampling study groups.

* To whom correspondence should be addressed

Neuropsychology of Childhood Epilepsy, edited by Jambaqué et al.
Kluwer Academic / Plenum Publishers, New York, 2001.

Population based studies usually give a more optimistic view of outcome while clinical series based on patients who attend specialized clinics or hospitals tend to have more severely affected cases and a worse scenario.[2,41,42]

Even when comparing series based only on selected groups of patients attended in specialized hospitals, data may be contradictory if they rely on retrospective or prospective data. The earlier pediatric studies of Aicardi and Chevrie[1,2] have emphasized the high incidence of neurological sequelae after SE. In these studies, where only convulsive status lasting more than one hour were included, death occurred in 11% (in half of the cases death was attributed to the seizures themselves), 37% developed permanent neurological deficits (mainly motor), 48% were mentally retarded, and 20% of the neurological deficits were considered to be the direct result of SE. The authors concluded that the immature brain is more prone to SE-induced sequelae.

In a more recent study, based less heavily on retrospective data and using 30 minutes as the inclusion criteria for SE, Maytal *et al.* found a lower incidence of neurological deficits in the 193 children studied.[28] Mortality was around 4% and new neurological deficits were found in 17 (9%) of the 186 survivors. Although the authors have shown that neurological sequelae were higher if SE occurred at a younger age, they indicate that the outcome seems to be determined by the underlying neurological problem rather than the SE itself.

The causal relation linking SE and the further development of epilepsy seems to be related to the etiology of SE and has serious implications for the subsequent treatment of these patients.[16] Using a simplistic classification, taking together studies from both adults and children samples, idiopathic SE does not seem to increase the risk to epilepsy[36,37] while symptomatic SE may facilitate the development of seizure disorders at some point in the future.[4,5,6,17,20,30] Retrospective studies in patients with partial epilepsy subjected to surgery for temporal lobe epilepsy showed that approximately 50% of the cases reported a history of "severe" or "prolonged" infantile convulsions.[13,14,26] In a recent study in which only medial temporal lobe epileptics were evaluated (mass or tumor lesions excluded), 81% of patients had histories of convulsions during early childhood or infancy. Complicated febrile seizures comprised 94% of the patients in whom detailed descriptions of the febrile seizures were available.[15] On the other hand, prospective studies have shown that the risk for the development of epilepsy is similar for those presenting SE or an isolated seizure as the first unprovoked ictal event.[17,18,35,36,38] Once again, the etiology of SE seems to be determinant and substantial evidence indicates that febrile or symptomatic SE are those with the highest risk to later induce epilepsy.[10,27]

A common problem for all clinical studies in addressing the issue of whether neurological sequelae and chronic epilepsy can be accounted for by a previous episode of SE is the uncertainty of prior unrecognized brain abnormality that preceded both the SE and the presumed SE-induced neurological deficit.[24] In temporal lobe epilepsy, for example, it has been suggested that mesial temporal sclerosis may represent a sequel of disturbed embryogenesis.[33] Another limitation of human clinical-pathologic studies of refractory epileptic patients is the fact that surgically resected specimens always depict a frozen picture of an "end-point" process of epileptogenesis. For these reasons, experimental preparations are particularly useful because they allow the study of SE-induced epileptogenesis and neurological sequelae starting from a "non-pathologic" brain and the course of neuropathological changes can be assessed more dynamically.

1. AN ANIMAL MODEL OF CHRONIC EPILEPSY

During the last years we have developed an experimental model of chronic epilepsy using the cholinergic muscarinic agonist pilocarpine as the convulsant agent and that has been considered a good model for temporal lobe epilepsy in humans.[7] Acute effects of pilocarpine administration to adult rodents are characterized by long lasting limbic SE associated with sustained electrographic discharges in limbic structures. After a single application of pilocarpine, SE can last from 6 to 12 hours. After spontaneous remission of SE, animals are comatose and both hippocampal and cortical recordings are depressed with high-voltage spiking activity.[40] These events have been called the "acute period" of this model and metabolic studies performed during this period have revealed increased glucose utilization mainly in the hippocampus and other limbic structures, thalamus and substantia nigra.[34] The pattern of neuronal loss observed in these animals matches closely with the areas that are metabolically activated during SE. Animals surviving the acute period of SE proceed to a latent "seizure-free" period with an apparently normal behavior except for some aggressiveness upon manipulation and towards other animals if they are maintained in groups.[8] This period lasts 1–8 weeks depending on the animal strain and finishes with the occurrence of the first spontaneous seizure. The recurrence of spontaneous seizures during the "chronic period" ranges from 2–5 seizures per week and is associated with profound memory impairment.[8,21,22]

In accordance with these observations, we performed a study in order to investigate whether pilocarpine-induced SE in developing rats would lead to the appearance of spontaneous seizures later in life.

2. ACUTE PERIOD IN DEVELOPING RATS

The susceptibility of rats to pilocarpine-induced seizures showed to be clearly age-dependent.[9] Younger rats (post-natal day [PN] 3–9) resisted to higher doses of pilocarpine when compared to adult rats. This higher threshold to the pilocarpine effects began to decrease in 10–14 day-old rats reaching a completely inverse situation in 15–25 day-old animals. This increased susceptibility to the convulsant action of pilocarpine was characterized by a shortened latency for behavioral and electrographic signs, and increased severity of seizures and lethal toxicity relative to older and younger animals. The susceptibility to pilocarpine gradually decreased with age and reached the mature level in 35–45 day-old rats (Fig. 1).

Pilocarpine administered to developing rats induced a characteristic array of behavioral patterns. Hyper- or hypoactivity, tremor, loss of postural control, scratching, head bobbing and myoclonic movements of the limbs dominated the behavior in 3–9 day-old rats. No overt motor seizures were observed in this age group. More intense behavioral signs evolving to limbic seizures and status epilepticus occurred when pilocarpine was administered in 10–12 day-old rats. The electrographic activity in these animals progressed from low voltage spiking registered concurrently in the hippocampus and cortex during the first week of life into localized epileptic discharges in the hippocampus which spread to cortical leads during the second week of life. No morphological alterations were observed in the brains of 3–12 day-old rats subjected to the action of pilocarpine.

The adult pattern of behavioral and electrographic changes after pilocarpine was encountered in 15–21 day-old rats. Akinesia, tremor and head bobbing progressed to

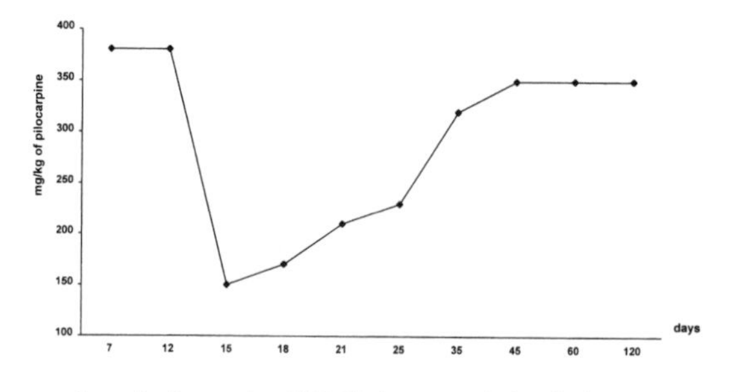

Figure 1. Graphic representation of pilocarpine (PILO) doses needed to induce status epilepticus in rats according to age. Notice the increased susceptibility to pilocarpine in 15–25 day-old rats.

motor limbic seizures and status epilepticus. High voltage fast activity superposed over hippocampal theta rhythm progressed into high voltage spiking and spread to cortical records. The electrographic activity became well synchronized and then developed into seizures and status epilepticus. Morphological analysis of frontal forebrain sections in 15–21 day-old rats which had status epilepticus after pilocarpine revealed an attenuated pattern of damage in the hippocampus, amygdala, olfactory cortex, neocortex and certain thalamic nuclei. An adult pattern of the damage to the brain, in terms of extent and topography, was present in 4–5 week-old rats.

More recently we have observed that, in contrast to adult rats, developing animals aged 7–17 days presented an evident hypothermia following pilocarpine administration that could last up to 5 hours (Fig. 2). When 7–11 day-old rats were maintained in a heating box at 35°C following pilocarpine administration and during the whole period of status epilepticus, their body temperature was maintained in the same level as that of non-treated rats (Fig. 3a). In rats with 12–17 days of age, in the same heating box, an increase in body temperature during pilocarpine-induced status epilepticus was observed, although this increase could not be considered as a sign of hyperthermia (Fig. 3b). A characteristic hyperthermia similar to that observed in adult rats during pilocarpine-induced status epilepticus was observed in rats older than 24 days of age even if they are maintained at room temperature (Fig. 3c).

Developing animals (7–24 days) that were maintained at 35°C following pilocarpine administration presented more severe status epilepticus with longer duration and

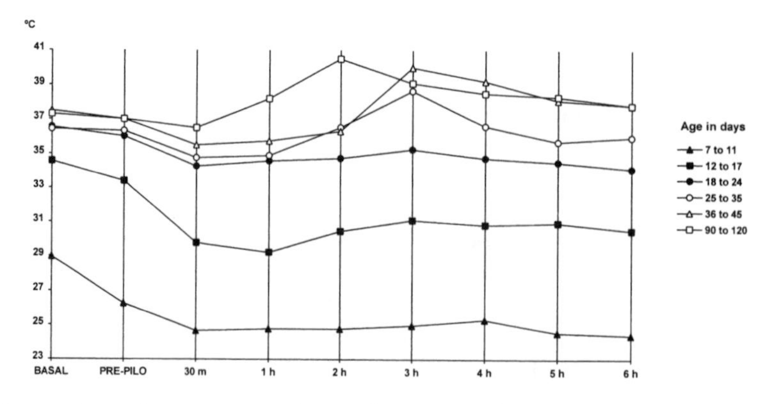

Figure 2. Changes in rectal temperature during the development of pilocarpine-induced status epilepticus in rats according to age. The following is valid for this and Fig. 3: Basal temperature was taken before the sc administration of scopolamine methylnitrate (1 mg/kg). Pre-PILO temperature was measured immediately before the ip injection of pilocarpine.

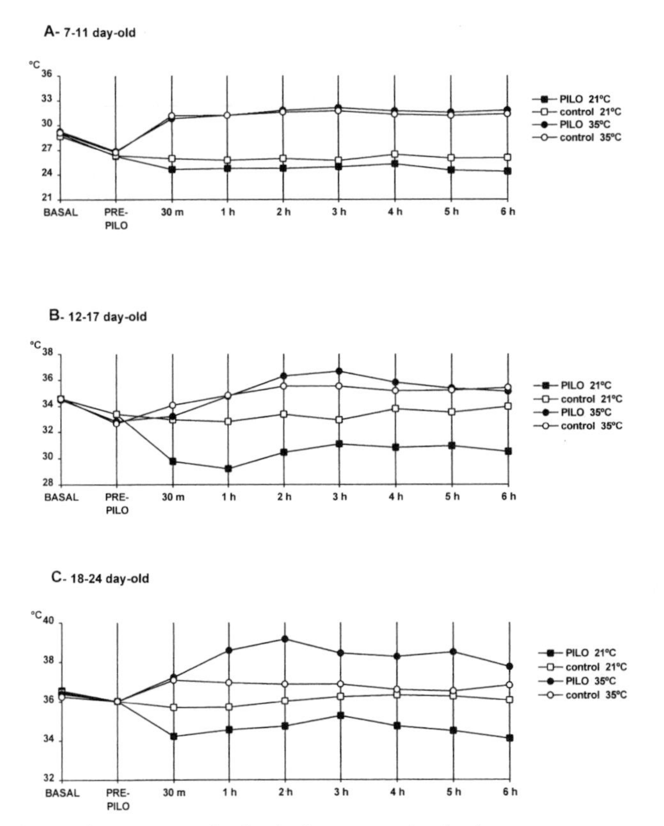

Figure 3. Changes in rectal temperature in developing rats maintained at room temperature (21°C) or in a heated box (35°C) after systemic pilocarpine or saline administration.

increased mortality rate when compared with age-matched rats maintained at room temperature.

3. CHRONIC PERIOD IN DEVELOPING RATS

Chronic seizures following pilocarpine-induced status epilepticus could be observed if the status was induced after the 18th day of life (Table 1), even if some degree of brain

Table 1. Frequency of spontaneous recurrent seizures during 4 months of observation in rats subjected to pilocarpine-induced status epilepticus at different periods of postnatal development

Age (days)	1st month	2nd month	3rd month	4th month
7–11	—	—	—	—
12–17	—	—	—	—
18–24	0.5 ± 0.5	1.6 ± 0.7	1.8 ± 0.8	2.5 ± 1.3
25–35	1.3 ± 0.8	2.6 ± 1.3	3.5 ± 1.9	3.6 ± 2.0
36–45	2.7 ± 2.1	4.6 ± 3.1	7.3 ± 5.2	7.7 ± 4.1
50–60	4.5 ± 3.8	8.7 ± 4.1	9.2 ± 3.8	9.1 ± 4.2
90–120	6.6 ± 4.5	10.5 ± 8.3	12.3 ± 5.8	10.2 ± 6.2

Data expressed as mean ± SD.

Table 2. Number of rats that survived pilocarpine-induced status epilepticus over the number of injected animals, number of animals that developed spontaneous recurrent seizures later in life and the latency (in days) for the appearance of the first spontaneous seizure according to age

Age (days)	Surviving animals/total	Animals with SRS	Latency for the first seizure (days)*
7–11	29/35	00	00
12–17	61/70	00	00
18–24	36/95	08	36.5 ± 24.8
25–35	26/66	14	23.2 ± 10.3
36–45	42/74	39	19.1 ± 7.4
50–60	45/62	45	17.8 ± 6.4
90–120	32/41	32	14.2 ± 4.6

*Data expressed as mean ± SD.

damage could already be detected when status epilepticus was induced in earlier stages of life. In contrast to adult rats, the latency (silent period) for the appearance of the first spontaneous seizure was longer and seizure frequency in the chronic period was smaller in 18–24 day-old rats (Table 2). In addition, hilar cell loss and density of Timm staining was less prominent in these animals suggesting a positive association between mossy fiber sprouting, de novo recurrent excitation of granule cells[11] and the development of spontaneous seizures in this epilepsy model.[32]

Although animals younger than 18 days of life did not develop spontaneous seizures as a consequence of pilocarpine-induced status epilepticus they presented electrographic and behavioral alterations similar to those observed in rats with genetic absence epilepsy[25] when studied 90–120 days later. When the youngest group was submitted to three consecutive episodes of status epilepticus at PN days 7, 8 and 9, electrographic and behavioral changes observed in the adult life were much more severe, including episodes of continuous complex spiking activity, and a small percentage (10–15%) of these rats presented spontaneous seizures. The examination of brains of these animals did not reveal any major pathological signs.

The age-related differences in the susceptibility of young animals to develop chronic epilepsy following pilocarpine-induced status epilepticus reflect the complexity of seizure activity in the immature brain and provide evidence for an apparent distinction between the epileptogenesis in the mature and developing nervous systems. Our observations show that developing rats submitted to pilocarpine-induced status epilepticus are able to develop typical manifestations of partial limbic epilepsy with secondary generalization later in life if status epilepticus occurs during or after the 3rd week of life and is followed by morphological changes such as neuronal loss mainly in limbic structures and dentate mossy fiber sprouting. On the other hand, although pilocarpine-induced status epilepticus does not induce neuropathological changes in younger rats, it may lead to important plastic changes in critical periods of brain maturation that can become apparent through other types of epileptic features. The functional state of several pathways involving the hippocampal and thalamic formations, substantia nigra, etc., key structures in the generation and spread of epileptic activity, should be taken into account in the interpretation of these age-related discrepancies.

ACKNOWLEDGMENTS

The authors wish to thank M. R. Priel, N. F. Santos, B. L. C. Ferreira, C. Timo-Iaria and L. Turski for their participation in the experiments reported in this article. This work has been supported by research grants from FAPESP, CNPq, FINEP and PRONEX (Brazil).

REFERENCES

1. Aicardi J and Chevrie JJ (1970): Convulsive status epilepticus in infants and children: a study of 239 cases. Epilepsia 11:187–197.
2. Aicardi J and Chevrie JJ (1983): Consequences of status epilepticus in inafnts and children. In Delgado-Escueta AV, Wasterlain CG, Treiman DW and Proter RJ (eds): "Status Epilepticus". New York: Raven Press, pp 115–125.
3. Aminoff MJ and Simon RP (1980): Status epilepticus. Causes, clinical features and consequences in 98 patients. American Journal of Medicine 69:657–666.
4. Annegers JF, Hauser WA, Sirts SB and Kurland LT (1987): Factors prognostic of unprovoked seizures after febrile convulsions. New England Journal of Medicine 316:493–498.
5. Annegers JF, Blakely SA, Hauser WA and Kurland LT (1990): Recurrence of febrile convulsions in a population-based cohort. Epilepsy Research 5:209–216.
6. Berg AT and Shinnar S (1991): The risk of seizure recurrence following a first unprovoked seizure: a quantitative review. Neurology 41:965–972.
7. Cavalheiro EA (1995): The pilocarpine model of epilepsy. Italian Journal of Neurological Science 16:33–37.
8. Cavalheiro EA, Leite JP, Bortolotto ZA, Turski WA, Ikonomidou C and Turski L (1991): Long-term effects of pilocarpine in rats: structural damage of the brain triggers kindling and spontaneous recurrent seizures. Epilepsia 32:778–782.
9. Cavalheiro EA, Silva DF, Turski WA, Calderazzo-Filho LS, Bortolotto ZA and Turski L (1987): The susceptibility of rats to pilocarpine-induced seizures is age-dependent. Developmental Brain Research 37:43–58.
10. Cendes F, Andermann F, Dubeau F, Gloor P, Evans A, Jones-Gotman M, Olivier A, Andermann E, Robitaille Y, Lopes-Cendes I, Peters T and Melanson D (1993): Early childhood prolonged febrile convulsions, atrophy and sclerosis of mesial structures, and temporal lobe epilepsy. Neurology 43:1083–1087.
11. Cronin J, Obenaus A, Houser CR and Dudek FE (1992): Electrophysiology of dentate granule cells after kainate-induced synaptic reorganization of mossy fiber. Brain Research 573:305–310.
12. Dunn DW (1988): Status epilepticus in children: etiology, clinical features and outcome. Journal of Child Neurology 3:167–173.
13. Falconer MA (1971): Genetic and related aetiological factors in temporal lobe epilepsy. A review. Epilepsia 12:13–21.
14. Falconer MA, Serafetinides EA and Corsellis JAN (1964): Etiology and pathogenesis of temporal lobe epilepsy. Archives of Neurology 10:233–248.
15. French JA, Williamson PD, Thadani VM, Darcey TM, Mattson RH, Spencer SS and Spencer DD (1993): Characteristics of medial temporal epilepsy: I. Results of history and physical examination. Annals of Neurology 34:774–780.
16. Gross-Tsur V and Shinnar S (1993): Convulsive status epilepticus in children. Epilepsia 34:S12–S20.
17. Hauser WA (1982): Status epilepticus: epidemiological considerations. Neurology 40:13–22.
18. Hauser WA, Anderson VE and Lowenson RB (1982): Seizure recurrence after a first unprovoked seizure. New England Journal of Medicine 307:522–528.
19. Hauser WA (1983): Status epilepticus: frequency, etiology, and neurological sequelae. In Delgado-Escueta AV, Wasterlain CG, Treiman DG and Porter RJ (eds): "Status Epilepticus". New York: Raven Press, pp 3–14.
20. Hauser WA and Kurland LT (1975): The epidemiology of epilepsy in Rochester, Minnesota, 1935 through 1967. Epilepsia 16:1–66.

21. Leite JP, Bortolotto ZA and Cavalheiro EA (1990): Spontaneous recurrent seizures in rats: An experimental model of partial epilepsy. Neuroscience and Biobehavioral Reviews 14:511–517.
22. Leite JP, Nakamura EM, Lemos T, Masur J and Cavalheiro EA (1990): Learning impairment in chronic epileptic rats following pilocarpine-induced status epilepticus. Brazilian Journal of Medical and Biological Research 23:681–683.
23. Leppik IE (1993): Status epilepticus. In Wyllie E (ed): "The Treatment of Epilepsy: Principles and Practice." Philadelphia: Lea & Febiger, pp 678–685.
24. Lothman EW and Bertram EH (1993): Epileptogenic effects of status epilepticus. Epilepsia 34:S59–S70.
25. Marescaux C and Vergnes M (1995): Genetic absence epilepsy in rats from Strasbourg (GAERS). Italian Journal of Neurological Science 16:113–118.
26. Margerison JH and Corsellis JAN (1966): Epilepsy and the temporal lobes: a clinical, electroencephalographic and neuropathological study of the brain in epilepsy with particular reference to the temporal lobes. Brain 89:499–530.
27. Mathern GW, Pretorius JK and Babb TL (1995): Influence of the type of initial precipitating injury and at what age it occurs on course and outcome in patients with temporal lobe seizures. Journal of Neurosurgery 82:220–227.
28. Maytal J, Shinnar S, Moshé SL and Alvarez LA (1989): Low morbidity and mortality of status epilepticus in children. Pediatrics 83:323–331.
29. Meldrum B, Vigouroux R and Brierley J (1973): Systemic factors and epileptic brain damage. Archives of Neurology 29:82–87.
30. Nelson KB and Ellenberg JH (1978): Prognosis in children with febrile seizures. Pediatrics 61:720–727.
31. Phillips SA and Shanahan RJ (1989): Etiology and mortality of status epilepticus in children. A recent update. Archives of Neurology 46:74–76.
32. Priel MR, Santos NF and Cavalheiro EA (1996): Developmental aspects of the pilocarpine model of epilepsy. Epilepsy Research 26:115–121.
33. Scheibel AB (1991): Are complex partial seizures a sequela of temporal lobe dysgenesis? Advances in Neurology 55:59–77.
34. Scorza FA and Cavalheiro EA (1996): Metabolic study in the pilocarpine model of epilepsy. In Delgado-Escueta AV, Wilson W, Olsen RW and Porter RJ (eds): "Basic Mechanisms of the Epilepsies III", p 568.
35. Shinnar S, Berg AT and Moshé SL (1988): Recurrence risk after a first unrpovoked seizure in childhood. Annals of Neurology 24:315–317.
36. Shinnar S, Berg AT and Moshé SL (1990): The risk of seizure recurrence following a first unrpovoked seizure in childhood: a prospective study. Pediatrics 85:1076–1085.
37. Shinnar S, Berg AT, Ptachewich Y and Alemany M (1993): Sleep state and the risk of seizure recurrence following a first unprovoked seizure in childhood. Neurology 43:701–706.
38. Shinnar S, Berg AT, Moshé SL, O'Dell C, Alemany M, Newstein D, Kang H, Goldensohn ES and Hauser WA (1996): The risk of seizure recurrence after a first unprovoked afebrile seizure in childhood: an extended follow-up. Pediatrics 98:216–225.
39. Treiman D (1983): General principles of treatment: Responsive and intractable status epilepticus in adults. In Delgado-Escueta AV, Wasterlain CG, Treiman DG and Porter RJ (eds): "Status Epilepticus." New York: Raven Press, pp 377–384.
40. Turski WA, Cavalheiro EA, Schwarz M, Czuczwar SJ, Kleinrok Z and Turski L (1983): Limbic seizures produced by pilocarpine in rats: behavioral, electroencephalographic and neuropathological study. Behavioural Brain Research 9:315–335.
41. Verity CM (1998): Do seizures damage the brain? The epidemiological evidence. Archives of Disables Childhood 78:78–84.
42. Verity CM and Ross EM, Golding J (1993): Outcome of childhood status epilepticus and lengthy febrile convulsions: findings of national cohort study. British Medical Journal 307:225–228.
43. Wasterlain CG and Shirasaka Y (1994): Seizures, brain damage and brain development. Brain Development 16:279–285.

4

EFFECTS OF FOCAL LESIONS, EPILEPSY, AND ANTIEPILEPTIC DRUGS ON MEMORY FUNCTIONS

Mohamad A. Mikati* and Amal Rahmeh

Adult and Pediatric Epilepsy Program
Department of Pediatrics
American University of Beirut
P.O. Box 113-6044/B52, Beirut, Lebanon

1. FOCAL LESIONS, EPILEPSY, AND MEMORY IN HUMANS

Memory is integral to cognition in that it is one of the four major classes of cognitive functions,[17] these being: receptive functions, memory and learning, thinking, and expressive functions. Broadly speaking, memory can be said to be served by two distinct neurophysiological/neuroanatomical systems, one system contributing to declarative memory and the second system contributing to procedural memory.

The bulk of memory research and theory has focused on declarative memory: this is the commonly understood meaning of the word "memory", involving the storage and conscious recall of facts and events. There are many elements of declarative memory including spatial memory, event memory, and semantic memory. Semantic memory is "timeless and spaceless", like the alphabet, and unrelated to individual experience (in contrast to event memory). As opposed to information the individual actively recalls, procedural memory involves motor and cognitive skills; it has been referred to as "a habit system". It is generally what is left over after amnesia; hence the assumption is that it is subserved by a separate neurophysiological/anatomic system. It can be further categorized as skill memory ("how to" memory), priming (cued recall), and classical conditioning.[17]

There is an abundance of theories that divide memory into stages and processing levels; for clinical purposes a good working model divides it into three stages. First, registration involves the perception of incoming stimuli and their retention for 1–2 seconds after which they are either further processed or they decay. Second, short term memory lasts for 30 seconds to several minutes. It is attention dependent and neurophysiologically has been described as a reverberating neural circuit; such a circuit may be made to

* To whom correspondence should be addressed

Neuropsychology of Childhood Epilepsy, edited by Jambaqué et al.
Kluwer Academic / Plenum Publishers, New York, 2001.

persist for hours by rehearsal or training, but will fade if not further processed. Third, long term memory is the ability to process/store information in such a way that it can be retrieved at a time distant from when it was stored. In order for this to happen consolidation must occur on information that is in short term memory; the process of consolidation may be quick or may continue for hours without the person's active involvement. Necessary also to successful long-term memory is successful retrieval of stored information after its consolidation.

Memory deficits following a specific insult to the brain such as head trauma, hypoxia, or ischemia may be caused by a disruption of any of the above-described stages of storage, consolidation, or recall. If the deficit affects information the individual is exposed to after the insult, it is called anterograde amnesia; anterograde amnesia may further be divided into a defect either primarily affecting the acquisition of new information (anterograde acquisition defect), or affecting the ability to maintain such information in long term memory (anterograde retention defect). Deficits that affect information processed before the insult cause retrograde amnesia.

The anatomic localization of memory functions has in the past greatly depended on case studies of patients with cerebral lesions (Table 1). For example H.M., the patient who became severely amnesic following bilateral temporal lobe resection[28] was so impaired that he was incapable of almost any new learning and forgot the events of daily life as quickly as they occurred (impaired anterograde acquisition). He also had severely compromised access to previously learned information (retrograde amnesia). This report and previous other reports implicated the hippocampus as a main site for formation of memory. Careful investigation of H.M. revealed that he also had, on measures of long term picture recognition and after being given an extra long time to learn new stimuli (15–20 seconds), normal memory on yes/ no tests after one day. This suggested that long-term storage and retention of such material occurs in extrahippocampal sites (anterograde retention).

Another more recently studied patient is R.B., who developed memory impairment following an ischemic episode affecting the hippocampus. During the five years until his death, R.B. exhibited marked anterograde amnesia and was found to have bilateral lesions involving the CA1 field of the hippocampus. He also had only mild retrograde amnesia. This emphasizes the importance of the hippocampus in the acquisition of new memories, and suggests that retrograde amnesia may be caused by injury to the CA1 region and not only to sites separate from that region.[33]

The focus of temporal lobe epilepsy may be in the left (LTLE) or right (RTLE) temporal lobe, and may therefore be associated with lateralizing cognitive defects. In general, these defects include difficulty with verbal memory tasks in patients with LTLE, and difficulty with spatial and figural tasks in patients with RTLE. Patients with other

Table 1. Localization of cerebral lesions and corresponding memory deficits (human studies)

Reference	Site of lesion	Memory impairment
28	Bilateral temporal lobe resection (patient H.M.)	Anterograde amnesia (acquisition but not retention defect), retrograde amnesia
33	Bilateral lesions of CA1 (patient R.B.)	Anterograde amnesia (acquisition but not retention defect), mild retrograde amnesia
3	Left temporal lobe epilepsy	Difficulty with verbal memory tasks
3	Right temporal lobe epilepsy	Difficulty with spatial and figural tasks

types of epilepsy can have distinct memory deficits. Breier *et al.*[3] examined preoperative memory performance to distinguish between patients who had been diagnosed as having extra-temporal lobe epilepsy (ETLE), RTLE, and LTLE. Analyses indicated that the ETLE group performed better than the RTLE group on non-verbal memory measures, and better than the LTLE group on verbal memory measures. However, not all patients with focal epilepsy have lateralizing cognitive signs. In fact, some studies have found that, in surgical patients, those with preoperative lateralizing cognitive deficits have better outcomes after resection in terms of seizure relief.

Temporal lobe epilepsy patients undergoing temporal lobectomy are expected to show modest increases in their IQ scores after surgery, especially if they undergo non-dominant resection and become seizure-free. However, these increments are similar to and sometimes smaller than those seen among nonsurgical patients retested over comparable or longer follow-up intervals. Some decrements from baseline can be forecasted for confrontation naming immediately following left temporal lobectomy. However, in some cases, specifically those that get relief from their seizures, language functions following left temporal lobectomy actually show significant improvements. Rausch[25] has described the specificity of numerous tests of memory for particular areas of resection, characterizing which tests are affected by left temporal resection versus those affected specifically by resection of the hippocampus. Deficits in intelligence and language generally resolve within a year after surgery. However, deficits in memory for verbal information are quite frequent and tend to persist even if a patient is seizure-free. Age of seizure onset and preoperative ability level are predictors of such postoperative deficits, with higher preoperative verbal memory index (on the Wechsler Memory Scale-Revised) predicting a larger percent drop in post operative performance. Some investigators have, thus, advocated that memory deficits remain a major complication of left hemisphere temporal lobectomy for epilepsy, and that the probability of such deficits may be sufficient to contraindicate surgery in patients who depend on memory for their livelihood.[4]

The above studies describe memory deficits in patients with partial epilepsy that were present at the time of the initial diagnosis, at the initiation of drug treatment, with the establishment of intractability, and before or after epilepsy surgery. These clinical studies are not able to distinguish whether the impairments that occur are secondary to the preexisting injury or lesion that caused the epilepsy, to the seizures or epilepsy *per se*, or due to the antiepileptic medications. Investigations in experimental models should allow us to answer those questions.

2. FOCAL LESIONS, EPILEPSY, AND MEMORY IN EXPERIMENTAL ANIMALS

Holmes *et al.*[12] used the Genetically Epilepsy Prone Rats and induced in them recurrent seizures by auditory stimulation. They subsequently found defective learning of spatial memory tasks as tested by the T-maze and the water maze in those rats. In another study,[11] prepubescent male rats were given kainic acid intraperitoneally. The kainate model is an extensively studied animal model of human temporal lobe epilepsy because of the ability of kainic acid to induce on parenteral, or local, administration localized paroxysmal discharges in the hippocampus that generalize to contiguous structures. The prepubescent male rats in that study underwent status epilepticus with bilateral forelimb clonus after kainic acid administration. They showed, when tested as adults, slower learning of water maze and T-maze as compared to control animals. No gross histological abnormalities were detected in these rats indicating that the spontaneous seizures and memory disturbances observed following systemic administration of kainic acid can occur without gross pathology.

In another study, animals receiving bilateral injections of kainic acid in the striatum showed deficits in learning a daily reward-alternation task[22] and exhibited impaired learning and memory in an avoidance task. In a model using kindling to evoke repeated brief seizures, Sutula *et al.*[31] found that these seizures induced a long-lasting deficit in a radial arm maze task that is a rodent test of memory known to be impaired by hippocampal damage. Kindled rats studied at one month after the last of 30 to 134 evoked generalized tonic-clonic seizures acquired competence in the performance of the radial arm maze task at a rate indistinguishable from controls (initial acquisition). However, these rats demonstrated a deficit in the ability to repeat the task on consecutive days, the severity of the deficit varying directly with the number of evoked kindled seizures (short-term retention). This shows that in animal models of epilepsy, the different stages of memory can (as in humans) be differentially impaired.

The type of memory deficits may differ from one epilepsy model to another. The type of memory deficits seen after kainic acid induced status epilepticus has recently been characterized by us.[20] In prepubescent P35 (post-natal day 35) rats, kainic acid induced status epilepticus is followed by a significant impairment in the acquisition of spatial information in the Morris water maze learning test. This impairment was present as early as ten days after kainic acid induced status. The long term retention of learned material in those rats, however, was not impaired since the time needed to find the platform at the beginning of one series was not longer than that at the end of the previous series completed 10 days earlier.[20] Also, the total capacity to eventually learn that task was not affected, since those rats eventually caught up with controls after four repeated series of trials. Previous investigations[1] demonstrated that prepubescent rats that undergo kainic acid induced status epilepticus have lesions in the limbic system, particularly in the CA1, CA3, polymorph, and granular areas of the hippocampus, the posterior cortical, lateral, and medial nuclei of the amygdala, the reuniens and medial dorsal thalamic nuclei, pyriform cortex, insular cortex, bed nucleus of the stria terminalis and lateral septal nucleus. To a lesser extent, the dorsal lateral thalamic nuclei, claustrum, and cingulate gyrus were involved, whereas the nucleus of the horizontal limb of the diagonal band was spared. This data is consistent with the hypothesis that acquisition of spatial information is dependent on limbic structures, such as CA1, fornix, and diagonal band, and that long-term retention of that material is dependent on extra-hippocampal sites, namely diagonal band and its extra-forniceal projections such as the entorhinal cortex (Table 2). Thus, lesions produced by kainic acid affect, mostly, circuits that are involved in acquisition of information, rather than those involved in anterograde long-term retention.

Our above findings of preserved anterograde retention in the kainic acid model, despite profound defects in its extent and in the rate of acquisition of memories, are supported by the work of Ridley *et al.*[26] Ridley *et al.*[26] tested the effects of lesions of the CA1 on learning and retention of visual-spatial information in marmosets. The lesions resulted in a severe retrograde amnesia, but no forgetting of the visual-spatial information acquired post-surgically (i.e., adequate anterograde retention). Conversely, lesions of the diagonal band did not cause retrograde amnesia, but did result in significant forgetting (impaired anterograde retention). Animals with fornix transection had severe

Table 2. Localization of cerebral lesions and corresponding deficits (animal studies)

Reference	Site(s)	Memory function	Corresponding human disease (and site of pathology)
1, 6,15, 16, 26, 27, 32	– Diagonal band – Cholinergic input through the fornix – CA1 region	Anterograde acquisition	– Korsakoff's syndrome and Alzheimer's disease (diagonal band and its input), – Patients H.M. and R.B. (CA1)
	– Diagonal band – Cholinergic input from the diagonal band through extraforniceal connections to extra CA1 sites like the entorhinal cortex – Extra CA1 sites like the entorhinal cortex	Anterograde retention	– Korsakoff's syndrome and Alzheimer's disease (diagonal band and its input)
	– CA1 – Fornix – Extra- temporal sites including frontal, and prefrontal, cortex	Long term retrograde retention	– Patients H.M. and R.B. (CA1)

anterograde acquisition defect, severe retrograde amnesia, and only some of them showed evidence of forgetting.

There is currently a significant body of literature that shows that learning and retention of visual-spatial information depend on a number of interconnected key areas. The diagonal band sends cholinergic projections to the hippocampus and entorhinal cortex. The major cholinergic input into the hippocampus and entorhinal cortex comes through the fornix (which also contains the major efferent pathways of the hippocampus). To a lesser extent some precommissural cholinergic fibers travel to the hippocampus and adjacent allocortical areas by way of the ventrofugal route or the cingulate bundle.[15] Cholinergic projections are involved in sustaining the contribution of the hippocampus to acquisition of memories as indicated by the observation that impairment of acquisition after diagonal band or fornix lesions can be ameliorated by systemic injection of a cholinergic agonist.

The hypothesis that the diagonal band projections are involved in acquisition as well as retention is supported by the type of memory deficits seen in patients with anterior communicating artery lesions, Korsakoff's syndrome and Alzheimer's disease. These patients have problems with both acquisition and retention of newly learned material and, at least initially, have little retrograde amnesia.[26] Patients with Korsakoff's syndrome have early degeneration of the cholinergic system although the disorder eventually involves other systems quite extensively. The type of acquisition memory defect after those lesions is similar, but not identical, to that occurring with hippocampal damage as evidenced, among other things, by five case studies[6] of patients with basal forebrain lesions involving diagonal band projections. These patients had acquisition memory deficits similar to those seen in Korsakoff's syndrome, and different from patients with bilateral CA1 lesions. They could remember newly taught separate stimuli (in contrast, for example, to H.M. and R.B. who could not), but could not associate them into a coherent memory of an episode in time, thus, manifesting a distinct type of acquisition memory defect.

Based on all of the above studies, a hypothesis for the circuits involved in learning and memory of visual-spatial material is summarized in Table 2. It must be emphasized here that other circuits are probably involved. Save and Moghaddam[27] showed that associative parietal cortex lesions interfered particularly with the training of mice to reach a submerged platform in the Morris water maze in total darkness, i.e., interfering with the learning of egocentric locomotion-generated (kinesthetic) information. Other circuits may be those underlying procedural rather than declarative memory. Vnek and Rothblat[32] suggest that learning of a visual discrimination task may contrast with declarative memory in that it involves a non-cognitive association. In their study, rats with aspiration lesions of the dorsal hippocampus, rats with neocortical control lesions, and normal controls were trained on three object discrimination problems and then retrained for 3 weeks to measure retention. All animals were at the same level of performance during the training (acquisition) phase, but those with dorsal hippocampus lesions fell below that of controls during retraining (retention). These data suggest a role for the hippocampus in retaining visual discrimination information, but not in acquiring it. The authors have also postulated a role for the corticostriatal system or the rhinal cortex in the acquisition of non-spatial visual discrimination material. LeDoux[16] showed that lesions of the amygdala's central nucleus interfered with conditioned heart rate responses of rabbits when the shock was paired to a sound, suggesting the importance of the prefrontal cortex and amygdala in classical conditioning.

An understanding of the neurophysiological mechanisms of learning can be used to selectively test different circuits in the brain: one such mechanism involves NMDA receptor activation by glutamate. This leads to long term potentiation, which is considered to be the cellular counterpart of learning. Blocking NMDA receptors in the amygdala impaired fear conditioning, a finding that is also consistent with a role for the amygdala in procedural memory. Rats given intraventricular infusions of the NMDA receptor antagonist, D-L-AP5, have an impairment of spatial, but not of visual discrimination learning,[21] suggesting that spatial memory is localized in areas with a high density of NMDA receptors. These data are also important in view of the reported reduction in NMDA receptor density in the kindling model of epilepsy as well as in human temporal lobe epilepsy in the CA3 region.[13]

3. EFFECTS OF ANTIEPILEPTIC DRUGS

Antiepileptic drugs obviously have the beneficial effect of controlling seizures and preventing status epilepticus, and, thus, may prevent seizure-related brain injury. However, these drugs do have numerous acute cognitive adverse effects including hyperactivity and attentional problems.[5,7,9,18,24,29] Recently, concern about the long-term effects of antiepileptic drugs, even after those drugs have been stopped, on cognition and intelligence have been raised.[8] If antiepileptic drugs impair brain development, then they could potentially worsen seizure-associated memory impairments seen in patients with epilepsy. Investigation of those questions in patients is complicated, and such studies may be unethical since they may imply withholding treatment from such patients. In order to answer the above questions, studies on experimental animal models of epilepsy are needed.

To investigate the potential role of drug therapy in preventing or exacerbating seizure-related brain injury in the prepubescent brain, we administered kainic acid to P35 rats.[19] Therapy with daily phenobarbital was started directly before (one experimental group), or 1 day after (another experimental group) kainic acid was administered, and was continued in both groups through postnatal day 153. Another group received kainic acid alone and no phenobarbital. Kainic acid resulted in status epilepticus on P35 in all rats that received it, but those receiving phenobarbital first manifested a shorter and less severe status epilepticus as compared to rats on whom phenobarbital was started after kainic acid. When tested after phenobarbital was tapered, rats started on phenobarbital immediately before kainic acid administration did not differ from control rats on behavioral testing and had no subsequent spontaneous recurrent seizures (SRS). As compared to the group that received kainic acid and no phenobarbital, rats receiving kainic acid followed by phenobarbital manifested, after phenobarbital taper, similar aggressiveness, similar histological lesions, similar frequency of SRS and had even greater disturbances in memory, learning, and activity level. This study thus showed that daily phenobarbital therapy (of SRS), started after kainic acid status epilepticus, exacerbated seizure-related brain injury (memory and hyperactivity) despite the fact that the rats experienced less SRS than the non-treated kainic acid group. Normal rats receiving similar phenobarbital therapy were similar to control rats. This shows that the epileptic brain is more vulnerable to the detrimental effects of phenobarbital than the normal brain.

In another study,[2] P35 rats receiving a convulsant dose of kainic acid followed by daily injections of valproate from P36-P75, had during valproate therapy no SRS, and after valproate was tapered, no SRS, no abnormalities in the handling test, and no deficits in visuospatial learning. They also had fewer histological lesions than animals receiving kainic acid alone. These results demonstrate that kainic acid-related injury can be prevented by a medication working through inhibitory mechanisms; that structural and functional damage in the prepubescent brain can be prevented through strategically timed pharmacotherapy; and that the long term consequences of antiepileptic drug treatment during development are related to the drug used.

In a study by Pitkanen *et al.*,[23] the efficacy of carbamazepine and Vigabatrin (20 mg/kg/day and 250 mg/kg/day respectively) in preventing hippocampal and amygdaloid damage in the perforant pathway stimulation model of status epilepticus in rats was investigated. Treatments with carbamazepine alone, Vigabatrin alone, or with the combination of the two drugs were equally effective in decreasing the number and severity of seizures during stimulation. In the Vigabatrin groups, the damage to the hilar somatostatin-immunoreactive neurons and hippocampal CA3 pyramidal cells was less severe than in the vehicle and carbamazepine groups. In the carbamazepine and combination groups, the severity of the neuronal damage in the hippocampus did not differ from that in vehicle-treated animals. The amygdaloid neurons were not protected by any of the treatments. Thus, Vigabatrin and carbamazepine treatments have similar anticonvulsant efficacy during the perforant pathway stimulation, but only Vigabatrin decreases seizure-induced neuronal damage. This demonstrates that different brain regions and neuronal networks may be protected unequally by different anticonvulsants (Table 3).

In addition to the potentially neuroprotective or neurotoxic effects of anticonvulsants or lack of either effect, there are potential acute effects of those medications. In rats treated with carbamazepine with plasma levels of 2.5 and 4.5 mg/ml, the learning and memory of the T-maze and passive avoidance tests were enhanced as compared to controls. In addition, acetylcholinesterase activity was decreased in the hippocampus and pyriform cortex (19%) in these groups.[30] These authors postulated that the decrease in acetylcholinesterase activity caused by carbamazepine in the therapeutic range might

Table 3. Effect of antiepileptic drugs on long term sequelae of status epilepticus

Reference	Animal model	Drug	Drug effect
19	Kainic acid	Phenobarbital after kainic acid	Reduction of number and severity of SRS* during therapy only. Worsening of memory impairment and behavioral problems even after phenobarbital was stopped.
2, 19	Kainic acid	Valproate after kainic acid	Elimination of SRS and behavioral abnormalities even after discontinuation of valproate. Reduction of histologic lesions.
23	Perforant pathway stimulation	Vigabatrin or Carbamazepine during perforant pathway stimulation	Decrease in the number and severity of seizures during pathway stimulation (either drug). Reduction in hippocampal lesions by vigabatrin only.

*SRS = Spontaneous recurrent seizures.

have lead to increased acetylcholine levels in the brain, thus producing improvement in learning and memory. Whether those effects are clinically significant or not, still remains to be seen, particularly that there is a large body of literature in humans showing significant impairments in neuro-cognitive function during the use of most antiepileptic drugs. Review of this literature is beyond the scope of this chapter.

4. EFFECTS OF OTHER AGENTS

The relationship between the brain cholinergic system and memory disorders prompted many researchers to develop drugs that specifically stimulate the central cholinergic system. Memory deficits similar, to some extent, to those present in aged and demented subjects can be induced in experimental animals by administering muscarinic receptor agonists such as scopolamine, or by lesions of the septohippocampal pathway and of the nucleus basalis magnocellularis. The aim is to increase the activity of cholinergic forebrain neurons projecting to the cortex and hippocampus with different proposed mechanisms of action such as acetylcholine precursors (choline, lecithin), cholinergic receptor agonists (arecoline, bethanechol), cholinesterase inhibitors (tetrahydroaminoacridine).[10] Phosphatidylserine and the nosotropic drugs (piracetam, aniracetam, oxiracetam) probably work through the same mechanisms too.[10] Other investigators used central dopaminergic stimulants such as apomorphine and amphetamine, and catechol-O-methyltransferase inhibitors[14] to indirectly activate the hippocampal cholinergic neurons. The beneficial effects of these drugs on different aspects of cognition and various stages of learning tasks in experimental animals, however, have often been followed by controversial clinical results. Memory systems in humans are bound to be different and more complex than those in animals. Thus, one should be careful not to over interpret findings documented only in experimental animals. These studies and the studies delineated above in this chapter emphasize the importance of investigating different factors that can impair or modify memory function, as well as the nature of this impairment in different clinical and experimental situations.

REFERENCES

1. Ben-Ari Y (1985): Limbic seizure and brain damage produced by kainic acid: Mechanisms and relevance to human temporal lobe epilepsy. Neuroscience 14:375–403.
2. Bolanos AR, Sarkisian M, Yang Y, Hori A, Helmers SL, Mikati M, Tandon P, Stafstrom CE and Holmes GL (1998): Comparison of valproate and phenobarbital treatment after status epilepticus in rats. Neurology 51:41–48.
3. Breier JI, Plenger PM, Wheless W, Thomas AB, Brookshire BL, Curtis VL, Papanicolaou A, Willmore LJ and Clifton GL (1996): Memory tests distinguish between patients with focal temporal and extratemporal lobe epilepsy. Epilepsia 37:165–170.
4. Chelune GJ (1991): Using neuropsychological data to forecast postsurgical cognitive outcome. Epilepsy Surgery 477–485.
5. Craig I and Tallis R (1994): Impact of valproate and phenytoin on cognitive function in elderly patients: results of a single-blind randomized comparative study. Epilepsia 35:381–390.
6. Damasio AR, Graff-Radford NR, Eslinger PJ, Damasio H and Kassell N (1985): Amnesia following basal forebrain lesions. Archives of Neurology 42:263–271.
7. Dodrill CB and Troupin AS (1991): Neuropsychological effects of carbamazepine and phenytoin: a reanalysis. Neurology 41:141–143.
8. Farwell JR, Lee YJ, Hirtz DG, Sulzbacher SI, Ellenberg JH and Nelson KB (1990): Phenobarbital for febrile seizures effects on intelligence and on seizure recurrence. New England Journal of Medicine 322:364–369.
9. Forsythe I, Butler R, Berg I and McGuire R (1991): Cognitive impairment in new cases of epilepsy randomly assigned to carbamazepine, phenytoin and sodium valproate. Developmental Medicine and Child Neurology 33:524–534.
10. Giovannini MG, Casamenti F, Bartolini L and Pepeu G (1997): The brain cholinergic system as a target of cognition enhancers. Behavioural Brain Research 83:1–5.
11. Holmes GL, Thompson JL, Marchi T and Feldman DS (1988): Behavioral effects of Kainic Acid on the immature brain. Epilepsia 29:721–730.
12. Holmes GL, Thompson JL, Marchi T, Gabriel PS, Hogan MA, Carl FG and Feldman S (1990): Effects of seizure, on learning, memory and behavior in the genetically epilepsy prone rat. Annals of Neurology 27:24–32.
13. Hosford DA, Crain BJ, Cao Z, Bonhavs DW, Friedman AH, Okazaki JVN and McNamara JO (1991): Increased AMPA-sensitive quisqualate receptor binding and reduced NMDA receptor binding in epileptic human hippocampus. Journal of Neuroscience 11:428–434.
14. Khromova I, Voronina T, Kraineva VA, Zolotov N and Mannisto PT (1997): Effect of selective catechol-o-methyltransferase inhibitors on single-trial passive avoidance retention in male rats. Behavioural Brain Research 86:49–57.
15. Kitt CA, Mitchell SJ, DeLong MR, Wainer BH and Price DL (1987): Fibre pathways of basal forbrain cholinergic neurons in monkeys. Brain Research 406:192–206.
16. LeDoux J (1994): Emotion, memory and the brain. Scientific American 270:50–57.
17. Lezak MD (1995): Theory and practice of neurophysiological assessment. In Lezak MD (ed): "Neurophysiological Assessment" (Third Edition), pp 7–16.
18. Meador KJ, Loring DW and Nichols ME *et al.* (1994): Comparative cognitive effects of phenobarbital phenytoin, and valproate in healthy adults. Neurology 44 (Suppl 2) A 322.
19. Mikati MA, Holmes GL, Chronopoulous A, Hyde P, Thurber S, Gatt A, Liu Z, Werner S and Carl E (1994): Phenobarbital modifies seizure-related brain injury in the developing brain. Annals of Neurology 36:425–433.
20. Mikati MA, Tarif S, Lteif L and Jawad M (1996): Time sequence of memory impairment after experimental status epilepticus. Epilepsia 37 (Suppl 5): 141.
21. Morris RG (1989): Synaptic plasticity and learning: Selective Impairment of learning in rats nad blockade of long-term potentiation in vivo by the N-Methyl-D-aspartate receptor antagonist AP5. Journal of Neuroscience 9:3040–3057.
22. Pisa M, Sanberg PR and Fibiger HC (1981): Striatal injection of kainic acid selectively impair serial memory performance in the rat. Experimental Neurology 74:633–653.
23. Pitkanen A, Tunnanen J and Halonen T (1996): Vigabatrin and carbamazepine have different efficacies in the prevention of status epilepticus induced neuronal damage in the hippocampus and amygdala. Epilepsy Research 29–45.

24. Pullrainen V and Jokelainen M (1995): Comparing the cognitive effects of phenytoin and carbamazepine in long-term monotherapy: a two year follow up. Epilepsia 36:1195–1202.
25. Rausch R (1991): Factors affecting neuropsychological and psychosocial outcome of epilepsy surgery. 487–493.
26. Ridley RM, Baker HF, Harder JA and Pearson C (1996): Effects of lesions of different parts of the septo-hippocampal system in primates on learning and retention of information acquired before or after surgery. Brain Research Bulletin 40:21–32.
27. Save E and Moghaddam M (1996): Effects of lesions of the associative parietal cortex on the acquisition and use of spatial memory in egocentric and allocentric navigation tasks in the rat. Behavioral Neuroscience 110:74–85.
28. Scoville WB (1954): The limbic lobe and memory in man. J Neurosurg 11:64.
29. Smith DB, Craft BR, Collins J, Mattson RH and Cramer JA, for the VA cooperative study group (1986): Behavioral characteristics of epilepsy patients compared with normal controls. Epilepsia 27:760–768.
30. Sudha S, Lakshmana MK and Pradhan N (1995): Changes in learning and memory, acetylcholinesterase activity and monoamines in brain after chronic carbamazepine administratin in rats. Epilepsia 36:416–422.
31. Sutula T, Lauersdorf S, Lynch M, Jurgella C and Woodard A (1995): Deficits in radial arm maze performance in kindled rats: Evidence for long-lasting memory dysfunction induced by repeated brief seizure. Journal of Neuroscience 15:8295–8301.
32. Vnek N and Rothblat LA (1996): The hippocampus and long term object memory in the rat. Neuroscience 16:2780–2787.
33. Zola-Morgan S, Squire LR and Amaral DG (1986): Human amnesia and the medial temporal region: enduring memory impairment following a bilateral lesion limited to field CA1 of the hippocampus. Neuroscience 6:2950–2967.

5

ONTOGENESIS OF GABAERGIC AND GLUTAMATERGIC SYNAPTIC TRANSMISSION

Jean-Luc Gaiarsa* and Yezekiel Ben-Ari

INMED, Institut National de la Santé et de la Recherche Médicale
Unité 29, Avenue de Luminy, B.P. 13
13273 Marseille Cedex 09, France

INTRODUCTION

Neonatalogists and neurobiologists have long been aware that the immature brain differs considerably from the mature brain in the development and propagation of seizures. The incidence of seizures is highest in the first decade of life and status epilepticus is more common in children than in adults. Similarly, animal studies have clearly demonstrated the existence of developmental periods of altered seizure susceptibility during postnatal life.

In order to document how development can influence seizure thresholds, several studies have investigated the physiological and morphological maturation of the mammalian brain. It clearly turns out that the immature brain does not grow linearly into the adult stage but instead that both qualitative and quantitative abrupt changes occur. Thus, differences in intrinsic membrane properties, receptor subunits composition or density, as well as transient exuberant axonal outgrowth have been reported, all of these changes being able to modify seizure thresholds during development. Perhaps, the most striking example of such abrupt changes comes from studies showing that the main inhibitory neurotransmitter in the adult brain, i.e., γ-aminobutyric acid (GABA), provides in fact the main excitatory drive in neonates.

Correlative ontogenetic studies are also required for clinicians to extrapolate to man the results obtained in the developing rat brain. In an early study, based on the comparison of brain weight, Dobbing and Sands[9] have proposed that 7-day-old rat pups correspond to full-term new-born humans. In more recent studies, Romijn and collaborators[41] compared four different parameters including numerical synapse formation, the development of key enzymes involved in the synthesis of GABA or acetylcholine, and the development of electrical activity of cerebral cortex. They draw the conclusion that the

* To whom correspondence should be addressed

Neuropsychology of Childhood Epilepsy, edited by Jambaqué et al.
Kluwer Academic / Plenum Publishers, New York, 2001.

cerebral cortex of the 12- to 13-day old rat is functionally comparable to that of the full-term newborn human.

In the present chapter we will review the recent progress made in understanding the functional maturation of the developing brain. Particular attention will be given to the ontogeny of inhibitory GABAergic and excitatory glutamatergic synaptic transmissions that are thought to be involved in the generation of epileptiform discharges. In the last section, we will address the relative contribution of developmental changes to the observed alterations in seizure susceptibility in neonates.

1. DEVELOPMENTAL CHANGES IN GABAERGIC SYNAPTIC TRANSMISSION

Studies on the mature hippocampus and neocortex have shown that local circuit inhibition mediated by GABA plays a central role in regulating neuronal excitability and preventing the generation as well as the spread of seizures. The GABAergic inhibitory drive is achieved through the activation of two types of receptors termed the GABAa and GABAb receptors (Fig. 1) (see 43 for review). GABAa receptors are exclusively located on the postsynaptic membrane and their activation leads to the opening of chloride-permeable channels. A low level of intracellular chloride concentration is maintained by extruding pumps and ions exchangers (see 25 for review) so that activation of GABAa receptors induces an influx of chloride ions leading to a membrane hyperpolarization of

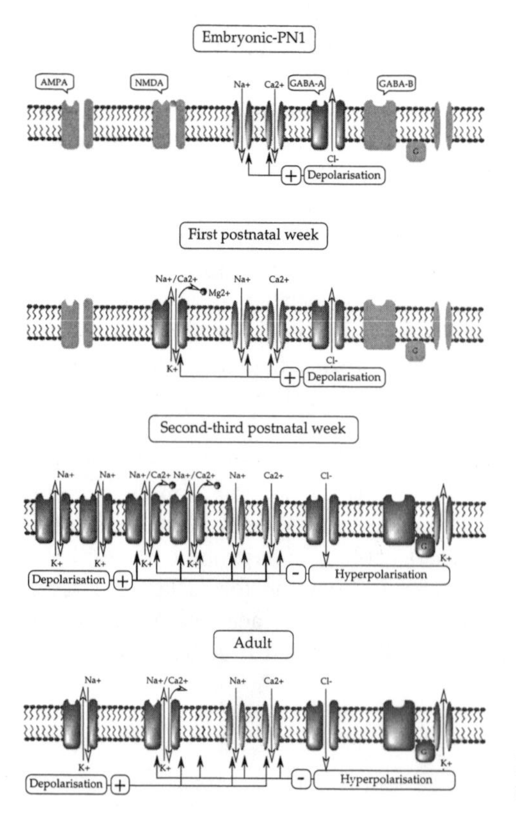

Figure 1. Sequential development of glutamatergic and GABAergic synaptic transmissions.

the cells. Unlike GABAa receptors, GABAb receptors are found both postsynaptically and presynaptically. At the postsynaptic level, GABAb receptors activate potassium channels through the activation of a pertussis toxin-sensitive G-protein. Activation of these receptors induces an efflux of potassium resulting in a hyperpolarization of the cells. At the presynaptic level, GABAb receptors inhibit transmitter release. Just as for postsynaptic receptors, this effect is mediated through the activation of a pertussis toxin-sensitive G-protein.

A completely different situation prevails at early stages of development. At embryonic and early postnatal stages, GABA operates like an excitatory neurotransmitter (Fig. 1). Indeed, activation of GABAa receptors, by exogenously applied GABA or endogenously released GABA, leads to a depolarization of neuroblasts and immature neurons in all regions of the CNS studied so far including the hippocampus, neocortex, cerebellum, hypothalamus, olfactory bulbs and spinal cord (see 4 for review). Interestingly, glycine that can be seen as the phylogenetically older inhibitory neurotransmitter, mainly present in the brain stem and spinal cord, also depolarizes embryonic and neonatal spinal neurons.[39] Even if GABAergic depolarization seems to represent a general feature of the developing neurons, this general pattern is subject to regional differences according to developmental gradients and heterogeneity in neuronal population. For instance, GABA depolarizes hippocampal CA3 pyramidal neurons until the fourth postnatal day of life (PN4), while hippocampal CA1 pyramidal and neocortical neurons are depolarized until PN9. This depolarizing action of GABA is not due to different properties or subunit composition of neonatal GABAa receptors[30] but rather to an inversion of the chloride driving force. In developing neurons, the intracellular chloride concentration is more elevated than in the adult so that activation of GABAa receptors leads to an efflux of chloride ions and thus to a depolarization of developing neurons.[46,49] The mechanisms responsible for the elevated intracellular chloride concentration are not fully understood yet. Recent studies have shown that functional potassium/chloride cotransport systems involved in the extrusion of chloride in the adult are not present at early stages of development.[46,49] In addition, some other pumps or ions exchangers responsible for intracellular chloride accumulation in dorsal root ganglion cells or sympathetic neurons (see 1 for review), may be present in developing neurons.

While in the adult, the GABAergic hyperpolarization prevents sodium action potential generation, in developing neurons the depolarization mediated by GABA reaches the threshold for sodium action potential generation (Fig. 1). GABA therefore operates as an excitatory neurotransmitter in the developing brain. Initially observed in hippocampal formation using intracellular recordings, the excitatory action of GABA has been confirmed and extended to other immature brain structures using techniques that do not affect intracellular chloride concentration (see 4 for review). GABAergic depolarization not only activates voltage-dependent sodium channels underlying the action potentials, but also voltage-dependent calcium channels (Fig. 1). Application of GABA or GABAa receptors agonits also leads to an increase in intracellular calcium concentration that is prevented by a further application of voltage dependent calcium channel blockers. This effect has been observed in culture and in acute brain slices, and more importantly following spontaneous synaptic release of GABA (see 4 for review).

The depolarizing and excitatory action of GABA in developing neurones is reinforced by the delayed maturation of postsynaptic GABAb receptor-mediated synaptic transmission (Fig. 1). Postsynaptic GABAb receptor-mediated inhibition is absent in both hippocampus and necortex from embryonic stages until the end of the first postnatal week of life.[12,13] The delayed maturation of postsynaptic GABAb receptor-mediated inhibition is unlikely due to the absence of receptors since GABAb binding has been reported to peak during the first postnatal week of life,[47] but is rather likely due to uncoupling between receptors, G-protein and potassium channels, at early stages of development.[13] Presynaptic GABAb receptors are however present and functional at early stages of development.[12,13] These presynaptic receptors provide a powerful inhibitory control on spontaneous network-driven activity in neonates.[34]

Starting from the end of the first postnatal week of life, GABA plays its well-documented role of inhibitory neurotransmitter. At that stage, activation of GABAa and GABAb receptors leads to a membrane hyperpolarization that prevents neuronal discharge.

From embryonic to immediately postnatal stages of development, GABA, acting on GABAa receptor, depolarizes developing neurons then activating sodium (Na+) and calcium (Ca2+) channels. At that stage, even if present and functional, AMPA, NMDA and GABAb receptors are not activated by synaptically released neurotransmitters. At early postnatal stages of development (during the first postnatal week of life), glutamatergic synaptic transmission is exclusively mediated by NMDA receptors. The depolarizing source required to remove the magnesium block from NMDA channels is provided by GABAa receptors. Starting from the end of the first postnatal week of life, GABA, as in adult, acts as an inhibitory neurotransmitter, preventing the activation of voltage-gated sodium and calcium channels. From the second to the third postnatal week of life, a functional over-expression of glutamatergic synaptic transmission takes place: a transient increase in the density of glutamate binding sites as well as exuberant axonal outgrowth have been reported. This general pattern of development is subject to regional changes according to the developmental gradients and heterogeneity in neuronal populations.

2. SEQUENTIAL DEVELOPMENT OF GLUTAMATERGIC SYNAPTIC TRANSMISSION

The importance of GABAergic excitation at early stages of development is reinforced by the fact that the establishment of GABAergic synaptic connections precedes the appearance of functional glutamatergic synapses. Thus, in the hippocampus, the spontaneous synaptic activity impinging on neonatal CA3 pyramidal neurons appears to be exclusively mediated by GABAa receptors until PN1.[23] Similarly, evoked glutamatergic synaptic potentials are not present before PN1,[11] while evoked GABAergic synaptic potentials are already present.[3]

When functional, glutamatergic synaptic transmission is mediated by NMDA receptors without any AMPA receptor mediated component.[11,24,48] This is in contrast to the adult situation where glutamatergic synaptic transmission is mainly mediated by AMPA and NMDA receptors (see 32 for review). The contribution of AMPA receptors to glutamatergic synaptic transmission increases as development progresses to reach adult levels by PN6 in the hippocampus[11] and by PN8–9 at the thalamo-cortical synapses.[24] It has been proposed that purely NMDA receptor mediated synaptic transmission could be due to the absence of functional AMPA receptors on the postsynaptic site.[11] Alternatively, purely NMDA mediated synaptic transmission might be attributed to a spillover of glutamate from a presynaptic bouton impinging on an adjacent cell[29] or from growth cone. Since the affinity of NMDA receptors for glutamate is about 500 times higher than that of AMPA receptors, NMDA receptors will be preferentially activated by glutamate spillover.

One of the key features of the NMDA receptor-gated channels is that they are blocked at resting membrane potential by magnesium so that a membrane depolarization is required to allow their activation. In the adult brain, this depolarization is provided by AMPA receptors (see 32 for review). In developing neurons, the voltage-dependent magnesium block of NMDA receptor-gated channels is as efficient as it is in adults. The depolarizing source required for their activation is provided by GABAa receptors when AMPA receptor mediated synaptic transmission is not yet functional.[26,31] Thus, in contrast to the adult brain where GABAa receptor-mediated hyperpolarization prevents the activation of NMDA-gated channels by strengthening their voltage-dependent magnesium block, GABAa receptor-mediated depolarization allows their activation (Fig. 1). This synergistic interaction between GABAa and NMDA receptors will have important implications in term of synaptic activity generated by the neonatal network[33,26] and in term of activity-dependent synaptic plasticity at early stages of development.[35]

The properties of NMDA receptor mediated synaptic transmission in developing neurons differ from their mature counterparts. Thus, it has been shown that the time course of both spontaneous and evoked NMDA receptors mediated synaptic potentials shortens with age in both neocortex and hippocampus.[7,21,27] This altered kinetic of NMDA receptor mediated EPSPs is presumably due to differences in receptor subunits composition as suggested by the developmental profile of the mRNAs encoding the different NMDA receptor subunits.[38]

3. BRAIN DEVELOPMENT AND EPILEPSY

How can the sequential expression of neurotransmitters affect seizures threshold? Both clinical and experimental results show that the immature central nervous system is particularly susceptible to seizures during restricted periods of development (see 22 for review). Similarly, animal studies have demonstrated the existence of periods of altered seizure susceptibility (see 45 for review).

3.1. The First Postnatal Week of Life

During the first postnatal week of life, that may correspond to prenatal stages in humans,[9,41] the rat neonatal brain is less prone to seizures both in vivo and in vitro. In vivo studies performed by Michelson and Lothman[36,37] have demonstrated that neonatal rats have the highest threshold for afterdischarges. On hippocampal as well as neocortical slices, application of several agents or procedures known to induce seizures in adults failed to generate any epileptiform discharges before the end of the first postnatal week of life in the hippocampus and before PN9 in the cortex.[3,6,20,28] Several factors may converge to reduce the sensitivity of immature circuits to seizures at a time when GABA acts as an excitatory neurotransmitter. These include i) different intrinsic properties of neonatal neurons:[3,6,49] action potentials are broader and smaller in amplitude compared to adults and the capacity of individual neurons to follow repetitive discharges as well as conduction velocity are reduced in the immature brain compared to the adult; ii) a delayed functional maturation of glutamatergic synaptic transmission (summarized above) and iii) a reduced network of recurrent axonal collaterals as documented in the hippocampus,[17] making the discharges, when elicited, less synchronized than in adults. In addition, at this stage persistent increase in glutamatergic synaptic efficacy, i.e., long term potentiation (LTP), which is believed to contribute to the maintenance and propagation of seizures,[5,44] is absent or poorly developed.[10] The delayed maturation of this phenomenon may be responsible for the low seizure sensitivity of immature tissue. In contrast, persistent decrease in glutamatergic synaptic efficacy, i.e., long-term depression (LTD) induced by repetitive stimulation, is easily induced at early stages of development.[10] The fact that LTD dominates in the immature brain will result in a downward regulation of network activity that could prevent seizure propagation. Finally, several studies have suggested that GABA may have a dual effect during this period of development. Thus, in addition to its excitatory effect due to membrane depolarization, GABA can exert an inhibitory shunting effect by locally increasing membrane conductance.[8,14] This inhibitory shunting action will also prevent excessive neuronal discharge.

3.2. The Second and Third Postnatal Weeks of Life

In contrast to the first postnatal week of life, during a period that extends from the second to the third postnatal week of life, the immature brain is particularly prone to seizures. Based on the criteria of Dobbing and Sands[9] and Romijin and collaborators,[41] this period corresponds to the immediate postnatal stage in man. During this period, immature rats show rapid generalization of seizures, with a quick progression through the early stages of kindling.[2,19] In vitro studies have also reported that, compared to adults, the hippocampus is more prone to develop epileptiform activity following repetitive orthodromic stimulation.[42] For instance, application of GABAa receptor antagonists leads to the appearance of unusually long epileptiform activity, on hippocampal as well as necortical slices (16, 20, see 45 for review).

A late onset of GABAergic synaptic inhibition could contribute to the enhanced seizure susceptibility in the developing hippocampus and neocortex. However, in several brain regions, including the hippocampus and neocortex, functional GABAergic synaptic inhibition is present at that stage (3, 12, 13, see 45 for review). Thus, as a possible mechanism underlying lower seizure thresholds, a transient functional over-expression of excitatory glutamatergic synaptic transmission has been proposed (Fig. 1). In agreement with this hypothesis, a transient increase in the density of glutamate binding sites has been reported in both the hippocampus and neocortex. This increase occurs during the second and third postnatal week of life (see 32 for review) thus coinciding with the period of enhanced seizure susceptibility. A similar transient increase has also been demonstrated using in situ hybridization to study the expression of mRNA coding for the AMPA receptor subunits during development.[40] In addition, as mentioned above, both spontaneous and evoked excitatory postsynaptic potential (EPSPs) mediated by NMDA receptors are unusually prolonged during the second postnatal week of life.[7,21,27] Due to the slower kinetic of EPSPs in developing neurons, the probability for summation will be higher in the immature brain thus resulting in a faster synchronization of the discharges. An increase in the density of glutamatergic synapses could also play a role in the enhanced seizure susceptibility in early postnatal life. Thus, during development, the projections from eye-specific layers of the lateral geniculate nucleus to the visual cortex overlap due to an excess of terminal axonal ramification. As development proceeds, axons from the LGN shorten to form the ocular dominance columns (see 15 for review). In a recent study, remodeling of recurrent excitatory collaterals that parallels changes in seizure susceptibility has been observed in the hippocampus.[16,17] Measurements of axonal length and varicosity density indicate that an excess of recurrent excitatory collaterals from CA3 pyramidal neurons exists during the second postnatal week of life. In addition with longer kinetic of EPSPs, this transient exuberant axonal outgrowth will favor synchronisation and thus the generation of epileptiform discharges.

Taken together, these observations show that both qualitative and quantitative changes occur at the level of the glutamatergic synaptic transmission leading to a functional over-expression during a restricted period of development. These alterations would promote excitatory synaptic transmission and thus would make the immature neonatal network more prone to generate seizures. However, during the same period of development, a functional over-expression of inhibitory GABAergic synaptic transmission has been observed. Indeed, during the second postnatal week of life, the synaptic activity in the rat hippocampus is characterized by the presence of spontaneous large hyperpolarizing potentials (LHPs).[3] These LHPs are network-driven events mediated by GABAa receptors and result from the synchronous activation of GABAergic interneurons by glutamatergic inputs.[33] This suggests that the glutamatergic functional over-expression that takes place at the level of pyramidal cells also occurs at the level of the GABAergic interneurons. This functional over-expression of inhibitory GABAergic synaptic transmission at this stage of development might represent a compensatory system to prevent excessive neuronal discharge in normal conditions.

CONCLUSION

In conclusion, the neonatal network shifts from one in which GABAa and NMDA receptors dominate and mediate most of the excitatory drive on developing neurons to an adult form in which glutamate mediates excitation through activation of AMPA and NMDA receptors and GABA exerts its well documented inhibitory action. This sequential pattern of development in addition with transient functional over-expression of neurotransmitter will affect seizure thresholds as well as clinical manifestation of the seizures in the developing brain.

In the mammalian CNS, maturation of neuronal network extends into postnatal life, and is often characterized by intense remodeling of connectivity. Initially demonstrated at the neuromuscular junction, remodeling of connectivity within the CNS also appears to be an activity-dependent process. The best-documented example of activity-dependent remodeling in the CNS comes from the study of axonal segregation within the visual system (see 15 for review). Similarly, it has been recently proposed that the appearance of functional postsynaptic AMPA receptors[11] or the shortening of NMDA receptor mediated EPSPs[18] are also modulated by neuronal synaptic activity. Thus, seizures in the developing brain may likely affect the pattern and/or the timing of appearance of functional synapses. Since learning and behavioral problems are over-represented in children with epilepsy, one of the main challenges in the future will be to determine whether and how synchronized epileptiform discharges can affect the sequential maturation of GABAergic and glutamatergic synaptic transmission, and more generally the brain development.

ACKNOWLEDGMENTS

This work was supported by the Institut National de la Santé et de la Recherche Médicale (INSERM).

REFERENCES

1. Alvarez-Leefmanns, FL (1990): Chloride channels and carriers in nerve, muscle and glial cell. In Alvarez-Leefmanns JF and Russel JM (eds): "Intracellular Chloride Regulation and Synaptic Inhibition in Vertebrate and Invertebrate Neurones. New York: Plenum Press, pp 109–150.
2. Baram TZ, Hirsch E and Schultz L (1993): Short-interval amygdala kindling in neonatal rats. Developmental Brain Research 73(1):79–83.
3. Ben-Ari Y, Cherubini E, Corradetti R and Gaiarsa JL (1989): Giant synaptic potentials in immature rat CA3 hippocampal neurones. Journal of Physiology 416:303–325.
4. Ben-Ari Y, Khazipov R, Leinekugel X, Caillard O and Gaiarsa JL (1997): GABA-A, NMDA and AMPA receptors: a developmentally regulated "menage à trois". Trends in Neurosciences 20(11): 523–529.
5. Ben-Ari Y and Represa A (1990): Brief seizure episodes induce long-term potentiation and mossy fibre sprouting in the hippocampus. Trends in Neurosciences 13(8):312–318.
6. Burgard EC and Hablitz JJ (1993): Developmental changes in NMDA and non-NMDA receptor-mediated synaptic potentials in rat neocortex. Journal of Neurophysiology 69(1):230–240.
7. Carmignoto G and Vicini S (1992): Activity-dependent decrease in NMDA receptor responses during development of the visual cortex. Science 258(5084):1007–1011.
8. Chen G, Trombley PQ and van den Pol AN (1996): Excitatory actions of GABA in developing rat hypothalamic neurones. Journal of Physiology (London) 494(2):451–464.
9. Dobbing J and Sands J (1979): Comparative aspects of the brain growth spurt. Early Human Development 3(1):79–83.
10. Dudek SM and Bear MF (1993): Bidirectional long-term modification of synaptic effectiveness in the adult and immature hippocampus. Journal of Neuroscience 13(7):2910–2918.
11. Durand GM, Kovalchuk Y and Konnerth A (1996): Long-term potentiation and functional synapse induction in developing hippocampus. Nature 381:71–75.
12. Fukuda A, Mody I and Prince DA (1993): Differential ontogenesis of presynaptic and post-synaptic GABAB inhibition in rat somatosensory cortex. Journal of Neurophysiolology 70(1):448–452.
13. Gaiarsa JL, Tseeb V and Ben-Ari Y (1995): Postnatal development of pre- and postsynaptic GABAB-mediated inhibitions in the CA3 hippocampal region of the rat. Journal of Neurophysiology 73(1):246–255.
14. Gao BX and Ziskind-Conhaim L (1995): Development of glycine- and GABA-gated currents in rat spinal motoneurons. Journal of Neurophysiology 74(1):113–121.
15. Goodman CS and Shatz CJ (1993): Developmental mechanisms that generate precise patterns of neuronal connectivity. Cell 72 (Suppl):77–98.
16. Gomez-Di Cesare CM, Smith KL, Rice FL and Swann JW (1996): Anatomical properties of fast spiking cells that initiate synchronized population discharges in immature hippocampus. Neuroscience 75(1):83–97.
17. Gomez-Di Cesare CM, Smith KL, Rice FL and Swann JW (1997): Axonal remodeling during postnatal maturation of CA3 hippocampal pyramidal neurons. Journal of Comparative Neurology 384(2):165–180.
18. Gottmann K, Mehrle A, Gisselmann G and Hatt H (1997): Presynaptic control of subunit composition of NMDA receptors mediating synaptic plasticity. Journal of Neuroscience 17(8):2766–2774.
19. Haas KZ, Sperber EF and Moshe SL (1990): Kindling in developing animals: expression of severe seizures and enhanced development of bilateral foci. Developmental Brain Research 56(2):275–280.
20. Hablitz JJ (1987): Spontaneous ictal-like discharges and sustained potential shifts in the developing rat neocortex. Journal of Neurophysiology 58(5):1052–1065.
21. Hestrin S (1992): Developmental regulation of NMDA receptor-mediated synaptic currents at a central synapse. Nature 357:686–689.
22. Holmes GL (1997): Epilepsy in the developing brain: lessons from the laboratory and clinic. Epilepsia 38(1):12–30.
23. Hosokawa Y, Sciancalepore M, Stratta F, Martina M and Cherubini E (1994): Developmental changes in spontaneous GABAA-mediated synaptic events in rat hippocampal CA3 neurons. European Journal of Neuroscience 6:805–813.

24. Isaac JT, Crair MC and Nicoll RA (1997): Malenka RC Silent synapses during development of thalamocortical inputs Neuron 18(2):269–280.
25. Kaila K (1994): Ionic basis of GABA-A receptor channel function in the nervous system. Progress in Neurobiology 42(4):489–537.
26. Khazipov R, Leinekugel X, Khalilov I, Gaiarsa JL and Ben-Ari Y (1997): Synchronization of GABAergic interneuronal network in CA3 subfield of neonatal rat hippocampal slices. Journal of Physiololgy (London) 498:763–772.
27. Khazipov R, Ragozzino D and Bregestovski P (1995): Kinetics and Mg2+ block of N-methyl-D-aspartate receptor channels during postnatal development of hippocampal CA3 pyramidal neurons. Neuroscience 69(4):1057–1065.
28. Kriegstein AR, Suppes T and Prince DA (1987): Cellular and synaptic physiology and epileptogenesis of developing rat neocortical neurons in vitro. Brain Research 431(2):161–171.
29. Kullmann DM, Erdemli G and Asztely F (1996): LTP of AMPA and NMDA receptor-mediated signals: evidence for presynaptic expression and extrasynaptic glutamate spill-over. Neuron 17(3):461–474.
30. Laurie DJ, Wisden W and Seeburg PH (1992): The distribution of thirteen GABAA receptor subunit mRNAs in the rat brain. III. Embryonic and postnatal development. Journal of Neuroscience 12(11):4151–4172.
31. Leinekugel X, Medina I, Khalilov I, Ben-Ari Y and Khazipov R (1997): Ca2+ oscillations mediated by the synergistic excitatory actions of GABA(A) and NMDA receptors in the neonatal hippocampus. Neuron 18(2):243–255.
32. McDonald JW and Johnston MV (1990): Physiological and pathophysiological roles of excitatory amino acids during central nervous system development. Brain Research Reviews 15(1):41–70.
33. McLean HA, Rovira C, Ben-Ari Y and Gaiarsa JL (1995): NMDA-dependent GABAA-mediated polysynaptic potentials in the neonatal rat hippocampal CA3 region. European Journal of Neuroscience 7(7):1442–1448.
34. McLean HA, Caillard O, Khazipov R, Ben-Ari Y and Gaiarsa JL (1996a): Spontaneous release of GABA activates GABAB receptors and controls network activity in the neonatal rat hippocampus. Journal of Neurophysiology 76(2):1036–1046.
35. McLean HA, Caillard O, Ben-Ari Y and Gaiarsa JL (1996b): Bidirectional plasticity expressed by GABAergic synapses in the neonatal rat hippocampus. Journal of Physiology (London) 496:471–477.
36. Michelson HB and Lothman EW (1991): An ontogenetic study of kindling using rapidly recurring hippocampal seizures. Developmental Brain Research 61(1):79–85.
37. Michelson HB and Lothman EW (1992): Ontogeny of epileptogenesis in the rat hippocampus: a study of the influence of GABAergic inhibition. Developmental Brain Research 66(2):237–243.
38. Monyer H, Burnashev N, Laurie DJ, Sakmann B and Seeburg PH (1994): Developmental and regional expression in the rat brain and functional properties of four NMDA receptors. Neuron 12(3):529–540.
39. Nishimaru H, Iizuka M, Ozaki S and Kudo N (1996): Spontaneous motoneuronal activity mediated by glycine and GABA in the spinal cord of rat fetuses in vitro. Journal of Physiology (London) 497:131–143.
40. Pellegrini-Giampietro DE, Bennett MV and Zukin RS (1991): Differential expression of three glutamate receptor genes in developing rat brain: an in situ hybridization study. Proceedings of the National Academy of Sciences U S A 88(10):4157–4161.
41. Romijn HJ, Hofman MA and Gramsbergen A (1991): At what age is the developing cerebral cortex of the rat comparable to that of the full-term newborn human baby? Early Human Development 26(1):61–67.
42. Schwartzkroin PA (1984): Epileptogenesis in the immature CNS. In Schwartzkroin PA and Wheal HV (eds): “Electrophysiology of Epilepsy”. London: Academic Press, pp 389–412.
43. Sivilotti L and Nistri A (1991): GABA receptor mechanisms in the central nervous system. Progress in Neurobiology 36(1):35–92.
44. Stelzer A, Slater NT and ten Bruggencate G (1987): Activation of NMDA receptors blocks GABAergic inhibition in an in vitro model of epilepsy. Nature 326:698–701.
45. Swann JW (1995): Synaptogenesis and epileptogenesis in developing neuronal network. In Schwartzkroin PA, Moshé SL, Noebels JL and Swann JW (eds): “Brain Development and Epilepsy”. New York: Oxford University Press, pp 195–233.

46. Takebayashi M, Kagaya A, Hayashi T, Motohashi N and Yamawaki S (1996): gamma-Aminobutyric acid increases intracellular Ca2+ concentration in cultured cortical neurons: role of Cl-transport. European Journal of Pharmacology 297(1–2):137–143.
47. Turgeon SM and Albin RL (1994): Postnatal ontogeny of GABAB binding in rat brain. Neuroscience 62(2):601–613.
48. Wu G, Malinow R and Cline HT (1996): Maturation of a central glutamatergic synapse. Science 274(5289):972–976.
49. Zhang L, Spigelman I and Carlen PL (1991): Development of GABA-mediated, chloride-dependent inhibition in CA1 pyramidal neurones of immature rat hippocampal slices. Journal of Physiology (London) 444:25–492.

6

ONTOGENESIS OF NEURONAL NETWORKS

Antoinette Gelot

Unité de Neuropathologie
Hôpital Saint Vincent de Paul
82 Avenue Denfert Rochereau
75674 Paris Cedex 14, France

INTRODUCTION

The brain is composed of trillions of neurons that are organized into complex neuronal circuits through a vast area of synaptic connections: an estimated 10^{11} neurons share 10^{14} synaptic connections. The appearance of the neuronal circuitry represents a critical stage of cerebral development, because it concludes the constructional stage of brain organization and initiates that of brain function. The emergence of synapses allows the ongoing development, so far exclusively controlled by genetic factors, to be modulated by extrinsic signals, by conferring to the brain the ability to receive, process, store and transmit information.

At the cellular level, neuronal network construction is related to the appearance of synapses and subsequent synaptic component differentiation. At the ultrastructural level, a mature synapse consists of the contact between the synaptic bouton (the pre-synaptic component, an axonal terminal filled with synaptic vesicles containing the neuromediator) and the dendritic membrane (the post-synaptic component, a focal differentiation of the dendritic membrane). The dendritic membrane contains the neuromediator receptors. When synapses are excitatory, the dendritic membrane can become folded and form an expansion, the spine.

At the tissular level, neuronal network ontogenesis consists on i) axonal growth (to the target neuron), ii) synaptogenesis (between the afferent axon and the target neuron dendrite) and iii) neuronal differentiation. The latter process implies that, following synaptic contact, the dendritic tree development of target neurons gives rise to the appearance of the pyramidal or granular neuronal type.

Consecutively, at the macroscopic level, neuronal network ontogenesis leads to visible brain morphogenetic changes that respectively include i) white matter expansion linked to an increased number of axons that constitute the associative, commissural or projecting pathways; ii) gyrification which results from the cortical surface expansion consecutive to the incoming of axons and the development of dendritic trees iii) emergence of the intracortical lamination that appears after the differentiation of cortical neurons

Neuropsychology of Childhood Epilepsy, edited by Jambaqué et al.
Kluwer Academic / Plenum Publishers, New York, 2001.

on granular and pyramidal types and their segregation into respectively, layers II and IV and layers III, V, VI .

Neuronal circuitry ontogenesis proceeds in four major steps which will be reviewed in this chapter. These include: i) axonal guidance and neuronal target recognition, ii) axo-dendritic contact and early synaptogenesis iii), neuronal network remodeling and iv) synaptic components differentiation. The first two stages of neuronal network genesis are activity-independent and occur exclusively under genetic control. By contrast, the later stages are synaptic activity-dependent and enable the brain to learn about and adapt to the world.

Epileptic activity can severely alter this developmental process either in the neonatal period by the disturbance of the late-occurring activity-dependent stages (neuronal network remodeling), or later in childhood by the induction in the mature brain of intracellular events (i.e., messengers expression, neurotrophic factors secretion) that allow the occurrence of plasticity (axon growth: sprouting, dendritic branching) and result in the construction of an abnormal neuronal network.

1. AXONAL GUIDANCE AND NEURONAL TARGET RECOGNITION

It has long been suggested that axons may be attracted to their targets by diffusible substances-chemotactic factors-.[3] Recent embryological experiments have confirmed this assumption.

During development, axons originating from neurons located at cortical layer V make contact with the pontine neurons by selective branching. If pontine neurons are lacking (by selective destruction) the cortical axons fail to branch in the pons. Moreover, if the pontine neurons lie at ectopic sites, axonal branching appears at these sites.[18] Axonal branch behavior is controlled by diffusible factors produced by pontine target neurons that promote cortical axon branching, influence initial axon branch orientation and stabilize the axon branches formed in the suitable site.

Upon reaching the target region, the axon growth cone has first to select the specific post-synaptic area, then the specific neuron target among a set of neuron types (cell-to-cell specificity) and finally the suitable contact site among the surface of the target neuron dendrite. Several experimental studies carried out both in invertebrates and vertebrates have demonstrated that selective cell surface protein expressions are involved at each level of selection. In the retino-tectal system, ligands are expressed by the target neurons in a graded fashion; simultaneously, the incoming axons display the corresponding receptors.[26] The resulting network of connections displays a high degree of somatotopy. The recognition of both the specific target cell and the site of axon apposition seems related to the expression of specific cell proteins at the surface of the target cell.[2,21,11]

2. EARLY SYNAPTOGENESIS OR INITIAL AXO-DENDRITIC CONTACT

This step is the continuation of the phase of target recognition. It is based on cell surface protein expression and it is activity-independent. It can be observed at the histological level.

Initially, the attracted axon endings are still growth cones and not yet synaptic boutons. When they make contact with the dendritic membrane, they undergo serial changes that will lead to the appearance of synaptic bouton features: the endoplasmic reticulum cisterns disappear while synaptic vesicles begin to accumulate. Post synaptic density or PSD, containing the neuromediator receptors, appears subsequently. The basic mechanisms underlying these morphological changes have been studied at the neuromuscular junction.[16,17]

At the presynaptic level, when the growth cone of the motor axon makes contact with the muscular membrane, an increase in intra-axonal calcium can be measured during the following minutes. If the surface proteins of either axonal or dendritic membranes are not accessible, no variation of calcium level is observed, suggesting the involvement of such proteins in this process. Increase in intra-cytoplasmic calcium level leads to immobilization of the apposed growth cone as well as to the recruitment of synaptic vesicles to the contact site. The axon ending displays spontaneous depolarization that progressively becomes coupled with neuromediator release.

At the post-synaptic level the post-synaptic components, as neuromediator receptors, that were so far randomly distributed, are recruited facing the apposed axon. During the post-synaptic reorganization, involvement of Agrin, an extra-cellular protein secreted by the motor neuron in synaptic vesicles, has been reported in several studies. Indeed, Agrin can bind to glycoproteins expressed at the dendritic surface and provokes the clustering of growth factors and adhesive molecules at the post-synaptic site.

The study of neuronal network formation in hippocampal formation under experimental conditions provides a typical illustration of these processes of selective and reciprocal recognition. Under physiological conditions, the axons of the granule cells in the fascia dendata, the so-called mossy fibers (MF) make synapse selectively with the pyramids cells in the CA3 sector of Ammon's horn and quite exclusively at the proximal part of their apical dendrite.[1,3,8] If the CA3 neurons are damaged, the MFs make synapse with the apical dendrites of the granule cells in the fascia dentata,[7] or with the CA1 neurons adjacent to the CA3 ones,[6] but in these cases, the target dendrites do not display the typical post-synaptic differentiation i.e., the thorny excrescence, which is observed along the proximal part of the CA3 apical dendrites.

Taken together, these molecular events support a progressive setting up of synaptic activity by optimization of both pre- and post-synaptic functions that leads to synaptic activity appearance. At this stage, the activity-dependent stage of neuronal network modeling can start.

3. REFINEMENT OF INITIAL SYNAPTIC NETWORKS

When the synaptic activity appears at the site of axo-dendritic contact, the established basic topographical map will be subject to an activity-dependent remodeling. This leads to the appearance of the adult configuration of the neuronal network by two opposite processes: either maintenance or elimination of both synaptic sites and of the corresponding axons.

The process of synaptic and axonal elimination has been extensively described in the occipital cortex[19] and is particularly observable in the corpus callosum where it leads to an obvious peri-natal transversal size decrease by elimination of commissural collaterals.[4]

Once more, the visual system has provided an appropriate support for experimental studies. In the adult lateral geniculate nucleus (LGN), the projections arising from each eye are segregated into distinct and alternating strips. During development, however, retinal inputs from both eyes are initially intermingled in the LGN. Their segregation into distinct layers occurs subsequently through a process that requires synaptic activity. This remodeling can be prevented by synaptic activity blockage or disturbed by inappropriate synaptic activity (for review: 20).

The molecular mechanism underlying this secondary refinement is thought to be related to the degree of synchronization of the projecting cells.[5] Inputs with correlated activity will be stabilized whereas inputs arising out of phase with the other inputs will be eliminated in such a way that "cells that fire together wire together". If such a model is applied to the ontogenesis of the LGN pathway, it appears highly probable that cells from the same field in the same eye are more likely to fire together than cells from a distinct field in each eye. Conversely, experimental alteration of the firing pattern (vision deprivation, bilateral silencing inputs) impairs pathway refinement and leads to the persistence of immature connections.

Similarly, callosal elimination is altered if an epileptic status is provoked during the critical period of intra-cortical circuit refinement: Grigonis[15] reported a persistent immature aspect of the corpus callosum in the rabbits when epileptic activity was induced during the neonatal period. From a neuropathological point of view, the observation of a thick corpus callosum, reminiscent of the aspect observed during the fetal stage in children suffering of a severe epileptic status since the neo-natal period is not rare and could be interpreted as a reflect of altered synaptogenesis (personal observations).

One possible mechanism for translating patterns of activity into patterns of synaptogenesis and growth could be the local release of neurotrophines by those post-synaptic neurons activated in a coordinated way. This induces a selected synaptogenetic response restricted to the activating presynaptic axons.[20]

The two events, synapse reinforcement and subsequent maintenance and synapse weakening and consecutive retraction occur in parallel and result in a fine somatotopy of the neuronal pathways. A similar correlated signaling is used in the mature brain for cognitive functions such as learning and memory.[5]

4. SYNAPTIC COMPONENT DIFFERENTIATION

This last step of neuronal pathway ontogenesis only occurs at the site of the maintained synapses. Experimentally, synaptic component differentiation is easily observed in a specific hippocampal synaptic site. This corresponds to the synaptic contact between the granule cell axons (the mossy fiber: MF) and the CA3 pyramidal neuron dendritic spine (the thorny excrescence: TE). Indeed, the MFs appear as very expanded presynaptic boutons and the TEs as complex branched spines, both representing the higher degree of pre- and post-synaptic component differentiation.

Their development proceeds following precise steps as demonstrated in several studies.[1,9,12] It is closely related to synaptic activity maturation. Once the synaptic activity can be recorded, the PSD can be discerned at the dendritic membrane. Simultaneously, numerous organelles involved in protein synthesis accumulate under the post-synaptic membrane and participate in the bulging of the dendritic membrane. While synaptic activity is acquiring mature features, the pre-synaptic bouton expands and the spine emerges progressively, increases in size and becomes more and more complex by branching and appearance of spinules. Once the activity at the MF/CA3 synapse reaches its adult morphology, the TE displays its characteristic features.

At the molecular level, two notable events are associated with these morphological changes: i) increased messenger expression of some cytoskeletal associated proteins such as Spectrin;[14] ii) transitory intradendritic Spectrin messenger.[13] Interestingly Spectrin is involved in both intra-cytoplasmic skeletal organization (neurite growth) and integral protein immobilization at specific membrane sites (membrane differentiation). At the MF/CA3 synapse, the period of intra-dendritic transport of Spectrin messengers coincides with the phase of synaptic activity maturation. Intra-dendritic transport of messengers has been reported during synaptogenesis.[10,25] It participates in focal protein synthesis,[23] a phenomenon classically involved in neuronal polarization acquisition (segregation between dendritic and axonal compartments). Taken together this data suggests a close control of post-synaptic component differentiation by synaptic activity through the modulation of structural protein synthesis.

In adult structures a similar intra-dendritic messenger transport can be induced by LTP.[22] This latter data is complementary with those observed in the MF/CA3 synapse after kindling, i.e., an experimental procedure that is an animal model of temporal lobe epilepsy (for review: 24). Under this condition, a MF sprouting can be observed (emergence of axonal collaterals) which results in the appearance of additional synapses on the CA3 field and in aberrant connections in the fascia dentata. Together (increase density of both the pre and the post-synaptic components) these events may promote further excitability of the hippocampal circuitry and contribute to the persistence of epileptic disorders.

CONCLUSION

Neuronal network ontogenesis proceeds in successive steps. The early stages (target recognition at regional, cellular and membrane levels) are activity-independent and appear to be exclusively controlled by intrinsic factors that determine the intracerebral gradients of chemotactic factors and recognition proteins. Morphogenetic genes, that orchestrate the initial stages of brain construction, appear to be good candidates for the control of these initial steps. However, by activation of the plasticity processes normally occurring during development, the epileptic activity could generate aberrant synaptic circuitry and further abnormal excitability.

Later stages consist of refinement of the initial neuronal circuitry under the control of synaptic activity. These events occur in humans during the perinatal period. As a consequence, occurrence of epileptic activity early during the neonatal period could alter the definitive neuronal pathway emergence. Moreover, the use of similar processes of activity-dependent construction of neuronal pathways during learning and memory could explain the frequent alterations of these cognitive functions in epileptic children.

In conclusion the developmental schedule of synaptogenesis represents a notable illustration of the complementary roles of both the expected and unexpected, the innate mechanisms and experience during brain construction.

REFERENCES

1. Amaral D and Dent T (1981): Development of the mossy fibers of the dentate gyrus: I. A light an electron microscopic study of the mossy fibers and their expansions. Journal of Comparative Neurology 195:51–86.

2. Broadie K and Bate M (1993): Innervation directs receptor synthesis and localization in drosophilia embryo synaptogenesis. Nature 361:350–353.
3. Cajal SR (1904): Capitulo XLIV vias efferentes y ganglios subordinados al asta de Ammon. In "Textura del Sistema Nervioso del Hombre y de los Vertebrados. Imprenta y Libreria de Nicolas Moya, Tomo II", Segunda Parte, pp 999–1065.
4. Clarke S, Kraftsik R, Vanderloos H and Innoncenti G (1989): Forms and mesures of adult and developing corpus callosum: is there sexual dimorphism? Journal of Comparative Neurolology 280:213–230.
5. Constantine-Paton M and Cline HT (1988): LTP and activity dependent synaptogenesis: the more alike they are the more different they become Current Opinions in Neurobiology 8:139–148.
6. Cook TM and Crutcher KA (1985): Extensive target cell loss during development results in mossy fibers in the regio superior (CA1) of the rat hippocampal formation. Developmental Brain Research 21:19–30.
7. Cotman CW and Nadler JV (1987): Reactive synaptogenesis in the hippocampus. In Cotman CW (ed): "Neuronal Plasticity". Raven Press: New-York, pp 227–271.
8. Frotscher M (1991): Target specificity of synaptic connections in the hippocampus. Hippocampus 1:123–130.
9. Gaiarsa J, Baudoin M and Ben-Ari Y (1992): Effect of neonatal degranulation on the morphological development of rat CA3 pyramidal neurons: inductive rule of mossy fibers on the formation of thorny excrescences. Journal of Comparative Neurolology 321:612–625.
10. Garner C, Tucker R and Matus A (1988): Selective localization of messenger mRNA for cytoskeletal protein MAP2 in dendrites. Nature 33:674–677.
11. Garrity P and Zipursky S (1995): Neuronal target recognition. Cell 83:177–185.
12. Gelot A (1996a): Expression de l'alpha-spectrine non erythrocytaire au cours du développement du système nerveux central. Thèse de science, Paris VI.
13. Gelot A, Moreau J, Ben-Ari Y and Pollard H (1996b): Alpha-brain spectrin belongs to the population of intra-dendritically transported mRNAs. Neuroreport 8:113–116.
14. Gelot A, Moreau J, Khrestchatisky M and Ben-Ari Y (1994): Developmental change of alpha-spectrine mRNA in the rat brain. Brain Research 81:240–246.
15. Grigonis A and Murphy E (1994): The effects of epileptic activity on the development of callosal projections. Developmental Brain Research 77:251–255.
16. Hall Z and Sanes J (1993): Synaptic structure and development: the neuromuscular jonction. Cell 72:99–121.
17. Haydon P and Drapeau P (1995): From contact to connection: early events during synaptogenesis. Trends in Neuroscience 18:196–201.
18. Heffner C, Lumsden A and O'Leary D (1990): Target control of collateral extension and directional growth in the mammalian brain. Science 247:217–220.
19. Innocenti G, Aggoun-Zouaoui D and Lehman P (1995): Cellular aspects of callosal connections and their development. Neuropsychologia 33:961–987.
20. Katz L and Schatz C (1996): Synaptic activity and the construction of cortical circuits. Science 274:1133–1138.
21. Keshishian H, Shiba T, Chang M *et al.* (1993): Cellular mechanisms governing synaptic development in drosophilia melanogaster. Journal of Neurobiolology 24:757–787.
22. Lyfort G, Yamagata K, Kaufmann C *et al.* (1995): Arc, a growth factor and activity regulated gene encodes a novel cytoskeletal associated protein that is enriched in neuronal dendrites. Neuron 14: 433–445.
23. Rao A and Steward O (1991): Evidence that protein constituants of post-synaptic membrane specializations are locally synthesized: analysis of proteins synthesized within synaptosomes. Journal of Neuroscience 11:2881–2895.
24. Represa A, Le Gall La Salle and Ben Ari Y (1989): Hippocampal plasticity in the kindling model of epilepsy in rats. Neuroscience Letter 99:345–350.
25. Steward O (1995): Targeting of mRNAs to subsynaptic microdomains in dendrites. Current Opions in Neurobiology 5:55–61.
26. Tessier-Lavigne M and Goodmann C (1996): The molecular biology of axon guidance. Science 274:1123–1132.

7

DEVELOPMENT OF SENSORY SYSTEMS IN ANIMALS

Franco Lepore*[,1] and Jean-Paul Guillemot[1,2]

[1]Groupe de Recherche en Neuropsychologie Expérimentale
Université de Montréal
C.P. 6128, Succ. Centre-Ville
Montréal, Qué. H3C 3J7 and
[2]Département de Kinanthropologie
Université du Québec à Montréal
C.P. 8888, Succ. Centre-Ville, Montréal
Qué. H3C 3P8

INTRODUCTION

When examining development, the most obvious signs of this process are the growth of the body, of the brain and of the sensory organs. Each influences in its own way the manner in which the animal ends up perceiving the world. This chapter aims at presenting a summary of some of our research concerning the development of sensory systems in animals. Given the space constraints, we have tried to select material that presents a few general principles of development, starting at birth. Moreover, we have selected for presentation data relating mainly to the visual system. Examples of pathological conditions that may interfere with normal sensory development are presented. Although these conditions are not directly related to the epileptic process, the general principles that are described can nevertheless be extrapolated to foresee the consequences that the abnormal firing discharges observed in epilepsy could have on the normal development of the brain.

1. DEVELOPMENT OF THE BRAIN

The most striking effects of changes in body size concern of course the somatosensory system. The brain must progressively process information from a much larger surface of skin, hair and muscles. The development of the eyes and cochlea also has important

* To whom correspondence should be addressed

Neuropsychology of Childhood Epilepsy, edited by Jambaqué et al.
Kluwer Academic / Plenum Publishers, New York, 2001.

consequences since the eye more than doubles in size and the optics improve tremendously over the first month. Similar growth takes place for the cochlea and the receptor organs. The increase in the size of the brain parallels the increase in the size of body structures.

However, there is a very important difference between the way the body increases and the way the brain increases. The larger body is due to an increase in the number of cells. The brain, on the other hand, around birth actually looses a large number of cells. Some areas of the brain loose more than 50% of their neurons and Innocenti[25] has shown that the callosum actually looses up to 70% of its fibers after birth. The fact that the brain looses a large number of cells around birth has given rise to two important concepts regarding development and brain function, namely, the notion of programmed cell death and that of exuberance. There is presumably a reason for the existence of programmed cell death. The developing brain generally follows a very precise and complex developmental schema according to genetically determined plans. It also has, however, an extensive amount of redundancy so that small errors in the execution of the genetic plan can be corrected or perfected by neurons serving similar functions. Excess neurons that are not required for building the final structure simply die and are eliminated.

So how does the brain increase in size if it looses many of its cells? It does so by increasing the amount of glial material,[1] the size of the neuronal soma[2] and the number and complexity of its processes.[22,23]

In the cat, the eyes open around the end of the first week. The visual neurons and their receptive fields (RFs) show many of the properties of adult animals, but with some limitations. Some of these inaccuracies, however, are not necessarily due to retarded brain development but possibly to the poor optics of the eyes, particularly during the first month. Thus: a) cells show responses which are sluggish and habituate rapidly; b) orientation selectivity exists at eye opening but it improves significantly during the first month; c) orientation columns, as shown by 2-deoxyglucose (2-DG) activity, appear around the third postnatal week; d) the spatial organization of receptive fields is poorly developed until about the third week and even then only a few neurons show simple and complex properties; e) directional selectivity is only present in 5% of the neurons at two weeks and 28% after one month; f) as for ocular dominance, during the first postnatal month, 25 to 33% of the neurons are activated by either eye. After 1 week, monocular neurons are activated essentially by the contralateral eye; g) many orientation specific neurons are also tuned to velocity in the low range $< 20°$.[35]

In the monkey, results are quite similar, although the RF properties are more adult-like, given that the optics of the eyes are somewhat better at birth. Acuity development at the behavioral level in both species seems to follow the development of the spatial resolution of the cells.[7,8,32,34,45]

In the somatosensory system, the barrel fields in the rat are already segregated after one week, although again they are not perfect.[3,4] There also appears to be a columnar organization. Somatotopic organization is well developed. Receptive field size is generally larger than in the adult.

In the auditory system, few studies have been carried out. In the cat, there is the problem of the poor functional and structural development at birth of the cochlea. However, the general organization of the system seems to be present at birth: a) evoked potentials can be recorded at birth with loud stimuli ($>$95 db); b) at post-natal day 4, evoked potentials can be distinguished from those of adults only by longer latencies; c) action potentials in response to sound can be evoked by the third week.[35]

2. EXUBERANCE

Exuberance refers to the fact that at birth, there exist an excessive number of neurons or of their projections that are then eliminated to give rise to the normal adult pattern. This elimination is achieved by a process of "pruning" whereby only the efficient neurons or synapses are maintained and the others disappear.

One system where exuberance has been extensively studied concerns the interhemispheric pathways through the corpus callosum. Pruning, or elimination of exuberant callosal connections, has been shown to be present in the somatosensory system,[27] in the auditory system and in the visual system.[35] As a result, at the end of the first postnatal month, callosal neurons in the visual and somatosensory systems interconnect those parts of the primary sensory areas that are involved with sensation at midline of the perceptual hemispaces. The elimination of exuberant callosal connections can take place through the death of callosal neurons, but more probably through retraction of collateral axons of cells that also send axons to ipsilateral structures.

One important consideration concerns the point to which such shaping of projections depends on experience or on hard wiring. The usual procedure for examining the influence of nature and nurture on the shaping of the callosal projection zone is to raise animals with some visual abnormality and to see how this influences the final pattern of projections. Animals are either rendered strabismic, are monocularly or binocularly deprived of vision (sutured) or are binocularly enucleated. Extensive studies have been carried out using this paradigm and the results show that the restriction of the callosal projection zone still occurs but some experience-dependent organization is also present. We carried out precisely one such experiment,[12] where neonatal kittens were monocularly deprived through eyelid suture of one of the eyes at the same time as the optic chiasm was surgically transected (see Fig. 1A). At adulthood, the sutured eye was opened and electrophysiological recordings were carried out in areas 17/18 of each hemisphere. The chiasm transection ensured that each eye projected only to the ipsilateral hemisphere.[43] Any cell activated through the contralateral eye must of necessity receive this input from the contralateral hemisphere via the corpus callosum. In normally raised cats whose chiasm is sectioned at adulthood, the majority of cells are mainly excited through stimulation of the ipsilateral eye. However, a significant proportion (about 1/3) can also be driven to various degrees by stimulation of the contralateral eye (see Fig. 1D). A remarkable difference with these results was observed in the ocular dominance distribution of cells in the neonatal monocularly deprived-chiasm transected animals (see Fig. 1E). When recording in the hemisphere ipsilateral to the visually exposed eye, almost all the cells were driven from this eye (see Fig. 1E). Stimulation of the contralateral eye had almost no effect on cell activity, showing that the callosal route could not be activated through this eye in these animals. When cells in the cortex ipsilateral to the previously sutured eye were examined, they were not only excited through this eye, but the vast majority also responded to stimulation of the contralateral eye, some actually being driven only from this eye (see Fig. 1F). This suggests that, because of the visual experience to which the animals had been subjected, the callosum had been modified from a balanced bidirectional pathway to mainly a unidirectional one, where information proceeded from the developmentally normally afferented hemisphere to the deprived hemisphere but not the reverse.

We examined whether this reorganization was mainly functionally determined or whether it also rested on anatomical modifications. A number of cats were prepared as

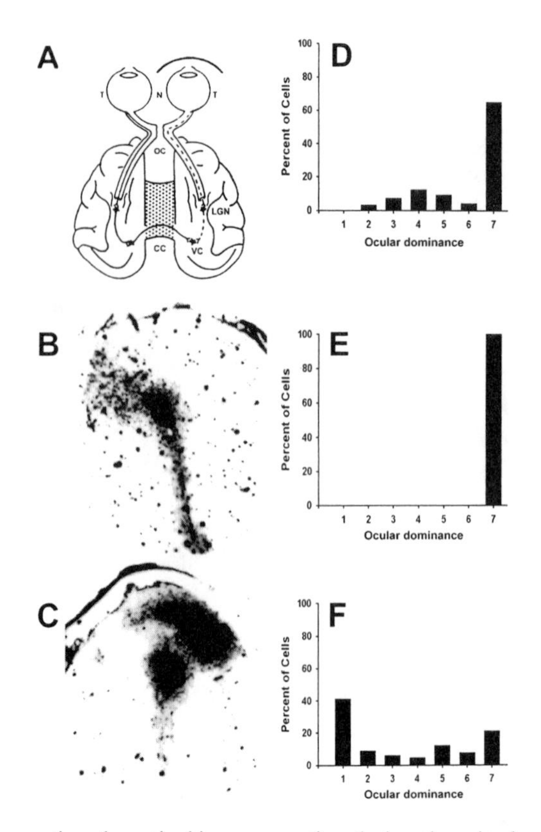

Figure 1. A—Animal preparation: the optic chiasm was sectioned when the animal was two to three weeks old and the eyelids of the right eye were sutured shut at the same time. T—temporal retina; N—nasal retina; OC—split optic chiasm; LGN—lateral geniculate nucleus; VC—visual cortex; CC—corpus callosum. B—autoradiograph of the left hemisphere of a neonatal chiasm split-right eyelid sutured animal whose eye was opened at adulthood and tritiated proline was injected in the visual cortex of the right hemisphere. Callosal fibers terminate mainly in the infragranular layers and very little marking is present in the supragranular layers. C—autoradiograph of the right hemisphere of a neonatal chiasm split-right eyelid sutured animal whose eye was opened at adulthood and tritiated proline was injected in the visual cortex of the left hemisphere. Callosal fibers terminate not only in the infragranular layers but also quite markedly in the supragranular layers. D—Ocular dominance distribution of cells recorded in the visual cortex of a normal cat whose optic chiasm was sectioned at adulthood. Extreme values 1 and 7 represent monocularly driven cells, where 1 indicates the cells driven by the contralateral eye and 7 those excited through the ipsilateral eye. Intermediate values represent relatively balanced ocular dominance, where category 4 stands for equal dominance between the two eyes. E—Ocular dominance distribution of cells recorded in the left hemisphere of a neonatal chiasm split-right eyelid sutured animal whose eye was opened at adulthood (conventions as in D). F—Ocular dominance of cells recorded in the right hemisphere of a neonatal chiasm split-right eyelid sutured animal whose eye was opened at adulthood (conventions as in D).

above, namely, one eye was sutured shut at two weeks of age, at the same time as the chiasm was transected. At adulthood, the eye was opened and 0.4 millicuries of tritiated proline was injected in the normally exposed hemisphere in some animals or in the hemisphere connected to the previously sutured eye in others. This radioactive protein is transported anterogradely and crosses to the other side. After waiting two days, to ensure that it was transported by callosal fibers to the visual cortex of the opposite hemisphere, the animals were sacrificed and their brains extracted. These were cut in coronal sections and the slices containing the posterior callosum (the splenium) and visual cortex were placed on photographic plates and stored in the dark to reveal the distribution of the radio-labeling. The photomicrographs confirmed quite elegantly the electrophysiological data reported above: proline was transported massively from the afferented to the deafferented hemisphere (see Fig. 1C) but only minimally in the other direction (see Fig. 1B). The supragranular layers were particularly devoid of radio-label. Again, it appears that the corpus callosum can develop essentially into a unidirectional pathway under appropriate developmental constraints. It is thus clear that nurture can influence both the functional and anatomical organization of sensory systems.

The preparations presented above all have in common the fact that input to one or both eyes is either abnormal (strabismus, monocular deprivation) or lacking (binocular deprivation, enucleation). We wanted to exam whether giving the animals relatively normal experience but reducing binocular activation would also affect the development of the callosal zone. We again used the split-chiasm preparation.[9] Kittens were lesioned at 10–15 days post-natal and allowed to develop. At adulthood, horseradish peroxidase (HRP) was injected at the 17/18 border of one hemisphere. After waiting for 48 hours to allow the HRP to retrogradly label the cells in the opposite hemisphere, the animals were

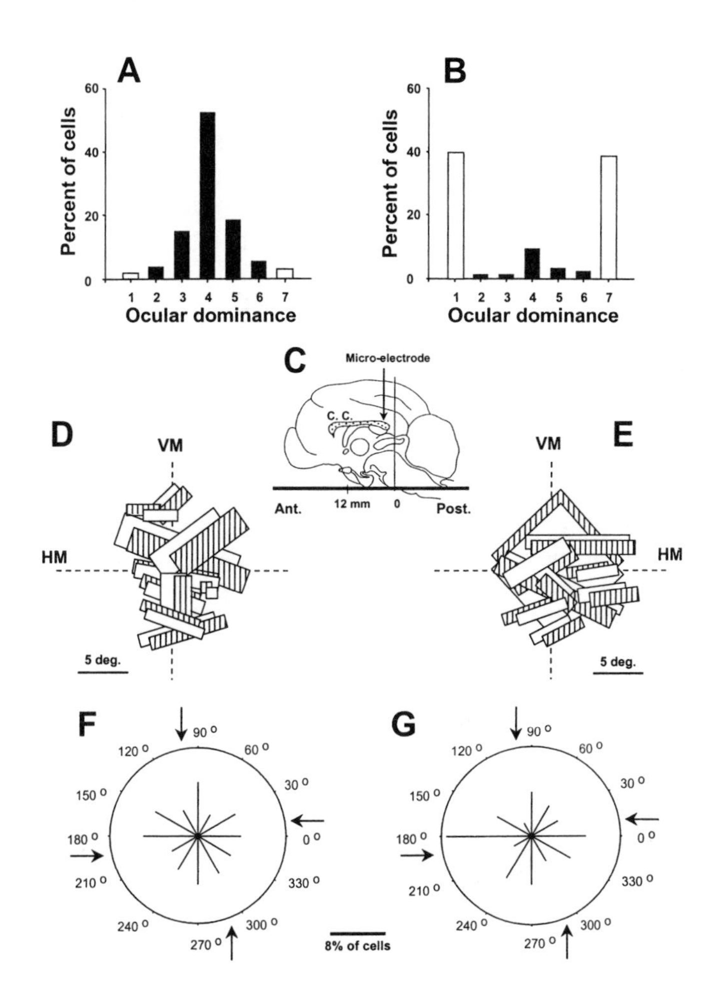

Figure 2.

sacrificed and their brains extracted and cut coronally. HRP-labeled cells were then counted in the contralateral hemisphere. Three outcomes were possible: a) a normal callosal zone might develop since vision was equal for both eyes; b) some abnormal organization would result from this manipulation due to the lack of binocular activation; c) the callosal zone might expand in these animals since all binocular activation now must, of necessity, take place through the corpus callosum. The results showed that the distribution of callosal neurons actually supported equally well the first two possibilities: the callosal zone was of appropriate size. However, there was a significant reduction of neurons in the supragranular layer whereas little or no change was observed in the infragranular layer.[9] These HRP results complement those obtained using proline autoradiograms described above: there is a reduction in the transport of radio-label in the supragranular layers because there are fewer callosal-projecting neurons to these layers. Taken together, the results show that the genetic plan was executed correctly in these abnormally innervated animals and that the lack of binocular input modified the nature of the callosal zone. However, they also demonstrate that there are limits to the amount of reorganization since the callosal zone did not expand despite the increased need for binocular activation, and since callosal activation to the infragranular layers was relatively normal.

Within the context of epilepsy, it is worth noting that only one study has addressed the issue of the influence that this pathological process may have on brain development. Indeed, one experimental model of epilepsy in the developing visual cortex of the rat has shown that maturational processes were impaired. Exuberant callosal connections connecting visual areas failed to be eliminated in the animals that experienced status epilepticus.[17] This study indicates that epilepsy may influence the anatomical organization of the visual system but it is uncertain at this point whether such anatomical malformations will be paralleled by functional disturbances in the immature epileptic brain.

3. CRITICAL PERIOD

Another important concept in development is that of the critical period. This concept proposes that there is a temporal window of opportunity during which experience can modify or shape the organization of a sensory system. After this period, experience has very limited influences in the final outcome.

In the visual system of the cat, where this has been studied extensively using such preparations as strabismus and monocular deprivation, it has been shown that these manipulations have very little effect if they occur before the third week or after about three months.[5,6,13,14,33,35,36] In the monkey, this period is somewhat extended, although the general results are the same.[18]

We wished to examine whether there also exists a critical period for callosal functional organization. We therefore studied interhemispheric transfer of a learned visual discrimination[36] in cats whose posterior two/thirds of the corpus callosum was cut at three different ages. It has been demonstrated that the corpus callosum is first noticeable between the two occipital regions by gestational age 56 days.[35] At birth, the callosum is quite complete, although many of the axons projecting to the contralateral hemisphere do not penetrate all the way up to the cortex. Myelination begins about post-natal day 16, when most of the exuberant cells have disappeared,[35] and is quite generalized by about 28 days, although myelination continues for several more months. The corpus callosum was thus sectioned in a first group of animals before it had undergone major myelination, namely at 19 or 20 days. For a second group, the same amount of callosum was sectioned after most of the myelination had taken place but before the end of the critical period for most visual functions, namely, 45 days. In the third group, the callosum was transected at adulthood. When the two young-sectioned animals reached adulthood, their optic chiasm was transected, as was that in the third group. This constitutes the classical split-brain preparation described by R. W. Sperry.[43] The animals were then evaluated in a visual discrimination task where they learned to discriminate a pattern using one eye and then tested for transfer using the other eye. Results indicated[36] that the post-natal day 19–20 sectioned animal showed significant transfer whereas the post-natal day 45 and adult-sectioned cats showed none. It was hypothesized that either the anterior callosum was reorganized in the earliest sectioned animals to allow transfer of visual information or that this function was taken over by reorganized sub-cortical commissures. These results, however, attest to the fact that there is a critical period for callosal functional organization and that it is somewhat different from that of other visual functions in being much more precocious.

4. USE IT OR LOOSE IT

A number of studies have examined the visual system and have shown that by the third post-natal week ocular dominance columns are present and can be demonstrated using the 2-DG technique. In the adult, they are quite well segregated, as originally illustrated from the classical results of Hubel and Wiesel.[18,21] These researchers also showed that raising an animal with abnormal visual experience in one eye, such as monocular deprivation, significantly affects the pattern of columnar organization, in that the column connected to the deprived eye shrinks whereas the one innervated by the opened eye expands.

A similar process can be demonstrated in the somatosensory system. It has been elegantly shown by Woolsey *et al.*[49] and Van der Loos *et al.*,[46] for example, that the destruction of one or a series of vibrissae results in the shrinkage or the disappearance of the appropriate barrel and the expansion of the remaining barrels within the deafferented cortex.

Although a series of hypotheses have been advanced to explain the expansion/contraction of the columns or barrels, the most popular is that, in the developing brain, synaptic space is at a premium, resulting in competition between the various inputs to a particular cell.[3,4] The axon that carries the strongest information, in terms of spike numbers, occupies increasing amount of synaptic space and the others are either eliminated or make fewer synaptic contacts.

In a series of experiments, we wanted to examine whether the deleterious effects resulting from a reduction or alteration of visual input from one eye can be modulated by manipulating the amount of interocular competition. Hubel and Wiesel,[20] as well as numerous other researchers[11,31,41] have shown that deviating one eye (strabismus) puts it at a competitive disadvantage in terms of its driving cortical cells. In fact, most cortical cells in the primary visual areas of these animals are driven by the non-deviated eye. The RFs of cells driven by the deviated eye show abnormal properties. At the behavioral level, the deviated eye becomes amblyopic and hence its visual acuity is reduced. This has also been shown to be the case in humans who develop strabismus before or during the critical period. It is generally postulated that these effects are not due to the development of some abnormality in the deviated eye but rather to changes in its cortical terminal fields. Thus, the lack of retinal correspondence of the inputs to the two eyes projecting to the same cortical cell cannot be integrated such that the one originating in the deviated eye is gradually eliminated. At the synaptic level, it is hypothesized that contacts of axon terminals from the deviated eye, which must compete for space with those of the normally aligned eye, are gradually withdrawn and post-synaptic sites become occupied by terminals from the normal eye.

We again used the split-chiasm preparation to examine interocular competition. Thus, animals underwent the transection of the optic chiasm at the same time as they were rendered strabismic by cutting the lateral muscle of one eye (esotropic strabismus). They were then raised in the animal colony and, at adulthood, the visual acuity of each eye was assessed behaviorally, using gratings of different spatial frequencies. Single-cell studies were also carried out to evaluate ocular dominance and receptive field properties.[21] Control groups consisted of either completely normal cats, or cats with normally aligned eyes whose chiasm was cut either neonatally or at adulthood and of neonatally deviated esotropic cats whose chiasm was sectioned at adulthood. The electrophysiological results seemed to support at least in part our working hypothesis, which stated that because of reduced competition between the eyes, cells innervated by the deviated eye would show normal receptive field properties. Thus, neurons situated in the cortex ipsilateral to the normal eye were, as expected, driven by this eye. Those in the opposite cortex were also, surprisingly, excited through the deviated eye. RF properties, in terms of sizes and responsiveness, were normal in both cortices. However, those in the cortex ipsilateral to the deviated eye were mostly situated in the "wrong" hemifield: the RFs appeared to be situated in the nasal hemiretina, that is, the portion of the retina disconnected by the chiasma transection.

Briefly summarized, the behavioral results show that the section of the optic chiasm greatly reduced visual acuity, especially in adult-sectioned animals. Deviating the eye inward soon after birth produced the expected amblyopia in the sense that visual acuity in this eye was extremely poor, some animals being unable to solve the test even at the lowest spatial frequency which we presented. The resolution of this eye was of course even worse when the animals underwent the additional adult chiasmatomy: in this case, all the animals failed to discriminate the spatial frequencies to which they were subjected. The critical test was the one where neonatal monocular esotropia was combined with early optic chiasm section. This preparation was expected to result in animals whose acuity in the deviated eye would be similar to that of normally aligned, neonatally chiasm sectioned animals. This is because the chiasmatomy was presumed to eliminate interocular competition and hence allow the deviated eye to develop as well as the normally aligned eye. This prediction was all the more probable given that, as we have seen above, cells in the ipsilateral cortex were driven from the deviated eye and had some normally appearing RF properties. The results did not fulfill these expectations: the deviated eye in the early split chiasm cats had extremely poor acuity, comparable to that of neonatally deviated, adult chiasmatomized animals. In conclusion, these results show that interocular competition, though certainly important, is probably not the only factor influencing the development of amblyopia in animals raised with abnormal visual afferents of at least one of the eyes.

One factor which might influence the amount of deficit resulting from abnormal visual experience concerns the extent to which retinal correspondence must be precise in order to coherently activate a single cortical cell. Neurons in primary visual cortex have very small RFs so that small deviations of one of the eyes results in non-corresponding

RFs. To avoid blurred vision or even diplopia, the system must somehow neglect one of the inputs and hence lead to the well-documented suppression of input from the deviated eye. Primary visual cortex is afferented by the X inputs in cats[44] or parvocellular system in primates[10,19,28,29,39,40] generally assumed to be at the basis of high-resolution spatial vision. In many higher order areas, receptive fields are larger and the grating acuities of the neurons are lower, possibly due to the fact that they are innervated by the Y (cat) or magnocellular systems (monkey) and participate in different functions, such as ambient vision and movement discrimination. The constituent cells can therefore tolerate somewhat larger discrepancies between the visual inputs from each eye and still show coherent (binocular) activation.

This hypothesis was examined in an electrophysiological study carried out on adult cats rendered strabismic at an early age, before the end of the critical period. Cells in areas 17/18 were, as expected, mainly monocularly driven through the normally aligned eye. Cells in higher-order lateral suprasylvian area[42] showed quite a different pattern of response. In normal cats, most of the cells in this area are generally binocularly driven and have fairly large receptive fields. In early stabismic cats, contrary to the results obtained in areas 17/18, over two-thirds of the cells were binocularly driven. It is thus clear that cells in this area can tolerate some retinal incongruities and be activated by both eyes. However, as we have seen above, similarly prepared cats show low grating acuities at the behavioral level. One must therefore presume that this binocular convergence is involved in functions other that fine spatial analysis.

5. CROSS-MODAL PLASTICITY

In the monkey, Rizzolati *et al.*[38] described cells in the periarcuate cortex which are exquisitely tuned to visual-somatosensory stimulations. Some cells, for example, would only discharge to the visual stimulus if the hand also touched the peri-oral area of the face. Graziano and Gross[16] described cells in the putamen with bimodal somatosensory and visual receptive fields. Similarly, Hyvarinen *et al.*[24] described cells in visual area 19 of visually-deprived monkeys which responded to the sensory-motor activity of the hand when it manipulated a box which contained a piece of raisin. In the cat, we recorded fibers in the corpus callosum which were excited by stimulation in the visual and/or auditory and/or somatosensory modalities. The spatial relationships among these modalities were not random but were essentially in register.[30] Thus, a fiber that responded to a sound presented on one side also preferred a visual stimulus presented to the same side and the stimulation of the skin ipsilaterally to these two. We[26] and others[47,48] also recorded cells in the anterior ectosylvian area which were sensitive to bimodal and even trimodal stimulation. Again, the RFs in our cats showed a large amount of spatial coherence. Rauschecker and Korte[37] also found visual-auditory receptive fields in cells recorded in this area but took the experiment a step further. They raised some of their cats with an experimentally induced blindness. At adulthood, they recorded in the anterior ectosylvian area and found that the tuning properties of the (presumed) bimodal cells to auditory stimulation was greatly enhanced. To confirm that this had behavioral relevance, they tested some of their cats in an auditory localization task[37] and found that blind-raised cats localized better than normal cats.

Through what mechanism is achieved this apparent cross-modal plasticity? Again, the most likely possibility is through synaptic reorganization. If one presumes that the two or three sources of input on a bimodal or trimodal cell more or less compete for synaptic space, the elimination of one of the sources frees space which is taken over by the remaining sensory systems.

Another approach to show that reorganization can take place across different modalities has been taken by Frost and Metin.[15] Their typical experiment consisted in lesioning the somatosensory thalamic relay, namely, the ventrobasal nucleus, and the terminal field of visual inputs, namely the occipital cortex (also resulting in a degeneration of the dorsal lateral geniculate nucleus) and the superior colliculi in hamsters. Results showed that visual fibers would project to the deafferented somatosensory cortex and that the recipient neurons responded to visual stimulation. Their multi-unit receptive fields had some retinotopic organization and showed some of the characteristics of cells recorded in visual cortex of normally raised animals. It is clear therefore that reorganization can take place across cortical areas under conditions where one of the sensory systems cannot innervate its normal target.

CONCLUSION

We have tried to illustrate that sensory systems, and especially the visual system, develop according to some well-defined plans. Thus, at birth, many of the structures are present but their constituent cells manifest a large amount of redundancy. If experience is normal, the genetic plan is executed whereby essential connectivity is maintained through a process of synaptic stabilization and superfluous cells and contacts are eliminated. If experience is abnormal or altered, reorganizations can take place that have profound effects on residual functions. Availability of synaptic space for which axon terminals compete or the existence of appropriate exuberant pathways determines how this restructuring takes place. The extent of reorganization is, however, limited. In many cases, the reorganized portion of the plan is co-existent with the genetically determined one. In addition, it may affect the system at one level, for example, at the electrophysiological level, but not at another, such as the behavioral level, probably because the latter depends on the coordinated activation of many structures. These changes, when they occur, can only take place within a narrow window of opportunity generally called the critical period. Moreover, they occur not only within the same sensory systems but can also manifest themselves across sensory systems.

Although we have only looked at the development of sensory systems of animals in the present chapter, there is no doubt that the same general principles apply also to humans. How these principles may help explain the epileptic process has yet to be examined, as there is only one description[17] of what consequences a pathological state simulating epilepsy may have on the development of sensory systems. The experimental model of epilepsy in the developing visual cortex has shown that maturational processes are indeed altered: immature pathways which should have normally disappeared were stabilized by the epileptic process. Obviously, further studies need to be carried out in order to assess whether this phenomenon applies to all sensory modalities and whether these observations have relevance to behavior in general and whether they constitute a useful model of epilepsy in particular.

ACKNOWLEDGMENTS

These studies were supported by grants from the Natural Science and Engineering Research Council of Canada (NSERC) and the Fonds pour la Formation de Chercheurs et l'Aide à la Recherche (FCAR) awarded to Franco Lepore and Jean-Paul Guillemot. We also wish to thank colleagues and graduate students who collaborated in the studies reported herein. Their names appear in the cited material.

REFERENCES

1. Altman J (1967): Postnatal growth and differentiation of the mammalian brain with implications for a morphological theory of memory. In Quarton GC, Melnechuk T and Schmitt FO (eds): "The Neurosciences". New York: Rockefeller University, pp 723–743.
2. Altman J (1976): Experimental reorganization of the cerebellar cortex. VII. Effects of late x-irradiation. Journal of Comparative Neurology 165:65–76.
3. Armstrong-James M (1975): The functional status and columnar organization of single cells responding to cutaneous stimulation in neonatal rat somatosensory cortex. Journal of Physiology (London) 246:501–538.
4. Armstrong-James M and Fox K (1988): The physiology of developing cortical neurons. In Jones EG and Peters A (eds): "Cerebral Cortex. Volume 7: Development and Maturation of Cerebral Cortex". New York: Plenum Press, pp 237–272.
5. Blakemore C (1974): Development of functional connexions in the mammalian visual system. British Medical Bulletin 30:152–157.
6. Blakemore C (1988): The sensitive periods of the monkey's visual cortex. In Lennerstrand G, von Noorden GK and Campos EC (eds): "Strabismus and Amblyopia". New York: Plenum Press, pp 219–234.
7. Blakemore C (1990): Maturation of mechanisms for efficient spatial vision. In Blakemore C (ed): "Vision Coding and Efficiency". Cambridge: Cambridge University Press, pp 254–266.
8. Blakemore C and Vital-Durand F (1986): Effects of visual deprivation on the development of the monkey's lateral geniculate nucleus. Journal of Physiololgy (London) 380:493–511.
9. Boire D, Morris R, Ptito M, Lepore F and Frost DO (1995): Effects of neonatal splitting of the optic chiasm on the development of feline visual callosal connections. Experimental Brain Research 104:275–286.
10. Casagrande VA (1994): A third parallel visual pathway to primate area V1. Trends in Neurosciences 17:305–310.
11. Chino YM, Smith EL 3rd, Yoshida K, Cheng H and Hamamoto J (1994): Binocular interactions in striate cortical neurons of cats reared with discordant visual inputs. Journal of Neuroscience 14: 5050–5067.
12. Cynader MS, Lepore F and Guillemot J-P (1981): Compétition interhémisphérique auc ours du développement postnatal. Revue Canadienne de Biologie 40:47–51.
13. Di Stefano M, Bédard S, Marzi CA and Lepore F (1984): Lack of binocular activation in area 19 of the Siamese cat. Brain Research 303:391–394.
14. Di Stefano M, Ptito M, Quessy S, Lepore F and Guillemot J-P (1995): Receptive field properties of areas 17–18 neurons in strabismic cats with the early section of the optic chiasm. Journal für Hirnforschung 36:277–281.
15. Frost DO and Metin C (1985): Induction of functional retinal projections to the somatosensory system. Nature 317:162–164.
16. Graziano MS and Gross CG (1993): A bimodal map of space: somatosensory receptive fields in the macaque putamen with corresponding visual receptive fields. Experimental Brain Research 97:96–109.
17. Grigonis AM and Murphy EH (1994): The effects of epileptic cortical activity on the development of callosal projections. Developmental Brain Research 77:251–255.
18. Hubel DH (1978): Effects of deprivation on the visual cortex of cat and monkey. Harvey Lectures 72:1–51.
19. Hubel DH and Livingstone MS (1987): Segregation of form, color, and stereopsis in primate area 18. Journal of Neuroscience 7:3378–3415.
20. Hubel DH and Wiesel TN (1965): Binocular interaction in striate cortex of kittens reared with artificial squint. Journal of Neurophysiology 28:1041–1059.
21. Hubel DH and Wiesel TN (1977): Ferrier lecture: Functional architecture of macaque monkey visual cortex. Proceedings of the Royal Society of London: Biological Sciences B198:1–59.

22. Huttenlocher PR and de Courten C (1987): The development of synapses in striate cortex of man. Human Neurobiology 6:1–9.
23. Huttenlocher PR, de Courten C, Garey LJ and Van der Loos H (1982): Synaptogenesis in the human visual cortex–evidence for synapse elimination during normal development. Neuroscience Letter 33:247–252.
24. Hyvarinen J, Carlson S and Hyvarinen L (1981): Early visual deprivation alters modality of neuronal responses in area 19 of monkey cortex. Neuroscience Letter 26:239–243.
25. Innocenti GM (1986): General organization of callosal connections in the cerebral cortex. In Jones EG and Peters A (eds): "Cerebral Cortex, Volume 5: Sensory-Motor Areas and Aspects of Cortical Connectivity". New York: Plenum Press, pp 291–353.
26. Jiang H, Lepore F, Ptito M and Guillemot J-P (1994): Sensory modality distribution in the anterior ectosylvian cortex (AEC) of cats. Experimental Brain Research 97:404–414.
27. Killackey HP and Chalupa LM (1986): Ontogenetic change in the distribution of callosal projection neurons in the postcentral gyrus of the fetal rhesus monkey. Journal of Comparative Neurology 244:331–348.
28. Lennie P (1980a): Perceptual signs of parallel pathways. Philosophical Transactions of the Royal Society of London: Biological Sciences B290:23–37.
29. Lennie P (1980b): Parallel visual pathways: a review. Vision Research 20:561–594.
30. Lepore F, Ptito M and Guillemot J-P (1986): The role of the corpus callosum in midline fusion. In Lepore F, Ptito M and Jasper HH (eds): "Two Hemispheres-One Brain: Functions of the Corpus Callosum". New York: Alan R Liss, pp 211–229.
31. Maffei L and Bisti S (1976): Binocular interaction in strabismic kittens deprived of vision. Science 191:579–580.
32. Mitchell DE (1990): Sensitive periods in visual development: insights gained from studies of recovery of visual function in cats following early monocular deprivation or cortical lesions. In Blakemore C (ed): "Vision Coding and Efficiency". Cambridge: Cambridge University Press, pp 234–246.
33. Mitchell DE (1991): The long-term effectiveness of different regimens of occlusion on recovery from early monocular deprivation in kittens. Philosophical Transactions of the Royal Society of London Biological Sciences B333:51–79.
34. Mitchell DE, Murphy KM, Dzioba HA and Horne JA (1986): Optimization of visual recovery from early monocular deprivation in kittens: implications for occlusion therapy in the treatment of amblyopia. Clinical Vision Sciences 1:173–177.
35. Payne B, Pearson H and Cornwell P (1988): Development of visual and auditory cortical connections in the cat. In Jones EG and Peters A (eds): "Cerebral Cortex: Development and Maturation of Cerebral Cortex", Volume 7. New York: Plenum Press, pp 309–389.
36. Ptito M and Lepore F (1983): Interocular transfer in cats with early callosal transection. Nature 301:513–515.
37. Rauschecker JP and Korte M (1993): Auditory compensation for early blindness in cat cerebral cortex. Journal of Neuroscience 13:4538–4548.
38. Rizzolatti G, Scandolara C, Matelli M and Gentilucci M (1981): Afferent properties of periarcuate neurons in macaque monkeys. II. Visual responses. Behavavioural Brain Research 2:147–163.
39. Schiller PH and Logothetis NK (1990a): The color-opponent and broad-band channels of the primate visual system. Trends in Neurosciences 13:392–398.
40. Schiller PH, Logothetis NK and Charles ER (1990b): Role of the color-opponent and broad-band channels in vision. Visual Neuroscience 5:321–346.
41. Sengpiel F and Blakemore C (1996): The neural basis of suppression and amblyopia in strabismus. Eye 10:250–258.
42. Spear PD (1991): Functions of extrastriate visual cortex in non-primate species. In Cronly-Dillon JR and Leventhal AV (eds): "Vision and Visual Dysfunction, Volume 4. The Neural Basis of Visual Function. "Boston: CRC Press Inc, pp 339–370.
43. Sperry RW, Myers RE and Schrier AM (1960): Perceptual capacity of the isolated visual cortex in the cat. Quarterly Journal of Experimental Psycholology 12:65–71.
44. Stone J (1983): Parallel Processing in the Visual System. New York: Plenum Press.
45. Teller DY, Regal DM, Videen TO and Pulos E (1978): Development of visual acuity in infant monkeys (Macaca nemestrina) during the early postnatal weeks. Vision Research 18:561–566.
46. Van der Loos H and Woolsey TA (1973): Somatosensory cortex: structural alterations following early injury to sense organs. Science 179:395–398.

47. Wallace MT, Meredith MA and Stein BE (1992): Integration of multiple sensory modalities in cat cortex. Experimental Brain Research 91:484–488.
48. Wilkinson LK, Meredith MA and Stein BE (1996): The role of anterior ectosylvian cortex in cross-modality orientation and approach behavior. Experimental Brain Research 112:1–10.
49. Woolsey TA and Van der Loos H (1970): The structural organization of layer IV in the somatosensory region (SI) of mouse cerebral cortex. The description of a cortical field composed of discrete cytoarchitectonic units. Brain Research 17:205–242.

8

CEREBRAL MATURATION AND FUNCTIONAL IMAGING

Catherine Chiron*[,1,2] and Isabelle Jambaqué[1]

[1]Service de Neuropédiatrie and INSERM U29
Hôpital Saint Vincent de Paul
82 Avenue Denfert Rochereau
75674 Paris Cedex 14, France and
[2]Service Hospitalier Fréderic Joliot
Departement de Recherche Médicale
Centre à l'Énergie Atomique, Orsay
France

INTRODUCTION

Normal cerebral maturation and therefore development of sensory and cognitive functions are sustained by neurobiological changes that are better known in animals than in humans. Postnatal developmental changes can now be approached in humans by means of non-invasive methods using external detection of brain signals. These cerebral functional imaging techniques include PET (positron emission tomography), SPECT (single photon emission computed tomography) and fMRI (functional magnetic resonance imaging). PET and fMRI recently experienced rapid development for stimulation studies because they can detect and localize the activation produced in the brain by selective cognitive tasks. Unfortunately, technical constraints make such studies almost impossible to realize in children. PET and SPECT can also measure at rest, respectively the local metabolism for glucose and regional cerebral blood flow, both parameters being proportionally close to the neuronal activity in the same region. These "at rest studies" are available in young children and are now being used in pathological conditions such as childhood epilepsy or developmental disorders. Most of the data available about postnatal functional brain maturation have been obtained by PET and SPECT "at rest studies". They have the advantage of providing measures at the regional level. It is therefore interesting to correlate the regional changes in metabolism or in cerebral blood flow during maturation with the development of sensory and cognitive functions in humans.

* To whom correspondence should be addressed

Neuropsychology of Childhood Epilepsy, edited by Jambaqué et al.
Kluwer Academic / Plenum Publishers, New York, 2001.

However, these "at rest" techniques also have some limitations: longitudinal studies are restricted because radioactive tracers are used, normal populations cannot be studied for ethical considerations, and one must bear in mind that the correspondence between cognitive functions and functional cerebral activity at rest remains indirect.

PET and SPECT "at rest" are also extensively used in the field of epilepsy. When epilepsy appears in very young children, it raises the question of the potential relationship between epilepsy and development since epileptogenic phenomena and developmental processes involve the same neuronal pathways within the brain. Some answers can be brought up by functional imaging, measuring cerebral blood flow and brain metabolism in infants.

1. TECHNIQUES

PET and SPECT "at rest" provide the two main types of functional imaging techniques applicable in very young children. PET uses positron emitters obtained from a cyclotron, namely ^{18}F-Fluodeoxyglucose (^{18}FDG) and PET cameras have less than 8 mm plane resolution that allow the study of cerebral metabolism. Multiple arterial samples are needed to measure absolute values of metabolism; this procedure, together with the high cost of cameras and cyclotron, causes PET not to be routinely used but dedicated only to research in children. SPECT tracers and SPECT cameras are not as expensive, and they are more accessible in nuclear medicine activity. SPECT studies the regional cerebral blood flow. Tracers are gamma emitters, ^{99}m-Technetium in static SPECT—the most often used in epilepsy studies—and 133-xenon in dynamic SPECT method—which is indicated for maturation studies—. Static SPECT-cameras provide high resolution images (8 mm), but absolute quantification of cerebral blood flow is not available. Using dynamic SPECT and 133-xenon, resolution of the images is not as good (12 mm) but absolute measures of cerebral blood flow are obtained, a necessary condition to study maturational changes.

There are particular difficulties in performing PET and SPECT in children. Sedation is needed under 6 years of age. When a barbiturate is administered before the injection of the tracer like in SPECT with 133-xenon, it produces a decrease in global CBF by about 15%, but the regional CBF distribution is not modified.[5] Another issue is related to the very stringent ethical limits to the administration of radioactive tracers in children. In adults, control populations for PET and SPECT studies are rather easily obtained from normal volunteers. Such a practice is ethically and legally prohibited in children so that normal control population *stricto sensu* is unobtainable in this age range. The only means to assess control values is to collect a population of children *a posteriori* considered normal, that means a series of patients exhibiting transient neurological or apparently neurological events but who later proved to develop normally.

Stimulation studies, still exceptional in children, are performed either with PET using H_2O^{15} as a positron emitter or with fMRI. Both lie on the comparison of images obtained during one or several stimulation tasks and a "control" task. PET assesses the increase of regional cerebral blood flow during the stimulation whereas fMRI measures the changes in the magnetic signal of hemoglobin in vessels during the stimulation, changes related with cerebral blood flow variations. Elementary and passive sensory stimulations as well as complex cognitive operations can be tested since a task mentally imagined induces the same changes as a task "truly" performed. Sophisticated statistical analyses are needed in order to detect, select, and localize the significant changes produced by the stimulation.

PET is limited by the radiation dose delivered by the isotope injected whereas fMRI does not require any injection and does not deliver any radiation. However maintaining complete immobility during the entire examination and performing the different tasks in order remain unsolved limitations for the use of fMRI in young children.

2. FUNCTIONAL IMAGING AT REST AND CEREBRAL MATURATION

2.1. Changes in Absolute Values of Metabolism and Cerebral Blood Flow During Childhood

The first set of data pertaining to childhood development[10] was gathered on 29 children aged 5 days to 15 years having suffered from "transient neurological events not significantly affecting normal neurodevelopment" such as facial angioma and partial or generalized controlled seizures due to acute pathology or epilepsy. Using PET, the authors showed that absolute values of local cerebral metabolic rate for glucose (lCMRGlu)—a measure of local cerebral metabolism for glucose—are low in newborns and increase progressively during infancy. Values reach the adult range by 2 years of age but continue to increase to about twice the adult values by 4–5 years and maintain this rate until about 9 years of age. Local cerebral metabolism subsequently decreases to once more reach adult levels by the end of the second decade. Similar results were subsequently reported in a study carried out during the first 6 months of life.[19]

Cerebral blood flow (CBF) data obtained using dynamic SPECT and 133-xenon in a population of 42 children aged from 18 days to 19 years confirmed this pattern. The children were considered normal *a posteriori*.[5] They were suffering from transient abnormal symptoms in which cerebral imaging was needed in order to confirm the integrity of the brain (facial angioma, sleep myoclonus, cerebello-opso-myoclonic syndrome, syncope, headache, etc.). All subjects had normal neurological examination, EEG and CT scan at the time of SPECT investigation; children with any evidence of cerebral lesion, epilepsy, or abnormal brain development were excluded from the study. SPECT examination was considered as part of their clinical evaluation and informed consent was obtained from the parents. Regional CBF was measured in 20 circular cortical regions of interest (ROI) on the 5 axial and 20 mm thick slices acquired, including 9 left, 9 right and 2 median ROIs. Global CBF was obtained by averaging all the values of the ROIs localized on the 3 medium slices and regional CBF was calculated in large cortical regions such as the sensori-motor cortex, Broca area, frontal cortex and various posterior associative areas. Global CBF followed the same developmental curve as absolute metabolic values (Fig. 1). At birth, its value was low, about 40 ml/mn/100 g, although not much lower than in adulthood. Global CBF rapidly increased from birth to 5–6 years to reach a plateau value 50% higher than in adults, about 80 ml/mn/100 g that remained stable until the age of about 10 years. At the end of the second decade, the global CBF slowly decreased to reach the adult value of about 45 ml/mn/100 g which thereon remained stable. All cortical regions showed the same non-linear CBF pattern.

Such a developmental pattern has been reported for most synaptic markers investigated in animals. In particular, a majority of neuroreceptors exhibit such a delayed postnatal overshoot.[22] In humans, the synaptic density calculated at different ages on postmortem brains also reproduces this non-linear pattern.[18] Several neurobiological

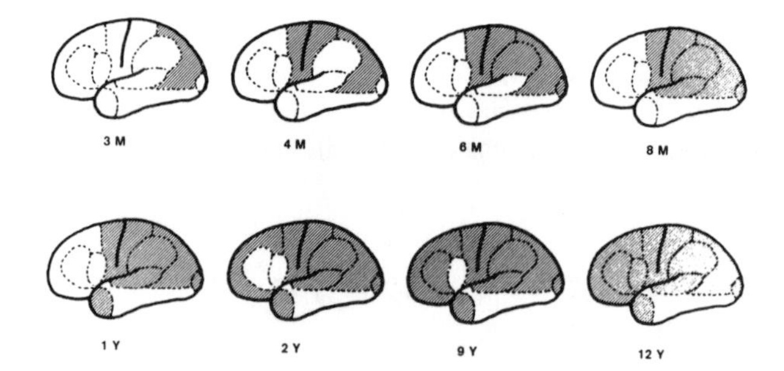

Figure 1. Changes in absolute values of global cerebral blood flow (mCBF) during childhood. The absolute value of the mean and standard deviation of cerebral blood flow, measured with SPECT using 133-xenon and expressed in ml/mn/100 g of cerebral tissue, is presented from birth to adulthood. See text for the description and interpretation of changes.

substrates have been suggested in order to support the developmental CBF curve. The most plausible is synaptogenesis, which follows a similar curve during the post-natal period, i.e., early increase and delayed decrease.[3] Moreover, synapses constitute the neuron component which has the highest level of activity in terms of CBF.[25] CBF may therefore be considered as an indirect marker of synaptic activity, which can be non-invasively quantified using cerebral functional imaging for the study of the human brain.

2.2. Changes in Relative Regional Values of Metabolism and Cerebral Blood Flow during Childhood

Relative regional changes also occur during postnatal brain maturation. Adjusting the absolute regional values for global values in the whole brain (the so-called "relative values") allows the comparison between one region and another. Different cortical regions will exhibit functional changes at different ages.

In neonates, metabolism is relatively elevated in the thalamus, brainstem, cerebellar vermis and sensori-motor cortex. It increases in the occipital and temporoparietal cortex as well as in the basal ganglia and cerebellar hemispheres by 3 to 5 months of life,[10,19] in the associative visual areas by 6 months, and finally in the frontal cortex at the end of the first year.[10] Because the metabolism increases less in the thalamus than in the cerebral cortex, there is a relative subcortical hypermetabolism during childhood. The same postero-anterior gradient is found studying cerebral blood flow through SPECT.[5] The earliest rCBF increase occurs in the sensorimotor cortex at 1 month followed by an increase in the occipital cortex at 3–6 months, the associative auditory and visual cortices at 4 and 7 months respectively, and the frontal cortex during the second year of life (Fig. 2).

These studies show that the main rise of functional cortical activity occurs within the first two years, the earliest regions to increase their activity being the sensorimotor and occipital cortices (before 3–6 months) and the latest, the frontal cortex (after 6–12 months). These developmental milestones are concordant with the age at which the corresponding cognitive functions develop in humans: motor control, particularly that of the face, rapidly develops during the first 2 months of life, visual abilities improve during the first 6 months and the so-called "frontal functions" appear after 1 year.

Using a different methodology based on a voxel by voxel analysis and not on an *a priori* choice of regions of interest as was the case in previous studies, Van Bogaert *et al.*[24] recently demonstrated maturational changes extending over a longer period. They studied glucose metabolism in 19 intellectually normal children aged from 6 to 15 years

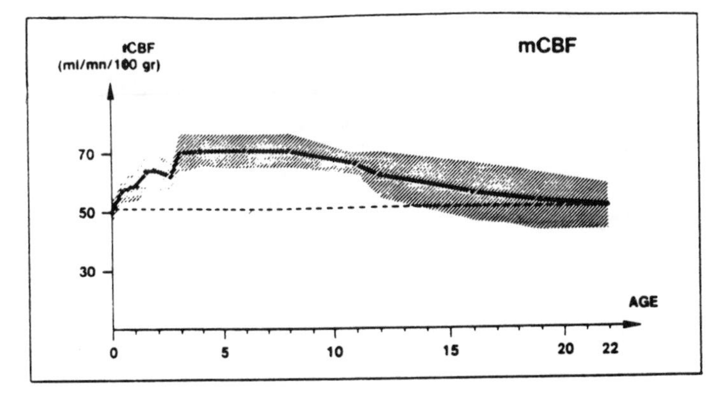

Figure 2. Changes in regional cerebral blood flow during childhood. Regional cerebral blood flow was measured with SPECT using 133-xenon. The region which experiences increase in cerebral blood flow at a given age is presented in gray. Cerebral blood flow increases in the posterior part of the brain as well as in the motor cortex during the first year of life (upper set of drawings) whereas the frontal cortex shows an increased blood flow later in life (lower set of drawings).

who suffered from idiopathic epilepsy. By means of a parametric statistical method currently used in stimulation studies (SPM, statistical parametric mapping), the most important changes were found in the thalamus and the anterior cingulate cortex, other changes were detected in the basal ganglia, mesencephalon, insula, posterior cingulum, frontal and postcentral cortex. Metabolism continuously rose during the entire childhood until the age of 30 years, before it decreased in these structures.

These results confirm the recent proposal made by Changeux[4] which considers the two-step model of brain development (overproduction of synapses, then elimination) as being overly simplistic. The neurobiological basis of Van Boagert *et al.*'s PET findings[24] could be the increase of synaptic activity in the thalamus due to a rise of corticothalamic connectivity during the first two (perhaps three) decades of life. The limbic system could also be involved in this prolonged developmental process.

2.3. Changes in Left and Right Cerebral Blood Flow during Childhood

Using dynamic SPECT and 133-xenon, CBF values were calculated, respectively, for the left and right sides in 39 children (19 males and 20 females) aged from 18 days to 19 years and who proved to be right-handed.[9] These children were reasonably representative of normal ones since they were recruited according to the same criteria as those used for our previous CBF study reported above.[5] The data were analyzed with a particular statistical method, a "mean change-point test", which can detect potential changes in the difference between left and right CBF values. This procedure not only provided a comparison between left and right CBF values, but also allowed a sequential follow-up of left-right asymmetry during childhood.

As expected, left CBF was significantly higher than the right one in these right-handed children. However, the mean left-right CBF difference appeared to be affected by age. Indeed, left CBF became predominant only after 3 years of age whereas right CBF appeared to be higher than the left one before age 3 (Fig. 3). Such a change in asymmetry was significant for the whole hemisphere and for two regions, the sensorimotor and the parieto-temporo-occipital cortex, the latter result being mainly due to a shift from right to left higher CBF in the posterior associative areas.

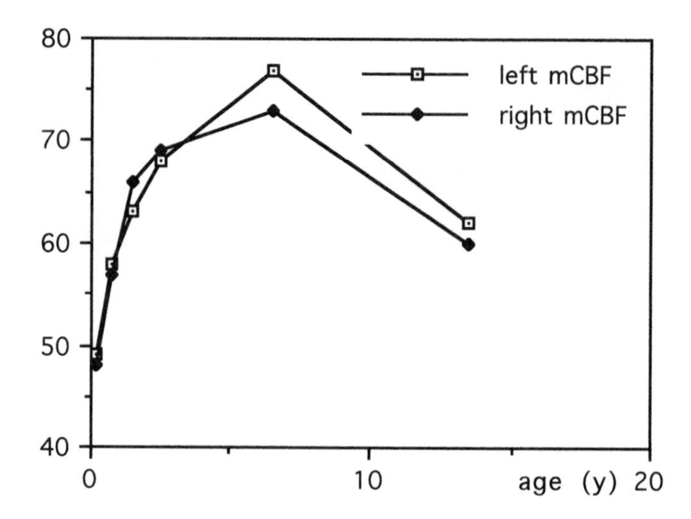

Figure 3. Respective changes in left and right hemispheric cerebral blood flow (left mCBF and right mCBF) during childhood. The absolute value of the mean cerebral blood flow, measured with SPECT using 133-xenon and expressed in ml/mn/100 g of cerebral tissue, is presented for the left and right hemispheres. See text for the description and interpretation of changes.

The hypothesis of an earlier development of the right hemisphere has sometimes been proposed.[12,16] This CBF study provides a quantitative demonstration that the right hemisphere matures earlier than the left one in human infants and shifts from right to left predominance during the fourth year of life. Moreover, the data show a previously unreported strong regional component for CBF asymmetry, predominant in the posterior associative multimodal cortex. The age at which the right and then the left CBF becomes predominant is concordant with the sequential emergence of functions sustained by the two hemispheres, dedicated first to visuospatial abilities (on the right) and then to language functions (on the left).

3. FUNCTIONAL CEREBRAL IMAGING WITH STIMULATIONS IN NORMAL CHILDREN

Only three studies have been published in this field because of the constraints of ethics and cooperation in this population. Two of them concern visual activation and one, working memory.

Born *et al.*[1] used fMRI and flickering light stimulation through the closed eyelids in 17 sleeping infants and children aged from 3 days to 4 years. The data from seven subjects had to be rejected because of movement artifacts or absence of responses. In the remaining subjects, the stimulation induced a decreased signal contrary to the classical increased signal observed in adults. Moreover, the changes in signal were restricted to the anterior and medial part of the calcarine sulcus in the younger infants. The functional organization of visual cortex thus seems to differ in the mature and immature brain.

Using PET and H_2O^{15} in 2-month-old alert infants having suffered from transient perinatal insult, Tzourio *et al.*[13] tested face recognition versus visual fixation of a circle. Whereas such a task induces a left activation in adults, a predominantly right hemisphere rCBF was observed in children. Once more, the neural basis of a cognitive function appears to differ in the developing and mature brains.

Using fMRI and tasks with sequences of letters, Casey *et al.*[2] studied nonspatial working memory in prepubertal children. At this age, the pattern of activation was similar to that seen in adults, involving the inferior and middle frontal gyri.

Finally, there are some interesting data based on behavioral studies in very young children. For instance, event-related potentials, a highly sensitive technique for temporal resolution but still lacking spatial accuracy, have recently been adapted to infants.[11] Three-month-old infants appear to be able to discriminate different phonemes and the discrimination process seems to involve first a temporal, and then a frontal treatment, with a moderate left-hemispheric advantage.

4. FUNCTIONAL CEREBRAL IMAGING IN INFANTILE EPILEPSY

4.1. The Potential Role of Maturation in the Determination of the Age of Onset of Epilepsy

It is interesting to note that seizures that are related to different cerebral sites usually emerge at different ages in epileptic children. Furthermore, epilepsy can begin a long time after birth even though the causal lesion is prenatal. For instance, neonates mainly exhibit motor seizures, occipital lobe seizures usually appear at about 2–3 months of age[20] and frontal lobe seizures are rarely observed in the first year of life.[21] This timing corresponds well to the regional sequence of development described from functional imaging during infancy. We therefore suspect that, as a consequence of maturational processes continuing in the brain after birth, the cortical location of a lesion could in part determine the age of onset of epilepsy during childhood. However, this age-lesion location relationship is also determined by other factors, such as the nature and size of the lesion.

4.2. The Potential Role of Maturation in the Expression of Epilepsy

In many childhood epilepsy syndromes, an age-dependent phenomenon of generalization may be observed. This seems to be the case for Infantile Spasms that develop during the first year of life[14] and for Continuous Spikes and Slow waves in Sleep that emerge during the second part of the first decade. In both conditions, the epileptogenic process involves the whole cortex, regardless of the focal or idiopathic origin of the epilepsy, and tends to disappear after a few years. Such a phenomenon could reflect transient cortical hyperexcitability during the two periods in which functional activity (CBF and metabolism) dramatically increases (the first year) or reaches particularly high values (the first decade).

4.3. The Potential Role of Epilepsy in Later Brain Development

In an immature yet previously normal cortex, persistence of epileptogenic activity could be deleterious to the later maturation of this cortex. Experimental models of epilepsy in the developing visual cortex have shown that maturational processes were impaired: immature pathways that should normally disappear were stabilized.[17] One can postulate that such a condition, which may favor the persistence of seizures, is also incompatible with the normal development of cognitive functions subserved by this part of the brain. This hypothesis is supported by data obtained longitudinally with SPECT.

In infantile spasms, CBF tends to remain abnormally high after 2 years of age if spasms persist but it returns to normal levels if spasms are controlled.[6] In the case of diffuse cortical malformation such as agyria-pachygyria, there is no tendency for spasms to be controlled and no tendency for CBF values to diminish with age.[8] In such extended abnormalities, the capacity for cortical maturation is limited and CBF remains at a fixed value during the entire childhood. One can therefore hypothesize that the recovery of normal maturational processes in the brain is a necessary condition for the cessation of infantile spasms.[15] Such a condition can be obtained by non-specific mechanisms, like decreasing diffuse hyperexcitability of the cortex with steroids,[6] the reference treatment for infantile spasms.

In patients with hemimegalencephaly, a particular malformation that affects a whole hemisphere unilaterally and induces very early seizures, CBF and metabolism are abnormal in the structurally normal hemisphere, except in the first 2–3 months of life. rCBF can get back to normal levels after hemispherectomy and seizure control, provided the patients are operated on before 2 years of age.[7] By contrast, when patients undergo surgery after 5 years of age, metabolism remains abnormal event though the patients are seizure free.[23] It can therefore be argued that frequent epileptogenic discharges rapidly produce functional disorders away from the lesion in the immature brain. Such disorders could produce irreversible dysfunctions in previously normal regions if seizures are not controlled early enough. These data suggest that, when cases are considered for surgery, it may be highly beneficial to operate the patients as early as possible, in order to ensure the normal development of the remaining brain.

CONCLUSION

Cerebral functional imaging techniques performed at rest, PET and SPECT, constitute interesting non-invasive means of studying brain development in humans, in normal as well as in pathological conditions. However, ethical constraints related to using radioactivity and performing research in pediatric populations are limiting the number of subjects reported in the studies. Stimulation studies using fMRI and PET with H_2O^{15} have a great theoretical interest but are very difficult to apply in young children. Developing techniques such as near infrared light scanning could be more adapted to study cerebral activation in this population.

In normal children, there are arguments in favor of a transient hyperfunctionality of the cerebral cortex at rest during the first decade, which disappears during the second decade, although some structures, like the cingulum, continue to mature later on. All cortical regions display such hyperactivity at various points in the development, in a manner that seems to correspond to the age at which cognitive functions sustained by the given regions develop. This provides additional arguments for the existence of critical periods of development in humans. These periods would be much longer than in animals, particularly in the frontal lobe.

Functional imaging also shows a developmental left/right asynchronism. The data support the hypothesis of an earlier development of the right hemisphere, the left one becoming predominant later at around 3 years of age. This is concordant with recent human behavioral studies.

In epileptic children, functional imaging may shed some light on the relationships existing between epileptogenic and maturational processes. The developmental stage of the brain plays a role in the expression of epilepsy in childhood, in terms of age of onset, type of seizures, EEG pattern and neuropsychological features. Conversely, because of the physisological and/or anatomical disorders induced by epileptogenic activity in the immature brain, an epileptic process that begins before two years of age may cause more severe motor or cognitive sequelae than a later one.

REFERENCES

1. Born P, Leth H, Miranda MJ, Rostrup E, Stensgaard A, Peitersen B, Larsson HB and Lou HC (1998): Visual activation in infants and young children studied by functional magnetic resonance imaging. Pediatric Research 44:578–583.
2. Casey BJ, Cohen JD, Jezzard P, Turner R, Noll DC, Trainor RJ, Giedd J, Kaysen D, Hertz-Pannier L and Rapoport JL (1995): Activation of prefrontal cortex in children during a nonspatial working memory task with functional MRI. NeuroImage 2:221–229.
3. Changeux JP and Danchin A (1976): Selective stabilization of developing synapses as a mechanism for the specification of neuronal networks. Nature 264:705–712.
4. Changeux JP (1997): Variation and selection in neural function. Trends in Neuroscience 20:291–293.
5. Chiron C, Raynaud C, Maziere B *et al.* (1992): Changes in regional cerebral blood flow during brain maturation in children and adolescents. Journal of Nuclear Medicine 33:696–703.
6. Chiron C, Dulac O, Bulteau C *et al.* (1994): Study of regional cerebral blood flow in West syndrome. Epilepsia 34:707–715.
7. Chiron C, Delalande O, Soufflet C, Plouin P and Dulac O (1994): Longitudinal evaluation by EEG and SPECT in infants with hemimegalencephaly. Epilepsia, 35:119.
8. Chiron C, Nabbout R, Pinton F, Nuttin C, Dulac O and Syrota A (1996): Brain functional imaging SPECT in agyria-pachygyria. Epilepsy Research 25:109–117.
9. Chiron C, Jambaque I, Nabbout R, Lounes R, Syrota A and Dulac O (1997): The right brain hemisphere is dominant in human infants. Brain 120:1057–1065.
10. Chugani HT, Mazziotta JC and Phelps ME (1987): Positron emission tomography study of human brain functional development. Annals of Neurology 22:487–497.
11. Dehaene-Lambertz G and Dehaene S (1994): Speed and cerebral correlates of syllable discrimination in infants. Nature 370:292–295.
12. De Schonen S and Deruelle C (1991): Visual field asymmetry for pattern processing are present in infancy: A comment at T. Hatta's study on children's performance. Neuropsychologia 29:335–338.
13. Tzourio N, De Schonen S, Mazoyer B, Bore A, Pietrzyk U, Aujard Y and Deruelle C (1992): Regional cerebral blood flow during visual stimuli processing in two-month alert infants. Society of Neuroscience 18:1121.
14. Dulac O, Chugani HT and Dalla Bernardina B (1991): Infantile Spasms and West Syndrome. London: WB Saunders.
15. Dulac O, Chiron C, Robain O *et al.* (1994): Infantile spasms: a pathophysiological hypothesis. Seminars in Pediatric Neurology 2:83–89.
16. Geschwind N and Galaburda AM (1985): Cerebral lateralisation. Archives of Neurology 42:428–459.
17. Grigonis AM and Murphy EH (1994): The effects of epileptic cortical activity on the development of callosal projections. Developmental Brain Research 77:261–265.
18. Huttenlocher PR (1989): Synaptic density in human frontal cortex. Developmental changes and effects of aging. Brain Research 163:195–205.
19. Kinnala A, Suhonen-Polvi H, Aarimaa T, Kero P, Korvenranta H, Riotsalainen U *et al.* (1996): Cerebral metabolic rate for glucose during the first six months of life: an FDG positron emission tomography study. Archives of Disease Childhood 74:F153-F157.
20. Lortie A, Plouin P, Pinard JM *et al.* (1993): Occipital epilepsy in neonates and infants. In Andermann F *et al.* (eds): "Occipital Seizures and Epilepsy in Children." London: John Libbey Eurotext, pp. 121–132.
21. Luna D, Dulac O and Plouin P (1989): Ictal characteristics of cryptogenic partial epilepsies in infancy. Epilepsia 30:827–832.

22. McDonald JW and Johnston MV (1990): Physiological pathophysiological roles of excitatory amino acids during central nervous system development. Brain Research Review 15:41–70.
23. Rintahaka P, Chugani H, Messa and Phelps ME (1993): Hemimegalencephaly: evaluation with positron emission tomography. Pediatric Neurology 9:21–28.
24. Van Bogaert P, Wikler D, Damhaut P, Sliwowski HB and Goldman S (1998): Regional changes in glucose metabolism during brain development from the age of 6 years. Neuroimage 8:62–68.
25. Wong-Riley MT (1989): Cytochrome oxydase: an endogeneous metabolic marker for neuronal activity. Trends Neuroscience 12:94–101.

9

EARLY SIGNS OF LATERALIZATION IN FOCAL EPILEPSY

Gail L. Risse* and Ann Hempel

Minnesota Epilepsy Group
P.A., 310 Smith Avenue N., Suite 300
St. Paul, MN 55102-2383, USA

INTRODUCTION

The development of lateralized functions in the human brain, and the factors that may alter this development, remain a mystery in contemporary neuroscience. This is due in part to the paradoxical necessity of inferring normal development from the study of abnormal brains. In adults, theories of hemispheric specialization have been based primarily on the study of behavioral-cognitive deficits that appear to result from acquired brain lesions. In these cases, it is assumed that normal development typically preceded the brain insult by a number of years. In children, however, the situation is quite different. When brain damage occurs early in development, it is superimposed on a dynamic, evolving neurodevelopmental process. The neuropsychological profiles of brain damaged children do not reflect the outcome of abnormal development but rather abnormal development in progress. It is the interaction between normal development and an acute neuropathological process or event that we believe may result in significant alteration of the genetic plan for lateralization of cortical functions.

In considering the cerebral lateralization of children with epilepsy, many of us begin with two common assumptions: one is that early brain damage lateralized to the left hemisphere frequently results in the reorganization of language and related verbal abilities to the right hemisphere. The second assumption is that this reorganization may help to minimize the cognitive deficits associated with this early damage. There are many unanswered questions about how this process occurs. What is the actual incidence of reorganization of language in children with evidence of early brain damage? In what age range is this reorganization most likely to occur? Are there differences between the sexes in language organization following an early brain insult? What effect does reorganization have on verbal intellectual development in general and other cognitive abilities compared

* To whom correspondence should be addressed

Neuropsychology of Childhood Epilepsy, edited by Jambaqué et al.
Kluwer Academic / Plenum Publishers, New York, 2001.

to children whose language did not reorganize? In the case of children with focal epilepsy, how similar are the neuropsychological profiles to those of adults with comparable seizure foci? This chapter will begin to address some of these questions using data from two sources: The results of language assessment with the Intracarotid Amobarbital Procedure (IAP) in children who are candidates for epilepsy surgery and the neuropsychological test profiles of children with focal epilepsy.

A number of previous studies have addressed the question of whether lateralized brain dysfunction in children results in neuropsychological profiles similar to those of adults with comparable lesions. Reports of language deficits in children who sustained left hemisphere injuries are common[5,28,36,39] and more severe impairment tends to be associated with later onset of injury.[28,39] Studies that have focused on measures of IQ have frequently reported lower overall scores in children with unilateral brain damage (regardless of side) compared to normal controls,[28,32,40] in some cases the effect correlating with lesion size, degree of hemiparesis or age at insult. However, selective impairment of verbal versus visual-spatial or perceptual skills in association with unilateral left or right lesions, respectively, have been reported less frequently.[2,16]

These trends also apply to children with epilepsy. Frequency and type of seizure activity have been linked to lower full-scale IQ,[13] and in some reports left hemisphere seizure onset has been specifically associated with lower verbal IQ as well as decreased performance on other verbal intellectual measures.[18,21,26] Other studies have failed to find left-right differences.[8,10,27] Investigators using more specialized or experimental cognitive measures have not necessarily been more successful in differentiating left from right hemisphere dysfunction associated with seizures. Beardsworth and Zaidel[3] reported greater impairment on delayed memory for faces in children with right temporal lobe epilepsy (TLE) compared to those with left TLE. However, in another study, measures of auditory, verbal and visual-spatial memory failed to differentiate children with left versus right TLE, although performance of the epilepsy sample fell significantly below that of normal controls.[11] Riva *et al.*[33] also failed to identify hemispheric specialization in an experiment using tachistoscopically presented verbal and spatial tasks lateralized to a single visual field. Verbal stimuli were processed less well in the hemisphere of seizure onset, regardless of side or presence of a structural lesion. For spatial stimuli, more errors were committed in the right field for left seizure focus patients, but were equally distributed in the two fields for right focus patients.

Some studies have suggested that failure to identify material specific cognitive deficits corresponding to unilateral hemispheric dysfunction may be attributable to gender differences.[23,35] Inglis and Lawson[23] re-analyzed data by gender for a total of 18 studies and concluded that male patients with left hemisphere pathology demonstrated relative deficits in verbal IQ, whereas a PIQ deficit was associated with right hemisphere lesions. This pattern was not consistent for females. Gender differences in cognitive patterns have been reported in adult epilepsy patients following temporal lobectomy.[17,38] However, clear differences between males and females in cognitive test performance among children with focal epilepsy have not been established.

Strauss *et al.*[35] examined data from 94 seizure patients who had undergone the IAP to determine if incidence of reorganization of language following cerebral injury differed between males and females. Their results suggest that reorganization in females is more likely to occur during the first year of life, whereas in males, some potential for reorganization may extend into puberty. This finding appears to contradict earlier reports suggesting greater plasticity or more diffuse representation of language functions in females.[24,30,31] In this chapter, we present additional data addressing some of these issues.

1. THE INTRACAROTID AMOBARBITAL PROCEDURE IN CHILDREN

The IAP employed was developed by Dr. Ann Hempel at the Minnesota Epilepsy Group specifically for use with children, and has been very successful with our population. The procedure involved a single catheterization of the femoral artery. The dose of sodium Amytal was based on age. Children in the youngest age group usually received an initial dose of either 50 or 75 mg, while older children typically received an initial dose of 75 or 100 mg and adolescents usually received an initial dose of 100 or 125 mg. The dose of sodium amobarbital was increased in prompt bolus increments of 25 mg if the initial injection failed to result in slowing in the ipsilateral hemisphere on EEG and complete contralateral hemiplegia. Injections of both sides were conducted on the same day with a minimum of 20 minutes between first and second injections. Suspected side of seizure origin was injected first in 89% of the cases.

Language testing consisted of counting forward or backward at the time of injection, naming, reading, repetition, responding to verbal commands and rote speech. Language testing of younger and lower IQ children involved rote speech, naming, repetition and responding to commands.

A hemisphere was considered dominant for language if language tasks were failed following ipsilateral injection and passed (in at least one modality) following contralateral injection prior to recovery of motor function. Recovery of motor function was defined as the first spontaneous movement of the contralateral hand observed post-injection. The patient was considered to possess bilateral language representation if language function in at least one modality was demonstrated following each injection prior to recovery of motor function.

1.1. Patient Population

This procedure has been completed successfully on a total of 80 children ranging in age from three to 17 years. Forty-one of the patients were male, while 39 were female. Fifty-three were right-handed, 22 were left-handed and five were indeterminate for handedness. The intellectual ability of this group was highly variable, the overall IQ ranging from profoundly impaired to above average (mean IQ = 79). In general, if a child was able to demonstrate consistent language behavior, even if quite limited in complexity, he/she was considered a candidate for the IAP, at least for the purposes of language lateralization. Most of the patients in the sample had a seizure focus lateralized to one hemisphere as determined by multiple EEG recordings. A total of 41 patients had a left-hemisphere seizure focus while 28 had a seizure focus on the right. Seizure focus was unknown for 11 patients.

1.2. Lateralization of Language by IAP

The number of patients with left, right, bilateral or unclassified language lateralization by IAP is presented by age group in Fig. 1. Note that the relative percentage of children clearly classified as left language dominant appears to increase with age, while the number of children in the unclassified category decreases. This could reflect the fact that it is easier to obtain valid results in older children. This finding may also reflect the somewhat higher incidence of a left seizure focus in the younger group.

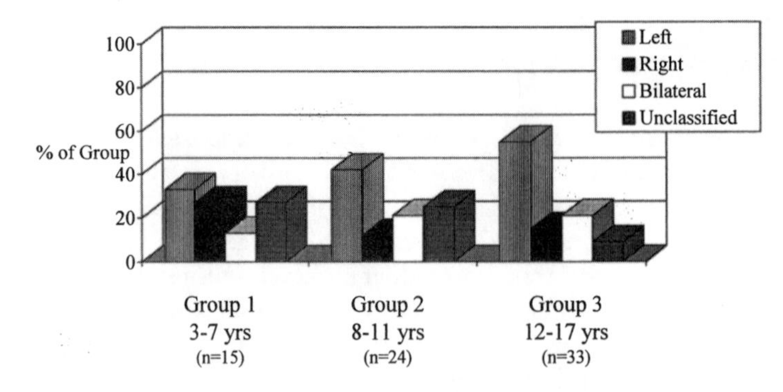

Figure 1. Language lateralization by IAP.

In addition, however, we must consider the possibility that very young children in the unclassified group may not have completed the process of language lateralization, especially considering that they have sustained early brain damage. In this sample, patients who could not be classified for language laterality often demonstrated speech arrest following both injections. We do not classify these patients as bilateral because we cannot be certain of the reason for unresponsiveness. It is likely that at least some of these patients in the youngest age group will develop a more lateralized language profile at a later age.

1.3. Factors Affecting Language Lateralization

1.3.1. Seizure Focus. Children with a left hemisphere seizure focus were much more likely to demonstrate atypical language lateralization than those with a right seizure focus (see Table 1). A total of 44% of the left seizure focus children were found to have right or bilateral language representation, while only 14% of right seizure focus children had atypical language lateralization. This difference was statistically significant ($\chi^2 = 9.59$, $p < 0.01$). Of the 18 cases with atypical language lateralization in the left focus group, 11 had right language dominance and 7 were bilateral.

Atypical language lateralization among children with a right seizure focus was limited to a total of four children with bilateral language. There were no right hemisphere dominant cases in the right seizure focus group.

1.3.2. Age of Onset. Age of seizure onset in relation to language lateralization was examined for the 41 patients demonstrating a left hemisphere seizure focus (Table 1). For purposes of this comparison, early onset was defined as 24 months or younger. All other cases were considered late onset. Among the 16 children in the early onset group, 10 (63%) were right hemisphere dominant for language, while only one child was left dominant and one was bilateral. Four children could not be classified for language lateralization. In the late onset group ($n = 25$), only one child (4%) was right dominant, while 13 children (52%) were left dominant, six were bilateral and five were unclassified. The difference in these distributions between early and late onset cases is statistically significant (Fisher Exact Test $p < 0.01$).

Table 1. Language lateralization

		Left	Right	Bilateral	Unclassified
Seizure Focus					
Left	(n = 41)	34%	27%	17%	22%
Right	(n = 28)	75%	0	14%	11%
Age of Onset (Left Focus)					
Early*	(n = 16)	6%	63%	6%	25%
Late	(n = 25)	52%	4%	24%	20%
Gender (Left Focus)					
Male	(n = 20)	45%	25%	5%	25%
Female	(n = 21)	24%	28%	29%	19%
Early Onset					
Male	(n = 9)	11%	56%	0	33%
Female	(n = 7)	0	72%	14%	14%
Late Onset					
Male	(n = 11)	73%	0	9%	18%
Female	(n = 14)	36%	7%	36%	21%
Handedness					
Left	(n = 22)	18%	50%	23%	9%
Right	(n = 53)	64%	0	15%	21%
Undetermined	(n = 5)	20%	20%	20%	40%

*<24 months.

1.3.3. Gender. The incidence of atypical language lateralization did not differ significantly between males and females with a left hemisphere seizure focus (Table 1). However, there was a trend suggesting females may be more likely to reorganize language. A total of 57% of the girls, compared to 30% of the boys, demonstrated atypical language lateralization. This trend is more pronounced among patients with later age of seizure onset. In the early onset group (n = 16), none of the seven girls and one of the nine boys were left dominant. A somewhat larger number of girls (86%) than boys (56%) demonstrated atypical language lateralization. However, it should be noted that 33% of the boys fell in the unclassified language category. Because bilateral speech arrest was a common reason for failure to classify language, and because it is possible that bilateral speech arrest may represent a form of atypical language, the difference in frequency of atypical language between males and females may not be as great as these numbers suggest. Among late seizure onset children, only one male (9%) vs. 6 females (43%) had atypical language lateralization. This difference approaches statistical significance (Fisher Exact test $p = 0.058$). Five of the six females displayed bilateral language while only one evidenced right hemisphere language. Thus, our data suggest that girls are more likely than boys to develop bilateral language in association with later left seizure onset.

1.3.4. Handedness. In this sample of 80 patients, 22 were left-handed. Sixteen of 22 (73%) left-handed patients displayed atypical language dominance, primarily in association with a left hemisphere seizure focus (Table 1). The incidence of left handedness among left language dominant children was 18%. There was a significant relationship between handedness and age of seizure onset, with early onset cases demonstrating a higher incidence of left handedness and late onset cases more frequently evidencing right handedness, but only for those with a left hemisphere seizure onset (Fisher Exact Test $p < 0.0001$).

2. NEUROPSYCHOLOGICAL PROFILES OF CHILDREN WITH FOCAL EPILEPSY

A total of 36 children with a seizure focus clearly lateralized to the right ($n = 21$) or left ($n = 15$) hemisphere were studied. Not all of these patients were in the sample of 80 patients who underwent the IAP. Cases were selected with an IQ of 70 or higher (range 70–114, $x = 92$). There were 15 males and 21 females. All but two children were right-handed and the left handers were left hemisphere dominant for speech and language based on IAP. Age ranged from five to 17 years with a mean of 11 years.

All children were administered the Wechsler Intelligence Scale for Children-III,[42] the Wechsler Adult Intelligence Scale—Revised[41] or the Wechsler Preschool and Primary Scale of Intelligence—Revised,[43] in combination with a comprehensive battery of more specialized neuropsychological tests including measures of language development, academic achievement, memory processing, visual-spatial and visual-motor performance, and executive functions.

Wechsler subtests are reported as scaled scores. Mean standard scores on all measures for the left and right seizure onset groups are presented in Table 2. Measures of verbal IQ discriminated the left from right seizure onset groups on three of five subtests, the children with left hemisphere focal seizures performing significantly below the right hemisphere group. There were no significant differences between the groups on performance subtests, with the exception of Picture Arrangement, on which the left hemisphere group performed significantly below the right.

Other aspects of verbal processing were assessed, including confrontation naming (Boston Naming Test),[25,45] repetition speech (Memory for Sentences subtest of the Stanford-Binet Intelligence Scale—4th Edition),[37] and language comprehension (Token Test).[6,12] Reading and written calculation skills were assessed with either the Woodcock-Johnson Tests of Achievement—Revised[44] or the Wide Range Achievement Test—Revised.[22] There were no significant differences in standard scores between the two groups, except in written calculation, and again the left hemisphere group was impaired (Table 2). There was an overall trend toward higher scores among children with right hemisphere seizure onset. Tests of nonverbal processing including visual-motor coordination (Developmental Test of Visual Motor Integration and visual-perceptual accuracy);[4] (Judgment of Line Orientation test)[7] were performed comparably by the two groups. Similarly, left and right onset groups did not differ on measures of executive function (Controlled Oral Word Association Test),[15,34] Wisconsin Card Sorting Test[20] and manual dexterity (Grooved Pegboard).[27]

Verbal memory was evaluated with a list learning measure, the Verbal Selective Reminding Test,[9,14,19] and the Story Memory subtest from the Wide Range Assessment of Memory and Learning (WRAML).[1] There were no significant performance differences on any aspect of verbal list learning including long term storage, consistent long term retrieval and delayed recall. Similarly, the two groups obtained equivalent scores on the immediate, delayed and recognition portions of the story memory measure. There were no differences between the groups on a measure of nonverbal memory, the Nonverbal Selective Reminding Test.[14] It is notable that delayed recall is impaired for both groups.

Table 2. Neuropsychological test performance by side of seizure onset

		Left (n = 15)	Right (n = 21)	p
IQ	VIQ	82	94	0.004**
	Vocab	5.9	8.5	0.003**
	Sim.	7.4	9.4	0.05*
	Arith	7.7	9.2	ns
	Comp	6.8	9.0	0.05*
	Dig Sp.	7.7	9.0	ns
	PIQ	94	97	ns
	Blk. Des.	10.7	10.7	ns
	Obj. Assem.	8.7	9.4	ns
	Pic. Comp.	9.3	9.5	ns
	Pic Arrang.	6.0	8.5	0.04*
	Coding	9.9	8.1	ns
	FSIQ	87	95	0.04*
Verbal Processing	Boston Naming	53	63	ns
	♣ Repet. Speech	88	86	ns
	♠ Token Test	77	87	ns
	♦ Reading	85	91	ns
	♦ Calculation	41	89	0.001 ***
Nonverbal Processing	VMI	82	91	ns
	JLO	65	69	ns
Executive and Motor Functions	COWAT	64	70	ns
	WCST (cats.)	60	66	ns
	Groov. Peg R	96	87	ns
	Groov. Peg L	87	75	ns
Verbal Memory	VSRT			
	LTS	85	87	ns
	CLTR	82	83	ns
	Delay %	64	70	ns
	WRAML			
	Story			
	Immed.	72	70	ns
	Delayed	73	75	ns
	Recog.	73	80	ns
Nonverbal Memory	NVSRT			
	LTS	82	83	ns
	CLTR	83	83	ns
	Delay %	55	56	ns

*$p < 0.05$.
**$p < 0.01$.
***$p < 0.001$.
♣ Memory for Sentences, Stanford-Binet.
♠ Token Test for Children or MAE Token Test.
♦ Woodcock-Johnson or WRAT-R.

2.1. Factors Affecting Cognitive Patterns in Children with Left or Right Focal Seizures

2.1.1. Early Seizure Onset. As with language lateralization, it is intuitively appealing to predict greater disruption of cognitive development in children who have suffered brain dysfunction from an early age. Indeed, the IQ differences reported earlier between left and right seizure onset groups appear to be primarily attributable to the early onset (<24 months) cases. For this reason, cognitive test performance was reconsidered for the early seizure onset group only (n = 14).

Consideration of individual IQ subtest scores reveals consistently lower scaled scores for the left early onset group (Table 3). However, left-right differences were only statistically significant for the Vocabulary, Similarities and Arithmetic subtests, most likely due to the small sample size. In contrast, performance on other measures of verbal processing failed to identify significant differences based on side of seizure onset. Once again, however, relative impairment for the left seizure onset group was strongly suggested on measures of confrontation naming, language comprehension and written calculation. Mean performance on these measures by the left seizure onset group fell in the clinically impaired range. There were no left-right differences on measures of nonverbal processing or executive functions, although the left seizure onset patients again were clinically impaired on measures of executive functions.

Measures of verbal and nonverbal memory processing also failed to differentiate early seizure onset children based on side of seizure focus. However, it is notable that both groups performed in the below average range on delayed recall conditions of the story memory and nonverbal memory measures.

2.1.2. Gender. Gender may also confound cognitive differences based on hemisphere of seizure onset. When patients were divided into male and female groups, left-right differences in full scale and verbal IQ were identified only for the males (Table 4). Significant left-right differences were noted on all verbal IQ subtests except Arithmetic in the male group, while none of these measures were performed differently by females based on side of seizure origin. In addition, left seizure focus males performed significantly worse than right focus males in the areas of language comprehension, repetition speech and written calculation, while none of these differences were significant for the females. Performance by the left onset males fell in the clinically impaired range on these tasks. On measures of nonverbal processing, left seizure focus males again performed in the impaired range on the Judgment of Line Orientation test and mean score for the group fell significantly below the right hemisphere males. Interestingly, there was an opposite trend for females, with somewhat greater impairment noted in the right seizure focus group (a pattern more consistent with the adult data). Performance by the left hemisphere females fell just below the average range on this task. The left-right difference for the females was not statistically significant.

CONCLUSION

In this group of children with chronic epilepsy, the role of the left hemisphere in early intellectual development appears to be primary. Onset of seizures from the left hemisphere at age 24 months or younger is associated with an increased incidence of right

Table 3. Neuropsychological test performance in patients with early seizure onset

		Left (n = 6)	Right (n = 8)	p
IQ	VIQ	80	99	0.01**
	Vocab.	5.3	9.4	0.007**
	Sim.	7.2	10.6	0.02*
	Arith.	6.3	10.5	0.05*
	Comp.	6.2	8.9	ns
	Dig. Sp.	6.5	8.5	ns
	PIQ	89	101	ns
	Blk. Des.	9.8	11.4	ns
	Obj. Assem.	8.2	9.9	ns
	Pic. Comp.	8.5	10.3	ns
	Pic. Arrang.	4.2	8.3	ns
	Coding	9.0	7.8	ns
	FSIQ	83	100	0.01**
Verbal Processing	Boston Naming	42	63	ns
	♣ Sentence Rep.	88	77	ns
	♠ Token Test	68	94	ns
	♦ Reading	75	93	ns
	♦ Calculation	43	90	ns
Nonverbal Processing	VMI	83	95	ns
	JLO	59	78	ns
Executive and Motor Functions	COWAT	65	92	ns
	WCST (cats.)	46	76	ns
	Groov. Peg R	93	85	ns
	Groov. Peg L	87	77	ns
Verbal Memory	VSRT			
	LTS	91	86	ns
	CLTR	94	81	ns
	Delay %	67	74	ns
	WRAML			
	Immed.	74	82	ns
	Delayed	64	70	ns
	Recog.	63	77	ns
Nonverbal Memory	NVSRT			
	LTS	70	80	ns
	CLTR	70	82	ns
	Delay %	50	60	ns

*$p < 0.05$.
**$p < 0.01$.
***$p < 0.001$.
♣ Memory for Sentences, Stanford-Binet.
♠ Token Test for Children or MAE Token Test.
♦ Woodcock-Johnson or WRAT-R.

hemisphere or bilateral language representation. Children with left hemisphere seizure onset also perform significantly below children with right hemisphere seizures on measures of verbal intelligence and verbal intellectual processing. This would appear to suggest that the reorganization of language may occur at the expense of verbal intelligence and other higher level verbal functions. However, a simple interpretation of these findings could be misleading. Gender also appears to be a factor in the developmental

Table 4. Neuropsychological test performance by gender

		Males (n = 15)			Females (n = 21)		
		Left	Right	p	Left	Right	p
IQ	VIQ	73	98	0.000 ***	91	93	ns
	Vocab.	4.4	9.0	0.000 ***	7.3	8.2	ns
	Sim.	6	10	0.027*	8.6	9.0	ns
	Arith.	6.1	9.4	NS	9.1	9.2	ns
	Comp.	4.7	9.4	0.05*	8.6	8.7	ns
	Dig. Sp.	5.1	9.6	0.04*	10	8.6	ns
	PIQ	84	92	ns	103	100	ns
	Blk. Des.	9.4	9.5	ns	11.9	11.4	ns
	Obj. Assem.	7.6	8.1	ns	9.8	10.2	ns
	Pic. Comp.	8.3	8.9	ns	10.3	9.9	ns
	Pic. Arrang.	5.4	8.0	ns	6.5	8.8	ns
	Coding	7.1	7.8	ns	12.3	8.4	0.03*
	FSIQ	77	94	0.006	96	96	ns
Verbal Processing	Boston Nmg.	46	59	ns	60	65	ns
	♣ Sentence Rep.	79	98	0.02*	97	78	ns
	♠ Token Test	51	96	0.02*	100	81	ns
	♦ Reading	73	90	ns	95	92	ns
	♦ Calculation	35	98	0.002**	47	84	ns
Nonverbal Processing	VMI	71	88	ns	61	57	ns
	JLO	53	96	0.03*	77	52	ns
Executive and Motor Functions	COWAT	66	91	ns	61	57	ns
	WSCT (cats)	55	75	ns	65	60	ns
	Pegs R	89	60	ns	102	104	ns
	Pegs L	79	59	ns	93	86	ns
Verbal Memory	VSRT						
	LTS	73	95	ns	97	82	ns
	CLTR	74	90	ns	90	78	ns
	Delay %	68	83	ns	88	79	ns
	WRAML						
	Immed.	69	75	ns	76	67	ns
	Delayed	63	66	ns	81	81	ns
	Recog.	58	72	ns	86	85	ns
Nonverbal Memory	NVSRT						
	LTS	88	87	ns	77	82	ns
	CLTR	87	85	ns	80	82	ns
	Delay %	54	46	ns	58	63	ns

* $p < 0.05$.
**$p < 0.01$.
***$p < 0.001$.
♣ Memory for Sentences, Stanford-Binet.
♠ Token Test for Children or MAE Token Test.
♦ Woodcock-Johnson or WRAT-R.

response to early left hemisphere dysfunction. In general, boys appear somewhat less likely to reorganize language, especially as age of seizure onset increases, compared to girls. At the same time, boys with a left seizure focus are more likely to demonstrate compromised verbal abilities, perhaps because verbal functions have not reorganized as effectively. Unfortunately we cannot determine the actual co-occurrence of these factors using the present data since the two samples described are not completely overlapping.

The failure to identify other lateralized cognitive deficits in this population may in part be attributable to the heterogeneous nature of the focal pathology in these groups. Although seizure onset has been localized to a single hemisphere, the sample size is too limited to subgroup patients by focal dysfunction of a given lobe. The possibility that lateralized or material specific memory deficits could be identified in a group of children with purely right or left temporal lobe epilepsy has not been excluded. It may be equally plausible, however, to speculate that lateralized or focal cognitive deficits similar to those demonstrated by adults with focal lesions may never be clearly delineated in a pediatric population due to the increased likelihood of cortical reorganization as a response to early brain insult.

REFERENCES

1. Adams W and Sheslow D (1990): Wide Range Assessment of Memory and Learning—Manual. Wilmington, DE: Jastak Associates.
2. Ballantyne A, Scarvie K and Trauner D (1994): Verbal and performance IQ patterns in children after perinatal stroke. Developmental Neuropsychology 10:39–50.
3. Beardsworth E and Zaidel D (1994): Memory for faces in epileptic children before and after brain surgery. Journal of Clinical and Experimental Neuropsychology 16:589–596.
4. Beery KE (1989): The Visual-Motor Integration Test. Administration, Scoring and Teaching Manual. Cleveland: Modern Curriculum Press.
5. Bendersky M and Lewis M (1990): Early language ability as a function of ventricular dilatation associated with intraventricular hemorrhage. Journal of Behavioral Pediatrics 11:17–21.
6. Benton AL and Hamsher K.deS (1989): Multilingual Aphasia Examination. Iowa City, Iowa: AJA Associates.
7. Benton AL, Varney NR and Hamsher K (1978): Visuospatial judgment: A clinical test. Archives of Neurology 35:364–367.
8. Blackburn L and Ritter F (1989): Effectiveness of cognitive and memory measures in lateralization of dysfunction in children with epilepsy. Epilepsia 30:688.
9. Buschke H (1973): Selective reminding for analysis of memory and learning. Journal of Verbal Learning and Verbal Behavior 12:543–550.
10. Camfield PR, Gates R, Ronen G, Camfield C, Ferguson A and McDonald G (1984): Comparison of cognitive ability, personality profile, and school success in epileptic children with pure right versus left temporal lobe EEG foci. Annals of Neurology 15:122–126.
11. Cohen M (1992): Auditory/verbal and visual/spatial memory in children with complex partial epilepsy of temporal lobe origin. Brain and Cognition 20:315–326.
12. DiSimoni FG (1985): Token Test for Children. Allen, TX: DLM Teaching Resources.
13. Farwell JR, Dodrill CB and Batzel LW (1985): Neuropsychological abilities of children with epilepsy. Epilepsia 26:395–400.
14. Fletcher JM (1985): Memory for verbal and nonverbal stimuli in learning disability subgroups: Analysis by selective reminding. Journal of Experimental Child Psychology 40:244–259.
15. Gaddes W and Crockett D (1975): The Spreen-Benton Aphasia Tests; normative data as a measure of normal language development. Brain and Language 2:257–280.
16. Gadian D, Issacs E, Cross M, Connelly A, Jackson G, King M, Neville B and Vargha-Khadem F (1996): Lateralization of brain function in childhood revealed by magnetic resonance spectroscopy. Neurology 46:974–977.

17. Geckler C, Chelune G, Trenerry M and Ivnik R (1993): Gender related differences in cognitive status following temporal lobectomy. Archives of Clinical Neuropsychology 8:226–227.
18. Griebel M, Williams J, Sharp G and Shema S (1995): Prediction of seizure focus from selected neuropsychological tests in pediatric presurgical patients. Epilepsia 36:110.
19. Hannay J and Levin H (1985): Selective reminding test: An examination of the equivalence of four forms. Journal of Clinical and Experimental Neuropsychology 7:251–263.
20. Heaton RK (1981): Wisconsin Card Sorting Test-Manual. Odessa, Fl: Psychological Assessment Resources, Inc.
21. Hempel A, Risse G, Frost M and Ritter F (1994): The utility of neuropsychological testing for identifying lateralization of the epileptogenic region in children. Epilepsia 35 (Suppl):80.
22. Jastak S and Wilkinson G (1984): Wide Range Achievement Test—Revised. Wilmington, DE: Jastak Associates.
23. Inglis J and Lawson JS (1981): Sex differences in the effects of unilateral brain damage on intelligence. Science 212:693–695.
24. Inglis J, Ruckman M, Lawson J, MacLean A and Monga T (1982): Sex differences in the cognitive effects of unilateral brain damage. Cortex 18:257–275.
25. Kaplan E, Goodglass H and Weintraub S (1983): Boston Naming Test. Philadelphia: Lea & Febiger.
26. Kasteleijn-Nolst Trenité D, Siebelink B, Berends S, van Strien J and Meinardi H (1990): Lateralized Effects of subcortical epileptiform EEG discharges on scholastic performance in children. Epilepsia 31:740–746.
27. Korkman M, Granström M-L and Lehto S (1996): Neuropsychological performance in children with lateralized epileptic foci. Paper presented: International Neuropsychological Society Annual Meeting, Chicago, Illinois.
28. Levine S, Huttenlocher P, Banich M and Duda E (1987): Factors affecting cognitive functioning of hemiplegic children. Developmental Medicine and Child Neurology 29:27–35.
29. Matthews CG and Klove H (1964): Instruction Manual for the Adult Neuropsychology Test Battery. Madison, WI: University of Wisconsin Medical School.
30. McGlone J (1977): Sex d ifferences in the cerebral organization of verbal functions in patients with unilateral brain lesions. Brain 100:775–793.
31. McGlone J (1980): Sex differences in human brain asymmetry: A critical survey. The Behavioral and Brain Sciences 3:215–263.
32. Riva D and Cazzaniga L (1986): Late effects of unilateral brain lesions sustained before and after age one. Neuropsychologia 24:423–428.
33. Riva D, Pantaleoni C, Milani N and Giorgi C (1993): Hemispheric specialization in children with unilateral epileptic focus, with and without computed tomography-demonstrated lesion. Epilepsia 34:69–73.
34. Spreen O and Benton AL (1969, 1977): The Neurosensory Center Comprehensive Examination for Aphasia. Victoria, B.C.: Neuropsychology Laboratory, University of Victoria.
35. Strauss E, Wada J and Goldwater B (1992): Sex differences in interhemispheric reorganization of speech. Neuropsychologia 30:353–359.
36. Thal DJ, Marchman V, Stiles J, Aram D, Trauner D, Nass R and Bates E (1991): Early lexical development in children with focal brain injury. Brain and Language 40:491–527.
37. Thorndike RL, Hagen, EP and Sattler JM (1986): Stanford-Binet Intelligence Scale, Fourth Edition. Chicago: Riverside Publishing.
38. Trennery M, Jack C, Cascino G, Sharbrough F and Ivnik R (1995): Gender differences in post-temporal lobectomy verbal memory and relationships between MRI hippocampal volumes and preoperative verbal memory. Epilepsy Research 20:69–76.
39. Vargha-Khadem F, Gorman A and Walters G (1985): Aphasia and handedness in relation to hemispheric side, age at injury and severity of cerebral lesion during childhood. Brain 108:677–696.
40. Vargha-Khadem F, Isaacs E, vanDerWerf S, Robb S and Wilson J (1992): Development of intelligence and memory in children with hemiplegic cerebral palsy. Brain 115:315–329.
41. Wechsler D (1981): Wechsler Adult Intelligence Scale—Revised. New York: Psychological Corporation.
42. Wechsler D (1991): Wechsler Intelligence Scale for Children—Third Edition. New York: Psychological Corporation.
43. Wechsler D (1989): Wechsler Preschool and Primary Scale of Intelligence—Revised. New York: Psychological Corporation.
44. Woodcock RW and Johnson MB (1989, 1990): Woodcock-Johnson Psycho-Educational Battery—Revised. Allen, TX: DLM Teaching Resources.
45. Yeates KD (1994): Comparison of developmental norms for the Boston Naming Test. The Clinical Neuropsychologist 8:91–98.

NEUROPSYCHOLOGY OF TEMPORAL LOBE EPILEPSY IN CHILDREN

Isabelle Jambaqué

Service de Neuropédiatrie
Hôpital Saint Vincent de Paul
82 Avenue Denfert Rochereau
75674 Paris Cedex 14, France

INTRODUCTION

Complex partial seizures are one of the most common seizure types encountered in both children and adults.[9] In adults, many studies have established that temporal lobe epilepsy may affect memory abilities,[20] auditory perception processing[25] and global visual processing.[11] However, few studies have focused on the deleterious impact that temporal lobe epilepsy may have during childhood. Temporal lobe epilepsy appears compatible with normal intellectual level but there is evidence that psychological difficulties of varying types and degrees are overrepresented in children suffering from such recurrent seizures. Temporal lobe epilepsy also appears to be compatible with a whole range of educational accomplishments. Nevertheless, epileptic children attending regular school are at greater risk of developing learning problems than other children.[21,35,38] Behavioral disturbances such as hyperactivity, affective and personality disorders, may also be present in children with temporal lobe epilepsy.[36]

1. TEMPORAL LOBE EPILEPSY IN CHILDREN

Temporal lobe epilepsy in childhood, with a typical spike focus in the anterior temporal areas, is both less common and more heterogeneous among children than in adolescents and adults.[3] Age of onset is usually during school years rather than early childhood. In addition, focal seizures of temporal lobe origin in infants may seem generalized clinically. In young children, the pathological abnormalities associated with seizures of temporal lobe origin mostly consist of dysplasias, migrational disorders, hamartomas and low-grade tumors, thus including the neocortex more than limbic areas. Mesial temporal sclerosis is seen more often in older children and its pathogenesis

Neuropsychology of Childhood Epilepsy, edited by Jambaqué et al.
Kluwer Academic / Plenum Publishers, New York, 2001.

remains a subject of controversy. The relationship between hippocampal sclerosis, febrile seizures and complex partial seizures in temporal lobe epilepsy continues to be the subject of a great debate in the literature. On the other hand, hippocampal sclerosis could be underdiagnosed in children and be the cause and not the consequence of temporal lobe epilepsy.[18]

2. COGNITIVE ABILITIES

Most children with temporal lobe epilepsy appear to have a normal IQ although IQ differences were found between children with bitemporal epilepsy or unilateral temporal epilepsy and those with idiopathic generalized epilepsy.[21] The picture is clearly different when temporal lobe epilepsy begins in infancy. Pascual-Castroviejo *et al.*[34] reported the 24-year preoperative evolution of a temporal astrocytoma in a patient suffering since 5 months of age from complex focal seizures that could not be controlled with medication. The patient developed severe psychomotor retardation.

When epilepsy begins at school age, there is evidence that global measures of intelligence provide little information regarding the specific nature of the cognitive deficits or educational needs that a child with temporal lobe epilepsy may display. Unfortunately, most studies on cognitive functions in epileptic children have focused on intelligence quotients (IQ) while handedness is only reported in a minority of studies. Nevertheless, in 1984, Camfield *et al.*[4] failed to identify any specific pattern of impairment as characteristic of children with temporal lobe epilepsy using an extensive battery of tests. However, schooling problems are a major concern of parents of children with temporal lobe epilepsy. More recently, Gadian *et al.*[16] found in 22 right-handed children with intractable temporal lobe epilepsy a loss of verbal cognitive functions associated with left-sided pathology whereas a loss of nonverbal functions was associated with right-sided pathology. These findings are consistent with the pattern of lateralization of brain function observed in adults.

3. LANGUAGE FUNCTIONS

Although temporal lobe epilepsy can produce ictal arrest of speech, patients with temporal lobe seizure foci are rarely aphasic. In the Landau-Kleffner syndrome, aphasia develops between 3 and 8 years of age. EEG is abnormal, usually with bilateral spikes and waves, more marked in the temporal regions, but seizures are usually rare.[27] Anterior left temporal lobe epilepsy appears to produce more subtle language disorders. In adults, it has been hypothesized that impairment of naming ability and word usage is a predominant symptom of complex partial seizures of left temporal origin.[30] Children with left temporal epilepsy can also exhibit naming disorders and a limited vocabulary.[14,21] In our experience, temporal lobe epilepsy of early onset can be associated with developmental language disorders in young children.

Relatively little is known about the cortical representation of language in patients with developmental pathology and epilepsy. Using electrical stimulation of chronically implanted subdural electrodes, Duchowny *et al.*[12] mapped the language areas of 34 patients evaluated for epilepsy surgery. They found that developmental lesions and early-onset seizures did not displace the cortical representation of language from prenatally determined sites. Lesions acquired before the age of 5 years, on the other hand, seemed to result in a transfer of language to the opposite hemisphere, provided the language areas had been destroyed.

Temporal lobe epilepsy is also found to be associated with poor reading development. Children with complex partial epilepsy have more difficulties than children with idiopathic generalized epilepsy although both exhibit poor reading performances.[36,38] Furthermore, children with left temporal discharges show lower reading performances than children with right temporal lobe epilepsy.[24,35] Reading problems appear unrelated to the level of intelligence.[32] There is, however, a correlation between reading performance and seizure frequency in children with temporal lobe epilepsy.[38]

4. MEMORY

Disorders of learning and memory have long been associated with temporal lobe dysfunction and they have therefore been studied in detail both prior to and following temporal lobectomy. In adults, left temporal lobe epilepsy is frequently associated with verbal memory impairments whereas right temporal epilepsy is associated with poor visual memory. Anticonvulsant drugs, particularly phenobarbital, phenytoin and primidone may exacerbate memory impairment.[37]

The assessment of memory in children is particularly challenging because until very recently few normative tests of memory were available for children. Moreover, most tests assess only learning and/or immediate recall but do not include measures of delayed recall. Furthermore, it is difficult to find a test that is suitable for a wide range of ages. Although the literature available on memory deficits during childhood is limited, memory deficits have been identified in epileptic children. Partial epilepsy has been found to cause more substantial memory impairments than idiopathic generalized epilepsy and this is particularly true for children with complex partial seizures emanating from the temporal lobe.[14,21]

Similarly, few studies aimed at studying the laterality effects associated with partial epilepsy have been conducted in children. In 1969, Fedio and Mirsky[14] found a dissociable pattern of memory deficits in epileptic children that resembled the one described in adults with temporal lobectomy. Children with left temporal epilepsy had relatively greater verbal memory deficits while children with right temporal seizures showed a greater impairment of visuo-spatial memory. In 1993, Jambaqué *et al.*[21] studied memory efficiency in 60 epileptic children with various types of epilepsy including temporal lobe epilepsies. They used Signoret's Memory Battery which quantitatively assesses new learning and delayed recall with both verbal and visual material. As expected, epileptic children frequently exhibited memory deficits. Poor memory efficiency was noted in both the learning and the recall phase. Strikingly, although IQ measures did not discriminate children with left and right temporal lobe epilepsy, the influence of hemispheric specialization was clearly evidenced in memory abilities. In particular, delayed story recall was lower than the immediate story recall in the left temporal group and a marked visual memory deficit was unraveled in the right temporal group.

In the same context, Beardsworth and Zaidel[1] investigated memory for faces in 29 children and adolescents suffering from temporal lobe epilepsy. Before surgery, the patients with right temporal epilepsy performed worse than either the left temporal epilepsy or control groups. The results suggest that, just as the adult population and in spite of early brain damage, children display a right-hemisphere specialization for face memory. By contrast, Elger *et al.*[13] failed to demonstrate material-specific memory deficits in children who were pharmacoresistant and candidates for surgery. However, they observed a frequent loss of visual memory after right temporal lobectomy and a lesser frequent loss of verbal memory after left temporal lobectomy in children than in adults.

In summary, temporal lobe epilepsy may contribute to memory deficits in children, as it does in adults. In a few studies, a material-specific dissociation in memory abilities has been demonstrated in children with left and right temporal lobe epilepsy. In general, visual memory seems to be particularly affected by right-sided lesions during childhood while the relationship between verbal memory and left hemisphere damage is more equivocal. This cognitive pattern can be related to the predominance of visual memory organization in young children[7] and/or to brain plasticity that enables a shift of verbal memory functions from the left to the intact right hemisphere.[12]

5. AFFECT AND SOCIAL BEHAVIOR

Emotional changes have been well recognized as ictal events in temporal lobe epilepsy both in children and in adults. These changes consist of various stages of fear and sadness, rage attacks, depression, embarrassment, joy or ecstasy. In fact, some early work related psychopathology to temporal lobe dysfunction.[17] In adults, Flor-Henry[15] found left temporal epilepsy to correlate with psychosis but Jensen and Larsen[25] failed to confirm this correlation. Other studies have reported a special relationship between affective disorders and right temporal epilepsy.[2]

In children, there is agreement that temporal lobe epilepsy is associated with higher rates of psychopathology than in the general population. Davidson and Falconer[8] reported aggressiveness and "bizarre behavior" in most children of his series before they underwent temporal lobectomy. Stores[35] found that boys with left temporal lobe foci were more often isolated, overactive and anxious than girls and boys with right temporal foci. Lindsay *et al.*[28] confirmed that children with a left temporal focus are particularly at risk.

Nevertheless, the influence of laterality on the development of behavioral disorders remains controversial. Elger *et al.*[13] reported rage outbursts, aggressiveness and maladjustment in children with temporal lobe epilepsy. These behavioral disturbances were more frequent in children with right than left temporal epilepsy but this difference did not reach statistical significance. However, the combination of an early onset of epilepsy, the presence of a structural lesion and high frequency focal spikes seemed to be associated with increased aggressiveness.

Hermann *et al.*[19] attempted to directly link the extent of behavioral problems to the neuropsychological impairments accompanying altered brain functions in epileptic children. Interestingly, behavioral improvements were found after surgery in childhood.[8,13,28,33] However, a number of factors could influence the development of behavioral problems in children with temporal lobe epilepsy, including psychological factors related to the psychosocial consequences of epilepsy *per se*, as well as recurrent epileptic discharges and prolonged treatment with anti-epileptic drugs.

A special mention must be made about the possible link between temporal lobe epilepsy and autism. Autistic disorder has occasionally been associated with refractory temporal epilepsy, suggesting a relationship between limbic dysfunction and autism.[5,9] Recently, Neville *et al.*[33] reported autistic regression in two children with partial epilepsy and subclinical seizure activity in both temporal lobes. Similarly, a dysfunction of the temporo-occipital region has been found in an epileptic child with autism[22] and

Kyllerman *et al.*[26] have reported transient psychosis in a girl who presented epileptiform activity over the temporal and parietal areas.

CONCLUSION

Temporal lobe epilepsy in children has not been studied as well as in adults. Nevertheless, there is evidence that temporal lobe epilepsy in childhood can produce specific cognitive deficits including mental retardation, language disorders, attention and memory disturbances. Neurobehavioral problems are also frequently encountered in children with intractable complex partial seizures. Left temporal lobe epilepsy is also considered to be a risk factor for the development of behavioral disturbances and learning disabilities. Academic underachievement could be the result of limited vocabulary abilities, poor reading development and verbal memory deficits. The early assessment of such cognitive impairments can guide neuropsychological rehabilitation.

REFERENCES

1. Beardsworth ED and Zaidel DW (1994): Memory for faces in epileptic children before and after brain surgery. Journal of Clinical and Experimental Neuropsychology 16:589–596.
2. Bear D and Fedio P (1977): Quantitative analysis of interictal behavior in temporal lobe epilepsy. Archives of Neurology 34:454–467.
3. Bourgeois BF (1998): Temporal lobe epilepsy in infants and children. Brain Development 20:135–141.
4. Camfield PR, Gates R, Ronen G, Camfield C, Ferguson A and Mc Donald GW (1984): Comparison of cognitive ability, personnality profile and school sucess in epileptic children with pure right versus left temporal lobe EEG foci. Annals of Neurology 15:122–126.
5. Carracedo A, Martin Mucrica F, Garcia Penas JJ, Ramos J, Cassinello E and Calvo MD (1995): Autistic syndrome associated with refractory temporal epilepsy. Revue Neurologique 23:1239–1241.
6. Cohen M (1992): Auditory/verbal and visual/spatial memory in children with complex partial epilepsy of temporal lobe origin. Brain and Cognition 20:315–326.
7. Cramer P (1976): Changes from visual to verbal memory organization as a function of age. Journal of Experimental Child Psychology 22:50–57.
8. Davidson S and Falconer M (1975): Outcome of surgery in 40 children with temporal lobe epilepsy. Lancet 1:1260–1263.
9. Deonna T, Ziegler AL, Despland PA and Van Melle G (1986): Partial epilepsy in neurologically normal children: clinical syndromes and prognosis. Epilepsia 27:241–247.
10. Deonna T, Ziegler AL, Moura-Serra J and Innocenti G (1993): Autistic regression in relation to limbic pathology and epilepsy: report of two cases. Developmental Medicine and Child Neurology 35:166–176.
11. Doyon J and Milner B (1991): Right-temporal lobe contribution to global visual processing. Neuropsychologia 29:343–360.
12. Duchowny M, Jayakar P, Harvey AS, Resnick T, Alvarez L, Dean P and Levin B (1996): Language cortex representation: effects of developmental versus acquired pathology. Annals of Neurology 40:31–38.
13. Elger CE, Brochaus A, Lendt M, Kowalik A and Steidel S (1997): Behavior and cognition in children with temporal lobe epilepsy. In Tuxhorn I, Holthausen H and Boenigk H (eds): "Paediatric Epilepsy Syndromes and their Surgical Treatment." London: John Libbey, pp 311–325.
14. Fedio P and Mirsky AF (1969): Selective intellectual deficits in children with temporal lobe or centrencephalic epilepsy. Neuropsychologia 7:287–300.
15. Flor-Henry P (1969): Psychosis and temporal lobe epilepsy: a controlled investigation. Epilepsia 10:363–395.
16. Gadian DG, Isaacs EB, Cross JH, Connelly A, Jackson GD, King MD, Neville BG and Vargha-Khadem F (1996): Lateralization of brain function in childhood revealed by magnetic resonance spectroscopy. Neurology 46:974–977.

17. Gibbs FA, Gibbs EF and Furster B (1948): Psychomotor epilepsy. Archives of Neurology 60:331–340.
18. Harvey AS, Berkovic SF, Wrennall JA and Hopkins IJ (1997): Temporal lobe epilepsy in childhood: clinical, EEG, and neuroimaging findings and syndrome classification in a cohort with new-onset seizures. Neurology 49:960–968.
19. Hermann BP and Whitman S (1984): Behavioral and personality correlations of epilepsy: a review: methodoloccal critique, and conceptual model. Psychological Bulletin 95:451–497.
20. Hermann BP, Seidenberg M, Haltenier A and Wyler AR (1992): Adequacy of language function and verbal memory performance in unilateral temporal lobe epilepsy. Cortex 28:423–433.
21. Jambaqué I, Dellatolas G, Dulac O, Ponsot G and Signoret JL (1993): Verbal and visual memory impairment in children with epilepsy. Neuropsychologia 31:1321–1337.
22. Jambaqué I, Mottron L, Ponsot G and Chiron C (1998): Autism and visual agnosia in a child with right occipital lobectomy. Journal of Neurology, Neurosurgery and Psychiatry 6:555–560.
23. Jensen I and Larsen K (1979): Mental aspects of temporal lobe epilepsy. Journal of Neurology Neurosurgery and Psychiatry 42:256–265.
24. Kastelejn-Nolst Trenité DGA, Smit AM, Velis DN *et al.* (1900): One line detection of transient neuropsychological disturbances during EEG discharges in children with epilepsy. Developmental Medicine and Child Neurology 32:46–50.
25. Kester DB, Saykin AJ, Sperling MR, O'Connor MJ, Robinson LJ and Gur RB (1991): Acute effect of anterior temporal lobectomy on musical processing. Neuropsychologia, 29:703–708.
26. Kyllerman M, Nyden A, Praquin N, Rasmussen P, Wetterquist AK and Hedstrom A (1996): Transient psychosis in a girl with epilepsy and continuous spikes and waves during slow sleep. European Child and Adolescent Psychiatry 5:216–221.
27. Landau WM and Kleffner FR (1957): Syndrome of acquired aphasia with convulsive disorder in children. Neurology 7:523–530.
28. Lindsay J, Ounsted C and Richards P (1984): Long term outcome in children with temporal lobe seizures. Developmental Medicine and Child Neurology 26:25–32.
29. Loiseau P, Signoret JL, Strube E, Broustet D and Dartigues JF (1982): Nouveaux procédés d'appréciation de la mémoire chez les épileptiques. Revue Neurologique 5:387–400.
30. Mayeux R, Brandt J, Rosen JN and Benson DF (1980): Interictal language and memory impairment in temporal lobe epilepsy. Neurology 30:120–125.
31. Milner B (1975): Psychological aspects of focal epilepsy and its neurosurgical management. Advances in Neurology 8:299–318.
32. Mitchell WG, Chavez JM, Lee H and Guzman BL (1991): Academic underachievement in children with epilepsy. Journal of Child Neurology 6:65–72.
33. Neville BG, Harkness WF, Cross JH, Cass HC, Burch VC, Lees JA and Taylor DC (1997): Surgical treatment of severe autistic regression in childhood epilepsy. Pediatric Neurology 16:137–140.
34. Pascual-Castroviejo I, Garcia Blazquez M, Gutierre Molina M, Carcelle F and Lopez Martin, V (1996): 24-year preoperative evolution of a temporal astrocytoma. Childs Nervous System 12:417–420.
35. Stores G and Hart J (1976): Reading skills of children with generalized or focal epilepsy attending ordinary school. Developmental Medicine and Child Neurrolog 18:705–715.
36. Stores G (1978): Schoolchildren with epilepsy at risk for learning and behavior problems. Developmental Medicine and Child Neurology 20:502–508.
37. Thompson PJ and Trimble MR (1982): Anticonvulsivant drugs and cognitive functions. Epilepsia, 23:531–544.
38. Williams J, Sharp G, Bates S, Griebel M, Lange B, Spence GT and Thomas P (1996): Academic achievement and behavoral ratings in children with absence and complex partial epiepsy. Education and Treatment of Children 19:143–152.

11

NEUROPSYCHOLOGY OF FRONTAL LOBE EPILEPSY IN CHILDREN

Maria Teresa Hernandez[1], Hannelore C. Sauerwein[1], Elaine de Guise[1], Anne Lortie[2], Isabelle Jambaqué[3], Olivier Dulac[3], and Maryse Lassonde*[,1]

[1]Groupe de Recherche en Neuropsychologie Expérimentale
Département de Psychologie
Université de Montréal
C.P. 6128, Succ. Centre-Ville
Qué., H3C 3J7, Canada
[2]Service de Neurologie
Hôpital Sainte-Justine
3175 Ch. Côte Sainte-Catherine
Montréal, Qué, H3T 1C5, Canada and
[3]Service de Neuropédiatrie
Hôpital Saint Vincent de Paul
82 Avenue Denfert Rochereau
75674 Paris, Cedex 14, France

INTRODUCTION

The frontal lobes are involved in motor function, speech and regulation and organization of behavior.[5,18,24,32,33] In adult patients, frontal lobe damage (e.g., 18, 24) and frontal lobe epilepsy (FLE)[13] have been associated with impairments of planning abilities, working memory, impulse control, attention and motor coordination. These deficits may also be present in children presenting neurological problems involving the frontal lobes.

Adequate functioning of the frontal lobes is crucial for the child's cognitive, affective and social development.[12] However, only a few studies have attempted to assess the manifestations of frontal lobe dysfunction in children suffering from FLE. The purpose of this chapter is to provide a general description of the neuropsychological characteristics of FLE children and to identify similarities and differences between children and adults with frontal lobe dysfunction. Furthermore, we present a number of neuropsychological tests usually employed to evaluate frontal lobe functioning in adults that have

* To whom correspondence should be addressed

Neuropsychology of Childhood Epilepsy, edited by Jambaqué et al.
Kluwer Academic / Plenum Publishers, New York, 2001.

proven to be useful in the assessment of FLE children when proper adaptations are made and adequate norms are established.

1. FRONTAL LOBE FUNCTIONS IN ADULTS

1.1. Functions Subserved by the Frontal Lobes

Anatomically, the frontal lobes are constituted of Brodmann's area 4 or the primary motor area which sends impulses to various muscle groups, the premotor or secondary area concerned with the organization and control of movements and the prefrontal cortex or tertiary area, a region of higher order integration associated with the regulation of motor and mental activity. Thus, the frontal lobes subserve a variety of functions including motor behavior at different levels of complexity, speech and executive functions involved in the regulation and organization of behavior.[5,18,32] The executive functions represent a cognitive construct referring to the ability to maintain an appropriate problem-solving set for the attainment of future goals.[8,18,31] Executive functions include planning, organized search, impulse control, execution of purposeful, goal-directed activities, monitoring and self-correction of behavior,[17,37] maintenance set, working memory, attention control and motor coordination.[29] All these functions allow the individual to direct his/her behavior, not only with respect to the immediate environment, but also according to a plan, a goal, or an abstract script.

1.2. Frontal Lobe Dysfunction in Adults

Several behavioral and information-processing disturbances have been described as a result of frontal lobe lesions in adults.[5,10,18,32] These disturbances affect primarily three aspects of human behavior: self-regulatory abilities, direction of attentional resources and the ability to make use of acquired knowledge.[20] Examples of frontal dysfunction include planning and organization difficulties, difficulties to keep "on line" all the necessary information during a problem-solving activity, physical and mental inertia, lack of spontaneity or inversely euphoria, disinhibition, impulsiveness, inadequate behavior, distractibility and perseveration. These impairments may be variously present in the cognitive, affective and/or social domain.

There is some evidence that the prefrontal cortex is not unitary in function. For instance, it has been reported that lesions of the medial and orbito-frontal portions in humans can produce affective and motivational disturbances as well as lack of inhibition of both external and internal stimuli which may interfere with purposeful behavior and provoke inappropriate motor acts. Dorsolateral lesions, by contrast, appear to affect primarily performance on tasks requiring working memory and cognitive functioning, including attention control.[10] Some of the cognitive and behavioral changes observed in adult frontal patients can also be demonstrated in non-human primates having undergone ablation of this area (e.g., 11).

1.3. Neuropsychological Findings in Adult Patients with Frontal Lobe Epilepsy

Studies concerned with identifying the neuropsychological characteristics of frontal lobe epilepsy in adult patients have provided useful information for the treatment and management of these behaviors.[3] However, cognitive functioning in frontal lobe epilepsy continues to be poorly understood.[34] The behavior pattern of patients with FLE is indistinguishable from that of patients with frontal lesions stemming from tumors or other brain disease, at least with regard to executive functions.[13] Data from epileptic patients having undergone frontal cortectomy have revealed deficits on tasks involving spatial and non-spatial processing, sensory delay, concept formation, response inhibition,[22,32] conditional associative learning,[27,28] attention,[13] working memory[34] and monitoring of a sequence of events (e.g., 9, 21, 24). A recent study[13] comparing the performance of FLE and temporal lobe epilepsy (TLE) patients, showed that the FLE group was significantly impaired on motor programming and coordination tasks as well as on complex tasks requiring response inhibition. Problems in concept formation, planning behavior and response inhibition were similar to those reported for patients with frontal lesions[18] Furthermore, the FLE group exhibited attention deficits. Verbal fluency was not affected. While impairments of mnesic functions have sometimes been reported in patients with frontal lobe lesions, Delaney *et al.*[13] did not find any deficits in this domain when comparing patients with unilateral frontal foci with healthy controls.

2. FRONTAL LOBE FUNCTION IN CHILDREN

2.1. Development of the Frontal Lobes

Being the latest brain regions to develop, the frontal lobes have been associated with the most elaborate aspects of human behavior. There is increasing evidence that the frontal lobes of the immature human brain are not functionally silent, as previously assumed. This realization has kindled a renewed interest in the study of frontal functions in the developing brain.[7] The frontal lobes grow significantly from birth to the second year of life. This growth spurt is followed by a slow-down in the growth rate between the 4th and 7th year. But the size of the frontal lobes increases gradually until young adulthood.[19] This growth is associated with significant changes, such as increases in the size and complexity of nerve cells,[6] progressive myelination which continues until puberty,[6] increased cortical fissuration required for refined control of behavior[6] and important changes in synaptic density. Synaptic density is highest during the first two years. This increase is followed by a period of synaptic elimination which continues throughout childhood and adolescence.[6] Furthermore, the frontal lobes, particularly the prefrontal cortex, develop rich connections with posterior and subcortical regions of the brain.[10,17,19] Because of these ongoing changes during ontogeny, the relationship between frontal lobe function and behavior is difficult to establish. There is evidence that the structural and physiological changes during frontal lobe development correlate with an increasing efficiency in information synthesis and activity modulation.[6] Although this integrative capacity develops gradually, it is accelerated between 8 and 12 years.[32]

2.2. The Emergence of Frontal Lobe Functions in Children

The construct "executive functions" has also been used to describe frontal activity in children.[36] In normal children, frontally mediated regulatory functions tend to emerge in the first year of life. They begin to mature between 4 and 7 years,[25] undergo major changes between 8 and 12 years[4,25] and continue to develop at least until puberty (ages 12 to 15 and beyond).[37] The performance of children younger than 8 years somewhat resembles that of adult patients with frontal lesions.[4,25] The emergence of executive functions is measured in terms of the child's ability to regulate his/her cognitive, affective and social behavior (e.g., 37).

2.3. Evaluation of Frontal Lobe Functions in Normal Children

The evaluation of the developmental aspect of frontal functioning is rendered difficult by the fact that few instruments have been conceived to measure these functions in children. Most of the tests that have been employed in the past have been adaptations of tests designed for the assessment of adults or tests that are based on developmental theory.[36] These include the Rey-Osterrieth Complex Figure,[17,36] the Stroop Test,[36] the Porteus Mazes,[36] the Thurstone Word Fluency Test,[22] and the Wisconsin Card Sorting Test (WCST)[17,21,22,36] to name but a few. Some of these tests have been standardized for children and used in the evaluation of executive functions in pediatric populations. Thus, Levin *et al.*[16] used the WCST, the Verbal and the Design Fluency Test, the children's version of the California Verbal Learning Test (CVLT), a word memory and learning test, a Go-no-go task, the Tower of London[31] and the Delayed Alternation Task (e.g., 11, 22) to study 52 healthy children. The children were divided into three age groups (7–8, 9–12 and 13–15 years). Major gains were noted in the 7 to 8 and 9- to 12-year-old children on the WCST, a test purported to measure mental flexibility and conceptual shift and on the Go-no go task, as evidenced by fewer perseverative and false-alarm responses. Further progress was observed in the 13- to 15-year-olds on the CVLT, a test of verbal learning and memory, the Design Fluency Test, a non-verbal production task, and the Tower of London, a test evaluating planning ability and speed of execution. No age differences were observed on a Verbal Fluency Test and on the Delayed Alternation Test.

In a similar study by Welsh *et al.*,[37] 10 adults (mean age: 22 years) and 100 healthy children ranging from 3 to 12 years in age, were submitted to a battery of "frontal" tasks, including a motor planning task,[37] a visual search task,[37] the Verbal Fluency Test,[22] the WCST, the Tower of Hanoi (TOH) and the Matching Familiar Figures Test (MFFT).[37] They were also given a Picture-recognition task[37] and, as a general cognitive measure, the Iowa Test of Basic Abilities.[37] The aim of the study was to identify the point in development at which adult-level performance was achieved. The result most relevant to the present discussion was that the age at which children attained adult-level performance differed according to the tasks used. Some of the tasks were performed efficiently by preschool children. Others were not mastered before the age of 12. The earliest appearing skill was picture-recognition memory which attained adult-level performance at age 4. Visual search and simple planning, assessed on the Tower of Hanoi, were performed at the adult level by the age of 6. Significant steps towards mastering these tasks were observed between the ages 3 to 4 and 5 to 6 years. Accuracy on the MFFT and conceptual shift on the WCST were achieved at the age of 10 years whereas motor sequencing, verbal fluency and complex planning (involving 4 disks or more on the TOH) appeared to continue beyond the age of 12 since the 12-year-olds in the study failed to attain adult-level performance on these tasks. We will return to some of these tests and their results in FLE children later in the chapter.

2.4. Frontal Lobe Dysfunction in Children

As mentioned before, frontal lobe lesions tend to affect three major categories of behavior: 1- regulatory activity which includes the ability to initiate, modulate or inhibit ongoing mental attention; 2- executive functions which refer to planning, goal-setting and modulating behavior with respect to the attainment of a goal[8] and 3- social behavior defined as the ability to adequately interact with others.[12] Consequently, frontal lobe dysfunction in children and adolescents would be expected to produce a number of specific deficits such as disinhibition, perseveration, failure to initiate appropriate activity, failure to maintain attention and effort over time, failure to independently modulate activity and impaired social behavior which may interfere with the child's environmental adaptation and general development. The inability of introspection and appreciation of the deficits represents an additional handicap for patients with frontal lobe dysfunction. A child with frontal damage may demonstrate combinations of these symptoms depending on the extent and site of the lesion, his/her premorbid personality and age at the time of the insult. It can further be expected that the deficits vary over time and in different situations. This has important implication for the child's reeducation which has to be constantly adapted to the specific pattern of behavioral and cognitive limitations at a given stage in development.

In a case study of four children with frontal injury, Mateer and Williams[20] found persistent cognitive and behavioral changes marked by impaired attention (i.e., distractibility), irritability and social problems in the absence of apparent intellectual, linguistic or perceptual deficits. Neuropsychological assessment of the four frontal cases in the postacute stage revealed a full scale IQ within the average range on the Wechsler Intelligence Scale for Children (WISC-R). However, attention problems were noted on the Arithmetic and Digit Span subtests. Similarly, deficits in sustained attention were evident on several tasks including the Trail Making Test, parts A and B[20] and two cancellation tasks.[20] Results on verbal memory tasks, including paragraph recall and sentence memory of the Detroit Test of Learning Aptitude of Hammill[20] and non-verbal memory tasks such as Reitan's Tactual Performance Test,[20] the spatial memory subtest of the Kaufman Assessment Battery for Children,[20] the Corsi Block Tapping Test[23] and Rey's Complex Figure[17] were highly variable. Difficulties on tasks involving planning and organization, like Mazes and Block Design (WISC-R) were observed in two of the children. Two other children presented difficulties with general academic skills (i.e., reading and comprehension, writing, arithmetic). One case displayed a reduction in verbal fluency (animal naming). All children were impaired with respect to reading speed and paragraph copy on the Diagnostic Reading Aptitude and Achievement Tests as well as in the word reading condition of the Stroop Test.[20] Finally all of them had difficulties with impulse control as evidenced by the presence of irritability, moodiness, impulsiveness. They were also impaired with respect to their social adjustment, their social interaction skills, coping strategies and daily living skills.[20]

3. THE NEUROPSYCHOLOGY OF FRONTAL LOBE EPILEPSY IN CHILDREN

3.1. Single Case Studies

Even though epilepsy is one of the most prevalent neurological disorders in childhood, few studies have attempted to establish a neuropsychological profile of children with epilepsy. Epilepsy is not only associated with recurring seizures but also with specific behavioral and cognitive alterations.[14] Identifying the nature of cognitive dysfunctioning in these children is important because the latter are in the process of acquiring skills for academic achievement and future psychosocial adjustment.[30] Since frontal lobe dysfunction is associated with impulsive behavior, frontal lobe epilepsy may interfere with the cognitive and affective development of the children.

To date, most of our knowledge about the neuropsychological consequences of FLE in children is based on single case studies. In this context, Jambaqué and Dulac[14] reported the case of an 8-year-old boy who displayed a number of frontal symptoms in conjunction with a focus in the right fronto-temporal region. The child had difficulties controlling his aggressiveness, exhibited strange behaviors and showed marked personality and affective changes. Initial neuropsychological evaluation revealed a global IQ in the high average range with the highest score in Block Design, and the lowest on the subtests Coding and Mazes. An important attention deficit was also present. Furthermore, the child was hyperkinetic and had difficulties inhibiting motor responses. His ability to reproduce a sequence of hand movements was impaired. His performance on this task was characterized by echopraxis. Manual dexterity was affected as well: his calligraphy was poor and he had difficulties drawing geometric figures. No language problems were noted. However, verbal fluency was reduced. When the seizures were controlled with pharmacotherapy the child's frontal symptoms abated and his scores improved accordingly.

In the same vein, Boone *et al.*,[2] reported a case of a 13-year-old girl with bilateral frontal foci who exhibited a reversible frontal syndrome characterized by sudden behavioral changes such as sexual disinhibition, loss of concern for personal hygiene as well as physical and verbal aggression. Neuropsychological testing revealed impaired performance on tasks requiring motor speed (Finger Tapping Test or FTT), attention (Digit Span of the WISC-R), alternation between two concepts (Trails Making Test, part B), planning ability (Mazes of the WISC-R) and response inhibition (Stroop Test). Motor performance on the FTT was depressed for either hand, particularly for the left hand. By contrast, no deficits were observed on tasks requiring categorization, abstract reasoning, conceptual shift (WCST) and sequential organization of visual information (Picture Arrangement of the WISC-R). Attention was inconsistent but the patient performed normally on Rey's Auditory Verbal Learning Test. Visual perceptual skills (Hooper Visual Organization Test), basic language skills (Reitan Indiana Aphasia Screening Exam) and fund of general knowledge (Information subtest of the WISC-R) were essentially normal. Visuo-motor integration and constructional skills were also preserved although impulsive responses were registered on the Beery Visual Motor Integration Test. Handwriting was poor and characterized by large overlapping letters. As in the case reported by Jambaqué and Dulac,[14] seizure control resulted in a general improvement of most functions.

3.2. Group Study

In this section, we describe the results of a selection of neuropsychological tests considered to be most sensitive to frontal dysfunction (e.g., 15, 36) that we administered to a group of FLE children and two control groups with other types of epilepsy. The FLE group consisted of 16 children (4 girls and 12 boys). The performance of this group was compared to that of a group of 8 children (4 girls and 4 boys) with temporal lobe epilepsy (TLE) and a second group composed of 8 children (4 girls and 4 boys) with generalized epilepsy whose principal epileptic manifestations were typical absences (GEA). The global IQ of all three groups, assessed on the WISC-III,[35] fell within the average range. Chronological age, age at seizure onset and duration of epilepsy were also comparable among the groups, as was the side of the epileptic focus in the FLE and TLE groups. Background information was obtained from the child's medical file with authorization from the parents and the attending neurologist. Furthermore, parents rated the behavior of the child on a questionnaire (the Achenbach Child Behavior Checklist).[1]

The FLE group displayed many deficits commonly seen in adults with frontal lobe epilepsy. Specifically, FLE children showed difficulties on tasks requiring sustained attention, programming of complex motor sequences, visuo-spatial organization, mental flexibility, response inhibition, planning ability, conceptual shift and problem-solving ability. Performance on memory tests was variable and depended on the aspect of memory evaluated. In the present study, all three epilepsy groups were impaired on the California Verbal Learning Test relative to the healthy controls but the FLE children did not differ from the TLE and GEA children with respect to the number of items learned and retained. However, they were more sensitive to interference. In keeping with the results reported in adults and children with frontal lesions, our FLE group had greater visuo-spatial difficulties than the other two epilepsy groups during the copy phase of the Complex Rey's Figure, which clearly affected their recall. In contrast, the TLE group was significantly more impaired in the immediate recall than in the copy of the design which reflects specific memory problems in this group. Finally, a behavior profile specific to FLE could be identified on the basis of parents' ratings. According to this profile, the FLE children had more attention problems, psychiatric (thought) problems and social problems than the other two epilepsy groups.

4. STRATEGIES FOR THE REEDUCATION OF CHILDREN WITH FRONTAL LOBE EPILEPSY

The reeducation of FLE children requires a multidisciplinary approach, including pharmacological management, tutoring, if necessary, and psychological support. The involvement of family and teachers is crucial. The recognition and the understanding of the FLE child's difficulties are important, especially since some of these problems may persist in spite of professional intervention. FLE children tend to be impulsive in their response style and have difficulties sustaining an effort throughout an activity. They function more adequately when goals are clearly stated and information is provided in a step-by-step fashion, insuring every step of the way that the task is well understood. They also need help in the organization and planning of their work and leisure activities. To overcome their various difficulties, FLE children require an external structure. A behavioral modification approach (based on a contingency-reward and response-cost system) can be helpful.[20] Training in self-management and positive feedback may also be used to encourage greater autonomy and raise self-esteem.

REFERENCES

1. Achenbach TM and Edelbrock C (1993): Manual for the Child Behavior Check List. Burlington VT: University of Vermont, Department of Psychiatry.
2. Boone KB, Miller BL, Rosenberg L, Durazo A, McIntyre H and Weil M (1988): Neuropsychological and Behavioral abnormalities in an adolescent with frontal lobe seizures. Neurology 38:583–586.

3. Botez MI (1987): Les syndromes du lobe frontal. In Botez MI (ed): "Neuropsychologie Clinique et Neurologie du Comportement" Montréal: Quebec: Les presses de l'Université de Montréal et Paris: Masson, pp 117–143.
4. Chelune GJ and Baer RA (1986): Developmental norms for the Wisconsin Card Sorting Test. Journal of Clinical and Experimental Neuropsychology 8:219–228.
5. Damasio AR and Anderson SW (1993): The frontal lobes. In Heilman KM and Valenstein E (eds): "Clinical Neuropsychology" New York: Oxford UP, pp 409–460.
6. Dempster FN (1993): Resistance to interference: Developmental changes in a basic processing mechanism. In Howe ML and Pasnak R (eds): "Emerging Themes in Cognitive Development Volumen I: Foundations." New York: Springer-Verlag, pp 3–27.
7. Dennis M (1991): Frontal lobe function in childhood and adolescence: a heuristic for assessing attention regulation, executive control, and the intentional states important for social discourse. Developmental Neuropsychology 7:327–358.
8. Duncan J (1986): Disorganization of behavior after frontal lobe damage.Cognitive Neuropsychology 3:271–290.
9. Frisk V and Milner B (1990): The relationship of working memory to the immediate recall of stories following unilateral temporal or frontal lobectomy. Neuropsychologia 28:121–135.
10. Fuster JM (1997): Human Neuropsychology. In Fuster JM (ed): "The Prefrontal Cortex." New York: Lippincott-Raven, pp 150–184.
11. Goldman-Rakic P and Friedman HR (1991): The circuitry of working memory revealed by anatomy and metabolic imaging. In Levin HS, Eisenberg HM and Benton AL (eds): "Frontal Lobes Function and Dysfunction." New York: Oxford UP, 72–91.
12. Grattan RM and Eslinger PJ (1992): Long-term psychological consequences of childhood frontal lobe lesion in patient D.T. Brain and Cognition 20:185–195.
13. Helmstaedter C, Kemper B and Elger CE (1996): Neuropsychological aspects of frontal lobe epilepsy. Neuropsychologia 34:399–406.
14. Jambaqué I et Dulac O (1989): Syndrome frontal réversible et épilepsie chez un enfant de 8 ans. Archives Françaises de Pédiatrie 46:525–529.
15. Lassonde M and Sauerwein H-C (1996): Tests spécifiques dans l'évaluation neuropsychologique pré- et post-chirurgicale de l'enfant atteint d'épilepsie. Approche Neuropsychologique des Apprentissages chez l'Enfant, Numéro Hors-Série, Septembre: 43–48.
16. Levin HS, Culhane KA, Hartmann J, Evankovich K, Mattson AJ, Harward H, Ringholz G, Ewing-Cobs L and Fletcher JM (1991): Developmental changes in performance on tests of purported frontal lobe Functioning. Developmental Neuropsychology 7:377–395.
17. Lezak MD (1983): Neuropsychological Assessment. New York: Oxford University Press.
18. Luria AR (1969): Frontal lobe syndromes. In Vinken PJ, Bruyn GW and Biemond A (eds): "Handbook of Clinical Neurology." Amsterdam: Holland: North-Holland Publishing Company, pp 725–757.
19. Luria AR (1973): The Working Brain: An Introduction of Neuropsychology. New York: Basic Books.
20. Mateer CA and Williams D (1991): Effects of frontal lobe injury in childhood. Developmental Neuropsychology 7:359–376.
21. Milner B (1963): Effects of different brain lesions on card sorting. Archives of Neurology 9:90–100.
22. Milner B (1964): Some effects of frontal lobectomy in man. In Warren JM and Akert J (eds): "The Frontal Granular Cortex and Behavior." New York: Mc Graw Hill, pp 313–334.
23. Milner B (1971): Interhemispheric differences in the localization of psychological processes in man. British Medical Journal 27:272–277.
24. Milner B (1982): Some cognitive effects of frontal lobe lesions in man. Philosophical Transactions of the Royal Society of London 298:211–226.
25. Passler MA, Isaac W and Hynd GW (1985): Neuropsychological development of behavior attributed to frontal lobe performance in children. Developmental Neuropsychology 1:349–370.
26. Peña-Casanova J (1990): Programa Integrado de Exploracion Neuropsicologica-Test Barcelona Manual. Barcelona, Spain: Masson.
27. Petrides M (1985): Deficits on conditional associative-learning tasks after frontal- and temporal-lobe lesions in man. Neuropsychologia 23:601–614.
28. Petrides M and Milner B (1982): Deficits on subject-ordered tasks after frontal and temporal lobe lesions in man. Neuropsychologia 20:249–262.
29. Roberts RJ and Pennington BF (1996): An interactive framework for examining prefrontal cognitive processes. Developmental Neuropsychology 12:105–126.

30. Seidenberg M (1989): Neuropsychological functioning of children with epilepsy. In Herman BP and Seidenberg M (eds): "Childhood Epilepsies: Neuropsychological, Psychosocial and Intervention Aspects" New York: John Wiley and Sons, pp 71–82.
31. Shallice T (1982): Specific impairments of planning. Philosophical Transactions of the Royal Society of London, Series B; Biological Sciences (London) 298:99–209.
32. Shue ML and Douglas VI (1992): Attention deficit hyperactivity disorder and the frontal lobe syndrome. Brain and Cognition 20:104–124.
33. Stuss DT and Benson DF (1986): The Frontal Lobes. New York: Academic Press Inc.
34. Swartz BE, Halgren E, Simpkins F, Fuster J, Mandelkern M, Krisdakumtorn T, Gee M, Brown C, Ropchan JR and Blahd WH (1996): Primary or working memory in frontal lobe epilepsy: An FDG-PET study of dysfunctional zones. Neurology 46:737–747.
35. Wechsler D (1991): Wechsler Intelligence Scale for Children-III. New York: Psychological Corporation and Harcourt Brace Jovanovich.
36. Welsh MC and Pennington BF (1988): Assessing frontal lobe functioning in children: views from developmental psychology. Developmental Neuropsychology 4:199–230.
37. Welsh MC, Pennington BF and Groisser (1991): A normative-developmental study of executive function: A window on prefrontal function in children. Developmental Neuropsychology 7:131–149.

12

NEUROPSYCHOLOGY OF PARIETO-OCCIPITAL EPILEPSY

Mary Lou Smith*[1,2] and Rebecca L. Billingsley[1,2]

[1]Department of Psychology
University of Toronto at Mississauga
3359 Mississauga Road North, Mississauga
Ontario L5L 1C6 Canada and
[2]Department of Psychology
The Hospital for Sick Children
555 University Avenue, Toronto
Ontario M5G 1X8 Canada

INTRODUCTION

Focal epilepsies of the parietal and occipital lobes are rare. In a retrospective study of 320 individuals who received surgical treatment for epilepsy at the Montreal Neurological Institute between 1929 and 1980, Rasmussen[12] found that 7% had epileptogenic cortex primarily in the sensorimotor area (Rasmussen regards the pre- and post-central gyrus as a single functional unit), and 6% had epileptogenic cortex in the parietal lobe behind the post-central gyrus. Based on 160 cases of focal epilepsy that could be classified, Manford *et al.*[11] found that 6.3% had seizures of parietal lobe origin.

The incidence of occipital-lobe epilepsy appears to be approximately 6–8% of individuals with focal epilepsy.[11] However, only 1% of the surgical cases of intractable epilepsy reviewed by Rasmussen[12] involved excisions primarily in the occipital lobe. Epileptogenic activity of occipital-lobe origin appears to be more common in children than in adults.[18] Benign epileptic syndromes of childhood, which tend to resolve with development, account in part for this discrepancy. These syndromes are described later in the chapter.

Seizures originating in the parietal lobe typically spread rapidly, either anteriorly to the frontal cortex or posteriorly to the occipital lobe.[19,20] Characteristics of a seizure originating in the parietal lobes therefore may change depending on the direction of

* To whom correspondence should be addressed

Neuropsychology of Childhood Epilepsy, edited by Jambaqué et al.
Kluwer Academic / Plenum Publishers, New York, 2001.

seizure propagation. Likewise, epileptogenic activity originating in the occipital lobes can spread to other brain regions. Seizure spread from the occipital lobes to the ipsilateral temporal lobe is especially common.[15,18] The rapid spread of epileptic discharge can make it difficult to pinpoint the seizure onset to the parietal or occipital lobes.

1. NEUROPSYCHOLOGY OF THE PARIETAL AND OCCIPITAL LOBES

There has been little systematic study of the neuropsychology of the parietal and occipital epilepsies, perhaps due to their relatively low incidence, and to the difficulty in localizing the seizures originating from these areas. The kinds of neuropsychological phenomena that could be expected to be associated with epileptic activity in the parietal and occipital lobes can, however, be understood in relation to the roles that these regions play in normal cognitive and sensory function. These roles are discussed briefly below.

1.1. Parietal-Lobe Function

The parietal lobes support numerous cognitive and sensory functions. Each parietal lobe can be divided into several separate neuroanatomical regions, based on histological differences and divergent neural responses. The primary somatosensory region, located in the postcentral gyrus (Brodmann's areas 1, 2, and 3), is involved in the reception of somesthetic information. The body surface is spatially represented in the primary somatosensory cortex, such that distinct cortical regions correspond to specific body parts. In addition to receiving somesthetic information directly from the ventral posterior thalamus, the primary somatosensory region plays a role in motor functioning. The primary motor and somatosensory areas are highly interconnected, allowing for precise motor responses based on somesthetic feedback. Damage to the primary somatosensory region of the parietal lobe can give rise to a loss of sensation contralateral to the damaged hemisphere, an inability to determine shape or texture, and/or an inability to recognize objects by touch alone (astereognosis).[2,10] Astereognostic deficits, however, may require specific damage to the cortical region representing the hand.[13] Damage to the primary somatosensory area can also result in motor problems, including paresis, inaccurate motor movements, and reduced motor speed.[8]

Adjacent to the primary somatosensory region is the somesthetic association area (Brodmann's area 5). This region receives input from both the ipsilateral and the contralateral primary somatosensory regions, as well as from the motor association areas of the frontal lobes. The somesthetic association area contributes to complex hand manipulation, the ascertainment of body position in space, and tactile discrimination. Damage to the region can result in disturbances similar to those after damage to the primary somatosensory cortex. One difference, however, is that an affected individual may be sensitive to touch, yet be unable to determine the location of the stimulus.[10] Disturbances in pain sensitivity also have been reported after damage to the somesthetic association area 5.

The superior-posterior aspect of the parietal lobe (Brodmann's area 7) is involved in the integration of visual, motor, and somesthetic input. The region helps individuals to perceive visual-spatial properties, including figure-ground relationships, depth, and spatial location. Epileptogenic activity in this region may lead to spatial problems, as

Salanova *et al.*[15] observed spatial deficits in 9 of 30 individuals with parietal epileptogenic foci. The superior-posterior parietal lobule also contributes to visual attention. Individuals with lesions in the superior aspect of the right parietal lobe often show left-sided neglect, suggesting that the right hemisphere in particular plays an important role in visual attention. Damage to the superior-posterior aspect of either parietal lobe can result in other visual-spatial disturbances, such as deficits in depth perception, figure-ground analysis, object manipulation, and spatial construction.

The inferior aspect of the parietal lobe is located at the junction of the parietal, temporal and occipital lobes. It includes the supramarginal and angular gyri, and contributes to mathematical, naming, reading, and writing abilities. The left inferior parietal lobe is also important for the accurate temporal sequencing of motor tasks. Individuals with lesions in the left inferior parietal region can suffer from apraxia, an inability to perform actions in an appropriate temporal order.[9] Lesions involving the left angular and supramarginal gyri can result in anomia, or word-finding problems, as well as agraphia (a deficit in the perception or production of written language). Salanova *et al.*[15] reported two individuals with parietal lobe epilepsy who had speech difficulty at seizure onset. The epileptogenic focus in each case was located near the left supramarginal gyrus. Damage to the left inferior parietal lobe can also lead to acalculia, an impairment in the ability to perform mathematical calculations.[1]

1.2. Occipital Lobe Function

Like the parietal lobes, the occipital lobes respond to numerous kinds of stimuli. The principal function of the occipital cortex, however, is the processing of simple and complex visual information. The primary visual cortex (Brodmann's area 17), the main receiving area for visual information, is located medially surrounding the calcarine sulcus and extends narrowly along the lateral convexity of the occipital pole. This region receives visual input from the lateral geniculate nucleus (LGN) of the thalamus and relays it back to the LGN, as well as to the visual association areas (Brodmann's areas 18 and 19) of the occipital lobe. Information is also passed to the parietal and temporal lobes, where more complex processing takes place. Connections between the occipital lobe and the parietal and temporal lobes are also reciprocal, allowing for complex interactions, such as the development of hand-eye coordination, to occur.[8]

Simple-cell neurons in the primary visual cortex are most sensitive to movement. These cells communicate with complex cells in the primary visual cortex that are involved in the initial stages of form, color, and orientational perception. Hypercomplex cells, which perform the next stage of visual analysis, are found primarily in the visual association areas and act to discern angles, corners, distance, and discontinuities, as well as to analyze further the movement and orientation of objects.

Damage to the occipital lobe can produce a variety of perceptual deficits. Destruction of the visual cortex of one hemisphere results in homonymous hemianopsia, that is, blindness in the contralateral visual field. Individuals suffering from homonymous hemianopsia often are unaware of their partial blindness, particularly if it results from damage to the right visual cortex. Lesions to the inferior or superior occipital lobe can result in quandrantanopsia, or blindness in one-fourth of a visual field.

As will be described later, seizures originating in the occipital lobe can produce simple and complex visual hallucinations. Electrical stimulation, tumors, and lesions in the primary visual cortex and the visual association areas can also lead to hallucinations. Simple hallucinations, such as those involving flashes of light or color, appear to arise from activity in the primary visual cortex.[18] Complex hallucinations are associated with abnormal activation in the visual association areas.[14,18] Complex hallucinations have also been reported by individuals with parietal-occipital, temporal-occipital, and inferior-temporal damage. Complex hallucinations appear to be more common after damage to the right hemisphere than the left hemisphere, but simple hallucinations appear to be just as likely after left or right occipital-lobe damage.[7]

Visual agnosia, an inability to recognize objects despite normal visual sensory functioning, can result from damage to the medial occipital lobe.[7,8] Individuals suffering from visual agnosia may recognize isolated aspects of objects or pictures, but they appear unable to integrate the separate aspects into a whole. Two other types of visual agnosia that can result from damage to the occipital lobes are simultanagnosia and prosopagnosia. Simultanagnosia is an inability to see more than one object or every aspect of an object at one time. It can result from lesions in the superior aspect of the left occipital lobe.[7,8] Individuals with bilateral damage to the medial and inferior visual association areas may suffer from prosopagnosia, an inability to recognize faces. They realize a face is a face, but they cannot determine a person's identity by looking at his or her facial features, even if the face is that of someone very familiar to them. An intact right hemisphere appears to be particularly important for accurate face recognition.[7]

2. CLINICAL MANIFESTATIONS OF PARIETAL-OCCIPITAL EPILEPSIES

The types of deficits described in the above sections on the functions of the parietal and occipital lobes are rarely reported in patients with seizure disorders. However, in some proportion of patients, the seizures themselves are characterized by clinical phenomena that illustrate the neuropsychological functions of these brain regions.

Subjective sensations are commonly associated with auras and seizures originating in the parietal and occipital lobes because both cortical regions are involved in sensory perception. Paresthetic sensations, typically consisting of tingling and numbness, are common manifestations of parietal epileptogenic activity. Individuals suffering from paresthesia may also report sensations of pins and needles or crawling under the skin. Such feelings are likely to start in an area of the body contralateral to the seizure focus and spread to adjacent regions (characteristic of a Jacksonian march), and may involve motor activity.[18] Ideomotor apraxia can also arise from parietal-lobe seizures. This involves a feeling of being unable to move a body part and, like parasthetic seizures, can begin in one region of the body and spread to other regions.

Several other auras and ictal sensations have been reported by individuals experiencing epileptogenic activity in the parietal lobe. Painful sensations, such as stabbing, cramping, or throbbing feelings, can accompany partial seizures of parietal-lobe origin. Thermal perceptions of hot or cold, as well as disturbances of body image and paresthesia can occur simultaneously with painful seizures.[15] Body image disturbances can involve the entire body, such as twisting or floating feelings, or may be associated with a single body part. Asomotagnosia, in which an individual experiences a feeling of an absence of a body part, is another kind of body image disturbance associated with parietal seizures. Sexual seizures, such as painful, spontaneous orgasm or sharp pains in the genitals, also can result from parietal seizures, but sexual seizures also may arise from temporal-lobe epileptogenic activity.[18]

Gustatory seizures are uncommon, but have been reported by individuals suffering from parietal seizures. The gustatory sensations usually involve both taste and smell.[18] Eye-deviation, staring, and contraction of the face are phenomena that have been associated with gustatory seizures.[18]

Occipital-lobe epileptogenic activity is often accompanied by visual disturbances. Of the 42 individuals with intractable occipital-lobe epilepsy who underwent surgery at the Montreal Neurological Institute between 1930 and 1991, 31 (73%) had visual auras and 12 (29%) experienced ictal blindness.[14] Ictal amaurosis, which involves blurred vision or intermittent visual loss with or without a loss of consciousness, is common in children suffering from occipital-lobe seizures, and is often associated with concurrent migraines.[18] Visual hallucinations can occur during or prior to seizure onset of occipital origin and can be simple or complex. Simple (or elementary) hallucinations usually consist of flashes of light, color, or flames, and may appear as stationary or moving forms. Complex hallucinations can include perceptions of people, animals, and other objects. The size of the perceived figures or scenes can be normal, abnormally large, or diminutive.

Visual illusions have also been reported by individuals suffering from occipital-lobe seizures and may be associated with secondary seizure spread to the parietal lobes. The simplest kinds of visual illusions are those that involve distortions of spatial perception, movement, and color vision. Individuals may perceive objects that are compressed, altered in shape, partially missing, or fragmented. Both visual illusions and visual hallucinations are likely to appear in a blind or defective visual field.[18]

Epileptic activity in the occipital as well as in the parietal cortex has been associated with vertigo and body image disturbances. Both regions have also been implicated in epileptic nystagmus, or oculoclonic seizures.[18] Rapid eye-blinking or fluttering has been associated with several other types of epilepsy, but may be a reliable indicator of occipital onset when the symptoms occur at seizure onset.[21] Head and eye deviation in children and adults with occipital epilepsy is usually contralateral to the seizure focus.[14,21]

3. SPECIFIC SYNDROMES INVOLVING SEIZURES OF THE PARIETAL OR OCCIPITAL LOBE

3.1. Reflex and Photosensitive Epilepsies

Rather than experiencing sensory and cognitive disturbances as a result of seizures in the parietal and occipital lobes, some children and adults experience seizures as a consequence of performing particular activities. In a review of 25 cases of primarily generalized reflex epilepsy, Goosens, Andermann, Andermann, and Remillard[4] found that seizure onset could be induced by playing cards in 60% of the individuals, performing calculations in 56%, playing chess in 44%, and playing checkers in 20%. The triggers led to generalized tonic-clonic convulsions in all except one person, but the mathematical and spatial activities that preceded the seizures were considered to be suggestive of parietal-lobe dysfunction. Goosens *et al.*[4] speculate that abnormal cortical function in the parietal lobes may lead to generalized seizures in these individuals, but the exact mechanism is unknown.

Reflex epilepsies triggered by mathematical and spatial activities are similar to photosensitive occipital epilepsy, which has been demonstrated in both humans and animals.

Photosensitive epilepsy is extremely uncommon. Of nearly 2500 individuals with epilepsy seen by Guerrini *et al.*[6] between 1988 and 1992, only 18 had occipital-lobe seizures induced by visual stimulation. In all 18 cases, the age of onset was before age 17. Television and video screens triggered seizures in each case. Other triggers include computer screens, sudden transitions from dark to light, and flickering sunlight. The initial seizure manifestations typically involve elementary visual symptoms, followed by cephalic pain, epigastric discomfort, nausea, and/or unresponsiveness.

3.2. Idiopathic, Localization-Related Epilepsies

Childhood epilepsy with occipital paroxysms (CEOP) is an idiopathic epilepsy.[3] The majority of children with CEOP develop normally, but there have been reports of a small number having developmental delay, behavioural disorders and visual defects. Age of onset ranges from 15 months to 17 years, although there are peaks in onset that appear to distinguish two variants. The early onset variation (with onset peaking at 3–5 years of age) is characterized by partial seizures, usually nocturnal, in which there is eye deviation and vomiting, although they frequently evolve to hemiconvulsions and generalized tonic-clonic seizures. The late onset form of CEOP (peaking at 7–9 years) consists of visual seizures with amaurosis, phosphenes, illusions or hallucinations, frequently followed by automatisms or hemiclonic seizures. Approximately 25% of children with this form experience migraine attacks after seizures.

Benign epilepsy with "extreme" somatosensory evoked potentials is another idiopathic syndrome which is associated with EEG abnormalities in the parietal and parietosagittal area.[3] This is a benign syndrome with infrequent seizures and spontaneous remission of symptoms. Neuropsychological development has been reported as normal.

3.3. Sturge-Weber Syndrome

Sturge-Weber syndrome is a congenital neurocutaneous syndrome. The most common neurological symptom of Sturge-Weber syndrome is epilepsy, which plays a major role in the prognosis and treatment of the syndrome. Although the brain anomalies in Sturge-Weber syndrome may extend throughout an entire hemisphere, or may involve both hemispheres, they frequently are confined to the occipital and parietal lobes. For this reason, a brief discussion of Sturge-Weber syndrome seems appropriate in a review of the parietal-occipital epilepsies.

The contribution of seizures to cognitive and psychological development in Sturge-Weber syndrome is well illustrated in studies by Sujansky and Conradi.[16,17] A review of 171 individuals ranging in age from 2 to 59 years (median age = 8 years), 80% of whom had seizures, revealed that developmental delay was reported in 71% of those who had seizures, and only 6% of those who did not. The earlier the age of onset of seizures, the more likely the individual was to have experienced developmental delay, and to require special education services. In examining outcome in adults with Sturge-Weber syndrome, Sujansky and Conradi[17] extended their findings to show that normal development, normal intelligence, and higher educational attainment were more likely to be associated with the absence of seizures or with later age of seizure onset. The strong influence of presence of seizures and age of onset suggest that the outcome in these cases may be related to seizure variables and/or medication effects rather than due to specific effects relating only to abnormalities in the parietal or occipital lobes. This conclusion is somewhat tempered, however, by the lack of information on the location and extent of the structural lesions in the patients studied by Sujansky and Conradi.

CONCLUSION

Extensive research on the neuropsychology of the parietal and occipital epilepsies has not yet been undertaken. The parietal lobes are known to be important for aspects of attention, spatial processing, spoken and written language, coordinated motor sequencing, and sensory function. The occipital lobes are known to be involved in visual processing. Although the clinical phenomena associated with seizures arising in the parieto-occipital areas provide clues to the functioning of these brain regions, systematic study is necessary to understand the ways in which epilepsy, particularly in the developing child, has an impact on these diverse and important functions.

REFERENCES

1. Boller F and Grafman J (1983): Acalculia: Historical development and current significance. Brain and Cognition 2:205–223.
2. Corkin S, Milner B and Rasmussen T (1970): Somatosensory thresholds: Contrasting effects of post-central gyrus and posterior parietal-lobe excisions. Archives of Neurology 23:41–58.
3. Gobbi and Guerrini (1997): Childhood epilepsy with occipital spikes and other benign localization-related epilepsies. In Engel J and Pedley TA (eds): "Epilepsy. A Comprehensive Textbook." Philadelphia: Lippincott-Raven, pp 2315–2326.
4. Goosens LAZ, Andermann F, Andermann E and Remillard GM (1990): Reflex seizures induced by calculation, card or board games, and spatial tasks: A review of 25 patients and delineation of the epileptic syndrome. Neurology 40:1171–1176.
5. Greenspan JD and Winfield JA (1992): Reversible pain and tactile deficits associated with a cerebral tumor compressing the posterior insula and parietal operculum. Pain 50:29–39.
6. Guerrini R, Dravet C, Genton P, Bureau M, Bonanni P, Ferrari AR and Roger J (1995): Idiopathic photosensitive occipital lobe epilepsy. Epilepsia 36:883–891.
7. Hecaen H and Albert ML (1978): Human Neuropsychology. New York: John Wiley.
8. Joseph R (1996): Neuropsychiatry, Neuropsychology, and Clinical Neuroscience. Baltimore, MD: Williams & Wilkins.
9. Kimura D (1993): Neuromotor Mechanisms in Human Communication. New York: Oxford University Press.
10. LaMotte RH and Mountcastle VB (1979): Disorders in somesthesis following lesions of the parietal lobe. Journal of Neurophysiology 42:400–419.
11. Manford M, Hart YM, Sander JWAS and Shorvon S (1992): The national general practice study of epilepsy. The syndrome classification of the International League Against Epilepsy applied to epilepsy in a general population. Archives of Neurology 49:801–808.
12. Rasmussen T (1987): Focal epilepsies of nontemporal and nonfrontal origin. In Wieser HG and Elger CE (eds): "Presurgical Evaluation of Epileptics: Basics, Techniques, Implications" Berlin: Springer-Verlag.
13. Roland PE (1976): Astereognosis. Archives of Neurology 33:543–550.
14. Salanova V, Andermann F, Olivier A, Rasmussen T and Quesney LF (1992): Occipital lobe epilepsy: Electroclinical manifestations, electrocorticography, cortical stimulation and outcome in 42 patients treated between 1930 and 1991. Brain 115:1655–1680.
15. Salanova V, Andermann F, Rasmussen T, Olivier A and Quesney LF (1995): Parietal lobe epilepsy: Clinical manifestations and outcome in 82 patients treated surgically between 1929 and 1988. Brain 118:607–627.
16. Sujansky E and Conradi S (1995): Sturge-Weber syndrome: Age of onset of seizures and glaucoma and the prognosis for affected children. Journal of Child Neurology 10:49–58.

17. Sujansky E and Conradi S (1995): Outcome of Sturge-Weber syndrome in 52 adults. American Journal of Medical Genetics 57:35–45.
18. Sveinbjornsdottir S and Duncan JS (1993): Parietal and occipital lobe epilepsy: A review. Epilepsia 34:493–521.
19. Williamson PD (1994): Seizures with origin in the occipital or parietal lobes. In Wolf P (ed): "Epileptic Seizures and Syndromes." London: John Libbey, pp 383–390.
20. Williamson PD, Boon PA, Thadani VM, Darcey TM, Spencer DD, Spencer SS, Novelly RA and Mattson RH (1992a): Parietal lobe epilepsy: Diagnostic considerations and results of surgery. Annals of Neurology 31:193–201.
21. Williamson PD, Thadani VM, Darcey TM, Spencer DD, Spencer SS and Mattson RH (1992): Occipital lobe epilepsy: Clinical characteristics, seizure spread patterns, and results of surgery. Annals of Neurology 31:3–13.

13

NEUROPSYCHOLOGICAL OUTCOME FOLLOWING PROLONGED FEBRILE SEIZURES ASSOCIATED WITH HIPPOCAMPAL SCLEROSIS AND TEMPORAL LOBE EPILEPSY IN CHILDREN

Daniela Brizzolara*, Paola Brovedani, Claudia Casalini, and Renzo Guerrini

Dipartimento di Medicina della Procreazione e dell'Età Evolutiva
University of Pisa, Italy
IRCCS Stella Maris, Via dei Giacinti 2
56018 Calambrone (Pisa) Italy

INTRODUCTION

Hippocampal sclerosis, defined as extensive neuronal loss in certain sectors of the hippocampus, is the principal pathological abnormality found in 50–70% of resected temporal lobes of patients with intractable temporal lobe epilepsy (TLE).[1] Magnetic Resonance Imaging (MRI) can reliably detect hippocampal sclerosis and atrophy if certain methodological constraints are met.[3]

The explicit memory deficits (both antero and retrograde amnesia) ensuing from selective damage to mesial temporal lobe structures in temporal lobectomy patients have originally highlighted the role of the hippocampal region (the hippocampus proper, the adjacent enthorhinal and perirhinal cortices) in declarative memory.[23]

Lesion data from both humans and animals, more recently supported by PET and functional magnetic resonance studies, have converged in showing the nodal role of the hippocampal region for memory functioning but have also underlined the importance of the interconnections and interactions of this region with other mesial temporal lobe structures and neocortical areas.[24]

As to the causes of hippocampal sclerosis, and its relation to TLE, there is as yet controversy. More than 25 years ago, Falconer, studying the neuropathology of

*To whom correspondence should be addressed

Neuropsychology of Childhood Epilepsy, edited by Jambaqué et al.
Kluwer Academic / Plenum Publishers, New York, 2001.

the temporal lobes resected in patients with TLE, found that mesial temporal lobe sclerosis was frequently associated with a history of early childhood febrile convulsions.[9]

The causal link between childhood febrile convulsions and hippocampal sclerosis is still controversial; a number of studies reporting a frequent association between the two are suggestive of a causative role of febrile seizures in hippocampal sclerosis and temporal lobe epilepsy.[5,6,15,20] Other studies espouse an opposing view, that hippocampal sclerosis is a consequence of repeated seizures in TLE.[21]

1. EVIDENCE FOR AN ASSOCIATION BETWEEN EARLY FEBRILE SEIZURES AND HIPPOCAMPAL DAMAGE

Febrile seizures are clonic, tonic-clonic or atonic attacks occurring with fever, in infancy or childhood (between 3 months and 5 years), with an incidence of 2–4%.[18] They occur without evidence of intracranial infection, antecedent epilepsy or other definable cause, have a benign long-term outcome in most cases, but can constitute a risk factor for subsequent epilepsy (2–7% or greater incidence of later afebrile seizures depending on the study). Factors related to the nature of febrile seizures (duration, frequency) and to patients' neurological status (central nervous system abnormalities) and genetic predisposition to epilepsy are important determinants of subsequent epilepsy. The duration of childhood febrile seizures has been shown to be the most significant determinant for subsequent TLE. Maher and McLachlan[16] found that patients who later developed epilepsy had mean duration of febrile convulsions of 100 minutes while those who did not, had a mean duration of 9 minutes. Total number of febrile convulsions, their daily frequency or age of their onset were not different between patients who did and did not develop epilepsy. Patients with TLE and prolonged febrile convulsions had mesial temporal lobe sclerosis as the pathological finding.

The causal role of early prolonged febrile convulsions (PFC) in mesial temporal lobe sclerosis and TLE has been suggested by Cendes *et al.*[6] Performing a volumetric MRI study measuring amygdala and hippocampal formation in patients with intractable TLE, focal extra-temporal and generalized epilepsy, they found that only patients with intractable TLE had a significant reduction of amygdala and hippocampal formation and that those with a history of childhood prolonged febrile convulsions (<30 minutes) had significantly smaller amygdala and hippocampus. While PFCs correlated with the degree of sclerosis of mesial temporal lobe structures, total duration of epilepsy did not. Sagar and Oxbury[20] found that in surgically resected temporal lobe tissue, cell counts of H1 zone, end folium pyramidal neurons and dentate gyrus granule cells in the hippocampal region, were all significantly lower in patients with the first febrile seizure occurring before three years of age. The convulsion was usually longer than 30 minutes, repetitive within the 24 hours or unilateral. Other data, although inconclusive as to the causal relationship between febrile seizures and hippocampal sclerosis, seem to support the notion that it is the duration of febrile seizures that is a key factor in determining hippocampal damage and not age at the first febrile convulsion, duration of epilepsy and estimated seizure frequency during life-time.[15]

The mechanisms relating early childhood febrile seizures, hippocampal damage and subsequent epilepsy are not completely understood. The hypotheses can be summarized as follows:

- early prolonged febrile seizures produce structural changes in previously normal hippocampus, which are the substrate for TLE;[10]
- pre-existing subtle structural abnormalities in the temporal lobe predispose to prolonged febrile seizures and subsequent TLE;[17]
- hippocampal sclerosis may be a secondary phenomenon as demonstrated by its presence in temporal lobe specimen which also show structural abnormalities including dysplastic lesions and tumors.[22]

2. HIPPOCAMPAL DAMAGE AND MEMORY DEFICITS IN ADULTS

The evidence of lateralized memory disorders in adults with intractable TLE and in patients with vascular damage to hippocampal region is compelling, although not univocal.[8,13] The crucial role of mesial temporal lobe structures in memory has been derived for the most part on the study of patients who have undergone temporal lobectomy for the relief of intractable seizures. However, the effects of recurrent seizures, long-term massive anti-convulsant therapy and the extent of surgical excisions on mnemonic outcome should not be underestimated. Moreover, and more importantly, the pathogenetic role of prolonged febrile convulsions in hippocampal damage and memory dysfunction cannot be parcelled out when epilepsy is severe.

We addressed the issue of whether children with prolonged febrile convulsions and mesial temporal lobe sclerosis are at risk of developing cognitive and memory impairments. It can be hypothesized that damage to the hippocampus putatively determined by early PFCs can lead to subsequent memory disorders unless other areas, such as contralateral mesial temporal lobe structures can compensate the specific memory functions subserved by the hippocampus as is true in other cases of early occurring damage. Thus, memory deficits deriving from early onset damage should not be as severe as those ensuing from acquired lesions (surgical or vascular).

In adult temporal lobectomy patients with large hippocampal excisions, long-term memory deficits have been reported after left resection for stimuli such as words, prose passages, while non verbal long-term memory deficits have been demonstrated after right resection on tasks such as recall of spatial location, abstract designs, faces, allocentric spatial representation. In some cases, degree of memory impairment correlated with the extent of hippocampal removal. Analogous verbal-non verbal memory impairments have been found in pre-surgical TLE patients who presented unilateral hippocampal volume loss at MRI.[13,14]

3. HIPPOCAMPAL DAMAGE AND MEMORY DEFICITS IN CHILDREN

There is little knowledge on the specific effects of hippocampal damage in children. Ostergaard[19] reported a case of a 10-year old boy (C.C.) who developed an amnesic syndrome after seizure-related respiratory arrest with anoxia. Neuropsychological assessment revealed below normal intelligence and impaired performance on verbal memory tasks (word list delayed recall, immediate and delayed story recall) and visual memory tasks (Rey-Osterrieth figure and Benton Visual Retention test). His memory difficulties were associated with extensive left mesial temporal and occipital lesions, demonstrated by CT scan, involving the hippocampus and parahippocampal gyrus and moderate lesions of the right anterior portion of parahippocampal gyrus and hippocampus.

A more recent case report (M.S.) presented severe anterograde amnesia at age 7 following the same type of brain lesions of case C.C.[4] MRI showed bilateral hippocampal atrophy. Testing when the patient was a young adult revealed deficient verbal and visual long-term memory both in recognition and recall. Vargha-Khadem *et al.*[25] reported a patient (J.F.) with seizures at 4 years of age who had subsequent severe declarative memory impairment. MRI demonstrated bilateral mesial temporal lobe sclerosis and hippocampal atrophy.

The evidence of childhood amnesia, at least in terms of anterograde memory impairment, seems to fit the adult pattern of hippocampal amnesia.

Material specific memory deficits have been documented in childhood following unilateral hippocampal damage. Incisa della Rocchetta *et al.*[11] described a 10-year old boy (S.R.) who underwent selective right amygdalo-hippocampectomy and who was selectively impaired in incidental spatial supraspan learning. Beardsworth and Zaidel[2] assessed memory for faces pre-and post-surgically in children and adolescents with right and left temporal lobe epilepsy most of whom having hippocampal sclerosis. The authors found that right TLE patients were significantly impaired with respect to left TLE patients in delayed memory for faces, both pre and postoperatively.

Dennis *et al.*[7] analyzed long-term verbal memory with a series of recognition and recall tasks, in children, adolescents and young adults with TLE after unilateral temporal lobectomy. Mesial temporal lobe sclerosis of variable degree was present in about 75% of the sample, was associated in most cases with other abnormalities and did not correlate with memory scores. Side of temporal lobectomy and extent of resection of mesial temporal lobe structures correlated with memory scores in some of the tasks.

Jambaqué *et al.*[12] studied verbal and visual memory with an extensive battery in children and adolescents with idiopathic generalized epilepsy, and partial temporal and extra-temporal lobe epilepsy. All children with epilepsy scored below normal controls on all measures. Partial epilepsy patients exhibited more pronounced memory impairments than idiopathic epilepsy children. Left TLE determined lower verbal memory scores, while right TLE lower visual memory performance, providing evidence for dissociable lateralized memory effects in children.

4. MEMORY ASSESSMENT OF CHILDREN WITH PFCS, HIPPOCAMPAL SCLEROSIS AND TLE

We assessed memory functioning in a group of 9 patients with a history of prolonged febrile convulsions and hippocampal sclerosis demonstrated with MRI. They were selected from a group of 86 consecutive patients who underwent MRI because of seizures as the presenting symptom in the absence of other neurological manifestations. Fifty-one of 86 patients presented with structural abnormalities of different origin. Of these, 17 had mesial temporal lobe structural changes compatible with hippocampal sclerosis, and had a history of PFCs. Eleven were referred for psychological testing. The 9 patients with normal or borderline intellectual functioning were submitted to neuropsychological testing including verbal and non-verbal memory tests. Our study suggests that structural mesial temporal lobe abnormalities correlate with material specific memory dysfunction even in the absence of severe epilepsy.

4.1. Methods

Six patients had right hippocampal sclerosis and 3 had left. The age range in the right hippocampal sclerosis group was 6;8 to 16;10 years (mean age 11;2 years). The age range of the 3 patients with left hippocampal sclerosis was 7;8–22;7 years.

Magnetic resonance imaging was performed with either a superconductive scan at a field strength of 0.5T (MRMAX, GE-CGR) or with a high intensity field scan (Signa Advantage 1.5 Tesla General Electric, Milwaukee). T1 and T2 PD were performed in the axial, coronal and sagittal planes.

Polygraphic video EEG recordings (21 channels) were performed in wakefulness and sleep with silver chloride electrodes applied to the scalp (International 10–20 system).

Neuropsychological testing included assessment of cognitive functioning with Wechsler Intelligence Scale for Children-Revised and Raven's Progressive Matrices. Visuo-constructive abilities were tested with a copying task, The Visual-Motor Integration Test. Verbal short-term memory was measured with digit span sequences and spatial short-term memory with Corsi's procedure. Verbal long-term memory was assessed, in some patients, with the California Verbal Learning Test. The test measures learning over trials, delayed recall (20 min) and recognition of a list of 16 words. Spatial long-term memory was measured with Corsi's supraspan sequences (span plus 2) requiring three correct successive repetitions to reach criterion (out of a maximum of 25 trials) and, in a restricted number of patients, with the Biber Figure Learning Test, which consists of 12 geometric designs to be copied and retained after a delay of 20 min. Patients' performance was referred to published Italian norms when available or to age matched controls.

4.2. Results and Discussion

The clinical findings are presented in Table 1. For the majority of the patients, lateralization of PFCs could not be determined. All patients had PFCs ranging from 20 to 60 minutes. Patients 1 and 2 did not present with epilepsy at the time of testing. For the rest of the sample, seizures were well-controlled in monotherapy or polytherapy except for patient 8. Seven patients had complex partial seizures, accompanied by secondary generalization in two.

Eight patients had normal IQ while one patient had borderline intellectual level (Table 2).

Verbal and nonverbal short-term memory data are presented in Table 3 which reports patients' verbal and spatial span and corresponding percentile values after adjustment for age and educational level. For patients older than 16 years of age, the span was converted to equivalent scores. Visual inspection of the data reveals that 4 out of 6 patients with right hippocampal sclerosis had deficient or clearly below normal spatial short term memory, while patients with left hippocampal sclerosis had normal or above normal scores. This evidence, although based on a small number of patients, suggests that the right hippocampus may be involved in spatial short-term memory functioning. The data seem to support a selective spatial short-term memory deficit in children with right hippocampal sclerosis, since verbal short-term memory performance was well above normal in the same patients. Patient 3 was an exception because both verbal and spatial short-term memory were below normal (12th and 23d percentile respectively). This was also true for verbal and spatial delayed recall.

Table 1. Clinical description of the patients

Patient	Sex	Side of hippocampal sclerosis	Age at Test (years)	Age at Onset of PFCs	Age at Onset of epilepsy	Duration of PFCs (min)	No. of Seizures	Seizure Semeiology	AEDs
1	f	right	6;8	2;6 yrs	—	20	—	—	—
2	m	right	9;1	14 mo	—	60	—	—	—
3	m	right	9;3	14 mo	6 yrs	30	5	Epigastric sensation, staring, secondary generalization	CBZ
4	m	right	12;2	6 mo	10 yrs	60	3	Motionless, prolonged unresponsiveness	PB
5	f	right	13;4	11 mo	3 yrs	30	seizure free >5 yrs	III defined feeling, unresponsiveness, oral and verbal automatisms	PB CLB
6	f	right	16;10	2 yrs	6 yrs	30	seizure free >5 yrs	Brief unresponsiveness, generalized tonic-clonic	CBZ
7	m	left	7;8	2 yrs	7 yrs	60	3	Epigastric sensation, oral automatisms, staring	CBZ
8	f	left	13;11	6 mo	6 yrs	20	1/month	Epigastric sensation, oro-alimentary automatisms, unresponsiveness.	CBZ LTG CLB
9	f	left	22;7	11 mo	12 yrs	30	seizure free >5 yrs	Epigastric sensation, gestural automatisms, secondary generalization	CBZ

Table 2. Intelligence Quotients (Wechsler Intelligence Scale for Children-Revised)

	VIQ	PIQ	FSIQ
Right hippocampal sclerosis			
1	87	120	102
2	108	102	105
3	123	115	122
4	98	90	93
5	88	72	79
6	106	94	101
Left hippocampal sclerosis			
7	131	111	126
8	113	92	103
9	75*		

*Raven's Progressive Matrices.

No conclusions can be drawn on the effects of left hippocampal sclerosis on verbal short-term memory given the small sample size.

A reduced short-term memory capacity following hippocampal sclerosis is an unexpected finding on the basis of the adult lesion data. An early damage to the hippocampus not accompanied by severe epilepsy, as is the case with our young patients, may constrain normal short-term memory development for side-specific material, as demonstrated by the marked discrepancy between spatial and verbal spans.

A stronger claim for memory dysfunction following hippocampal damage can be made for long-term memory, especially when learning over trials and delayed recall are assessed, as demonstrated by recent literature.[14] We hypothesized that spatial long-term memory would be impaired following right hippocampal damage, leaving verbal long-term memory intact. Spatial supraspan learning of five patients with hippocampal sclerosis, is presented in Fig. 1. All patients required a greater number of trials to learn the supraspan sequence with respect to controls matched for age and span.

Table 3. Verbal and spatial short-term memory

	Verbal		Spatial	
Patient	Span	Percentile/ Equivalent score*	Span	Percentile/ Equivalent score*
Right hippocampal sclerosis				
1	4	50	4	89
2	6	89	6	98
3	4	12	4	23
4	6	81	4	13
5	8	100	4	23
6	6	4*	3*	0*
Left hippocampal sclerosis				
7	5	82	4	60
8	6	82	7	100
9	7	4*	6	4*

*Equivalent Scores range from 0 to 4.
Zero and 1 deficient, 2 borderline, 3 and 4 normal performance.

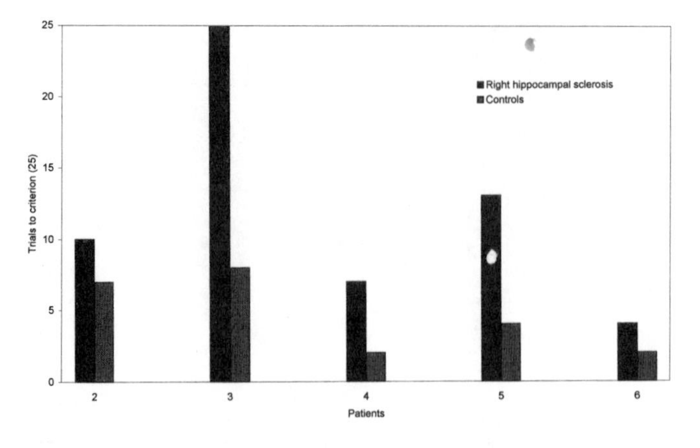

Figure 1. Spatial supraspan learning.

Verbal and visuo-spatial long-term memory was also tested in 3 patients with the California Verbal Learning Test and the Biber Figure Learning Test. According to recent adult temporal lobectomy evidence, one would expect an impaired delayed recall of words in left hippocampal sclerosis patients. Conversely, patients with right hippocampal sclerosis should be impaired in learning of abstract designs.[13] Two patients with right (patients 3 and 6) and one with left hippocampal sclerosis (patient 7) were submitted to the tests. Both patients with right damage showed difficulties in learning and retention of verbal material, and one of the two was impaired on delayed recall in the visuo-spatial task. The patient with left hippocampal sclerosis performed normally.

It is clear that no conclusions can be drawn on these last results given the limited number of patients studied. However, some methodological considerations can be addressed. The Biber memory task may induce dual coding of information since some of the stimuli are verbalizable. The intact left hippocampus could come into play in some of our patients. One of the patients with right damage who failed in the learning phase of the non-verbal memory task had low verbal skills, which could not be used for memorizing visuo-spatial information. We propose that "purer" tests of spatial memory, such as Corsi's or analogous spatial position memory tasks, are valid indicators of right hippocampal functioning. The sizeable discrepancy between verbal and spatial sequential memory in the majority of right hippocampal sclerosis patients, together with impaired spatial learning, are both suggestive of a specific, although mild, memory disorder in patients with early right hippocampal damage. The fact that the majority of our patients were either seizure-free or well controlled by anti-epileptic medication indicates that lateralized hippocampal sclerosis may be sufficient to determine mild memory impairment.

CONCLUSION

The evidence presented in this chapter demonstrates that hippocampal sclerosis is a substrate for TLE also in children.

The pathophysiological mechanisms responsible for hippocampal sclerosis are not completely understood. Given the high correlation between early PFCs and hippocampal sclerosis, there is a strong suggestion that the two may be causally related, although the basis for this relationship is not clear.

Hippocampal damage has been shown to produce long-term memory deficits of different severity; however, since most of the evidence relating memory impairment to hippocampal lesions has been derived from patients referred to neurosurgery centers for temporal lobectomy as relief of severe epilepsy, it is difficult to disentangle the effects of severe and recurrent seizures on memory from that of structural damage to the hippocampus.

We report evidence of mild memory deficits in a group of young patients with unilateral hippocampal sclerosis and normal intelligence who had either mild TLE or were seizure free.

Right hippocampal lesions were associated with reduced potential for learning spatial material. Our data support the idea that hippocampal sclerosis per se is sufficient to interfere with normal memory development.

Furthermore, early prolonged febrile convulsions may be considered a risk factor for neuropsychological outcome, as well as for subsequent epilepsy.

Further studies should be directed at corroborating our results with larger samples and at analyzing if and to what extent recurrent seizures specifically aggravate memory dysfunction.

REFERENCES

1. Babb TL (1991): Research on the anatomy and pathology of epileptic tissue. In Lüders HO (ed): "Epilepsy Surgery." New York: Raven Press, pp 719–727.
2. Beardsworth ED and Zaidel DW (1994): Memory for faces in epileptic children before and after brain surgery. Journal of Clinical and Experimental Neuropsychology 16:589–596.
3. Berkovic SF, Andermann F, Olivier A, Ethier R, Melanson D, Robitaille Y, Kuzniecky R, Peters T and Feindel W (1991): Hippocampal sclerosis in temporal lobe epilepsy demonstrated by magnetic resonance imaging. Annals of Neurology 29:175–182.
4. Broman M, Rose AL, Hotson G and Casey McCarthy C (1997): Severe anterograde amnesia with onset in childhood as a result of anoxic encephalopathy. Brain 120:417–433.
5. Cendes F, Andermann F, Gloor P, Lopes-Cendes I, Andermann E, Melanson D, Jones-Gotman M, Robitaille Y, Evans A and Peters T (1993): Atrophy of mesial structures in patients with temporal lobe epilepsy: Cause or consequence of repeated seizures? Annals of Neurology 34:795–801.
6. Cendes F, Andermann F, Dubeau F, Gloor P, Evans A, Jones-Gotman M, Olivier A, Andermann E, Robitaille Y, Lopes-Cendes I, Peters T and Melanson D (1993): Early childhood prolonged febrile convulsions, atrophy and sclerosis of mesial structures, and temporal lobe epilepsy: An MRI volumetric study. Neurology 43:1083–1087.
7. Dennis M, Farrell K, Hoffman HJ, Hendrick EB, Becker LE and Murphy EG (1988): Recognition memory of item, associative and serial-order information after temporal lobectomy for seizure disorder. Neuropsychologia 26:53–65.
8. Dobbins IG, Kroll NEA, Tulving E, Knight RT and Gazzaniga MS (1998): Unilateral medial temporal lobe memory impairment: Type deficit, function deficit or both? Neuropsychologia 36:115–127.
9. Falconer MA (1971): Genetic and related aetiological factors in temporal lobe epilepsy: A review. Epilepsia 12:13–31.
10. Gloor P (1991): Mesial temporal sclerosis: Historical background and an overview from a modern perspective. In Lüders HO (ed): "Epilepsy Surgery." Raven Press: New York, pp 689–703.
11. Incisa della Rocchetta A, Vargha-Khadem F, Connelly A and Polkey C (1992): Selective right mesiotemporal lesion and supraspan spatial learning in childhood. Journal of Clinical and Experimental Neuropsychology 14:371.
12. Jambaqué I, Dellatolas G, Dulac O, Ponsot G and Signoret JL (1993): Verbal and visual memory impairment in children with epilepsy. Neuropsychologia 31:1321–1337.
13. Jones-Gotman M. (1996): Psychological evaluation for epilepsy surgery. In Shorvon S, Dreifuss F, Fish D and Thomas D (eds): "The Treatment of Epilepsy. "Blackwell Science: London, pp 621–630.

14. Jones-Gotman M, Zatorre RJ, Olivier A, Andermann F, Cendes F, Staunton H, McMackin, D, Siegel AM and Wieser H-G (1997): Learning and retention of words and designs following excision from medial or lateral temporal-lobe structures. Neuropsychologia 35:963–973.
15. Kuks JBM, Cook MJ, Fish DR, Stevens JM and Shorvon SD (1993): Hippocampal sclerosis in epilepsy and childhood febrile seizures. Lancet 342:1391–1394.
16. Maher J and McLachlan RS (1995): Febrile convulsions: Is seizure duration the most important predictor of temporal lobe epilepsy? Brain 118:1521–1528.
17. Meencke HJ and Veith G (1991): Hippocampal sclerosis in epilepsy. In Lüders HO (ed): "Epilepsy surgery." New York: Raven Press, pp 705–715.
18. Nelson KB and Ellenberg JH (1976): Predictors of epilepsy in children who have experienced febrile seizures. New England Journal of Medicine 295:1029–1033.
19. Ostergaard AL (1987): Episodic, semantic and procedural memory in a case of amnesia at an early age. Neuropsychologia 25:341–357.
20. Sagar HJ and Oxbury JM (1987): Hippocampal neuron loss in temporal lobe epilepsy: Correlation with early childhood convulsions. Annals of Neurology 22:334–340.
21. Salmenperä T, Kälviäinen R, Partanen K and Pitkänen A (1998): Hippocampal damage caused by seizures in temporal lobe epilepsy. Lancet 351:35.
22. Shields WD, Duchowny MS and Holmes GL (1993): Surgically remediable syndromes of infancy and early childhood. In Engel J Jr (ed): "Surgical Treatment of the Epilepsies.", Second Edition. New York: Raven Press, pp 35–48.
23. Squire LR (1993): Memory and the hippocampus: A synthesis from findings with rats, monkeys and humans. Psychological Review 99:195–231.
24. Tulving E and Markowitsch HJ (1997): Memory beyond the hippocampus. Current Opinion in Neurobiology 7:209–216.
25. Vargha-Khadem F, Isaacs EB and Watkins KE (1992): Medial temporal-lobe versus diencephalic amnesia in childhood. Journal of Clinical and Experimental Neuropsychology 14:371–372.

14

NEUROPSYCHOLOGICAL ASPECTS OF SEVERE MYOCLONIC EPILEPSY IN INFANCY

Catherine Cassé-Perrot*[,1], Markus Wolf[2], and Charlotte Dravet[1]

[1]Centre Saint-Paul, 300 Boulevard Sainte Marguerite
13009 Marseille, France and
[2]Universitäts Kinderklinik
Neuropädiatrie, 72070 Tübingen, Germany

INTRODUCTION

Severe Myoclonic Epilepsy in Infancy (SMEI) was first described in 1982[12] and recognized as a syndrome in 1989.[7] This disorder is characterized by 1) a high familial incidence of epilepsy or febrile convulsions; 2) repeated convulsive seizures from the middle of the first year of life, that are clonic or tonic-clonic, either generalized or unilateral, and often prolonged; 3) the later occurrence of myoclonic jerks, atypical absences and partial seizures; 4) a progressive slowness of psychomotor development during the second year; 5) behavioral problems including hyperkinesia and poor interpersonal relationships.

To date, no underlying pathology or degenerative disease has been identified. This epilepsy is cryptogenic, affecting patients who were considered normal at the onset. Treatment is difficult and seizures persist into the adult age. We reviewed the neuropsychological course of 20 patients diagnosed as suffering from SMEI at the Centre Saint-Paul in Marseilles. In some cases, it was possible to correlate characteristics of the epilepsy (including seizure pattern and EEG features) to the neuropsychological pattern. Collecting detailed information on the neuropsychological development of these children may contribute to the understanding of mechanisms involved in both the deterioration and recovery of cognitive functions.

1. PATIENTS AND METHODS

All children with SMEI aged less than 12 years were selected to undergo the first neuropsychological evaluation. There were 12 boys and 8 girls whose ages ranged from

* To whom correspondence should be addressed

Neuropsychology of Childhood Epilepsy, edited by Jambaqué et al.
Kluwer Academic / Plenum Publishers, New York, 2001.

11 months to 16 years 7 months. Fifty evaluations were collected, patients being examined at a mean rate of once or twice yearly. In 10 cases the neuropsychological follow-up was over 3 years.

Neuropsychological evaluation consisted first of a discussion with the parents in order to collect information regarding early psychomotor and affective development, and to determine the psychological structure of the family. The child's behavior was observed in various conditions: together with the parents, alone with the neuropsychologist, in playing situations and when performing tests. Standardized test batteries were performed to assess, according to age, various cognitive abilities, including global intelligence, visuomotor coordination, language, space perception, drawing and handedness.[18] The tests included the developmental scales of Brunet-Lézine,[3] the Revised Wechsler Intelligence Scale for Children (WISC-R),[28] a French scale for language called "Epreuves pour l'examen du langage",[6] the McCarthy Scales of Children's abilities[21] and Harris' "Test of lateral dominance".[18]

The sample was divided in three groups according to age at testing, which also corresponded to different stages of the epileptic syndrome: 1) 4 children aged between 11 months and 2 years, 2) 12 children whose age varied between 2 years, 1 month and 6 years; and 3) 11 children aged between 6 years, 1 month and 16 years. Some of the children were tested several times and their data are therefore included in the second and third group.

2. RESULTS

2.1. Findings During the First Year of Life

In 17 patients, the first seizure occurred between 3 and 6 months of life and in 3 others, between 7 and 9 months. EEG and neuro-imaging were normal except in 3 cases, one who exhibited external hydrocephalus (patient 1) and two who showed brain atrophy (patients 6, 18). With regard to psychomotor development, only 11 children walked before the age of 15 months. The remaining 9 walked between 18 and 30 months and often displayed ataxia, hypotonia and pyramidal signs. Four children exhibited slowing of psychomotor development between 8 and 9 months of age, either after the second convulsive seizure or after an increase in seizure frequency, or even without any obvious reason.

2.2. Course During the Second Year of Life

Seizure frequency and severity increased and drug regimens became heavy, with frequent adjustments of type and dose of medication. Parents became aware of the severity of the epilepsy and concentrated their worry on the seizures. It often appeared to them that seizures were the only cause of retardation and behavioral problems, and that the child's condition would return to normal if seizures were brought under control. Retrospectively, none of the 20 children was able to construct a sentence at the end of the second year of life. Only two children presented with a quiet behavior and normal interest for people.

2.3. Analysis of the First Group

Four infants were first evaluated between 11 and 21 months of life (see Fig. 1). A slowing down in learning abilities began at the end of the second year and behavior was

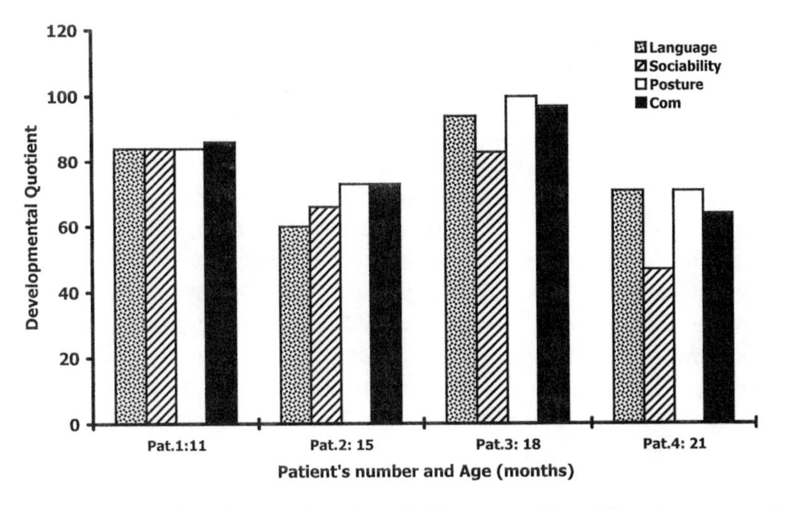

Figure 1. Group 1—Developmental quotient (D.Q.) expressed for all functions measured.

marked by poor motor coordination with hyperkinesia. Patients could use single words, but even their jargon was relatively poor. They exhibited little tendency to repeat syllables and words or name objects. Instability appeared with poor interpersonal relationship abilities that were often undifferentiated or restricted to adults. For a single patient examined at the age of 11 months (patient 1), DQ was homogeneous and in the normal range. For the girl who had the best development (patient 3) seizures were infrequent and of brief duration, with seizure-free periods. The most retarded boy (patient 4) had presented regression periods starting at the end of the first year of life. He lost the ability to walk after two episodes of status epilepticus that occurred in the second year of life.

2.4. Analysis of the Second Group

Twelve children were evaluated between 2 years 4 months and 5 years 10 months, for a total of 21 assessments. From the third year on, convulsive seizures were associated with other types of seizures (partial, absences, nocturnal). After 4 or 5 years of age, convulsive seizures tended to be shorter and less frequent, with a regular and important major decrease. Most DQ values spread from 60 to lower than 40 by five years of age.

One patient (nr 10) exhibited from the age of 1 year 10 months an increase in the frequency of generalized tonic-clonic seizures (GTCS) in spite of various drug combinations. At the age of 3 years and 2 months, the EEG showed major background slowing, multifocal spikes and spike-waves on both fronto-temporal regions and a few generalized bursts of spike-wave discharges. Her DQ was 61 with homogeneous deficiency. Her behavior was markedly unstable with an indifferent contact. At the age of 4 years 3 months, the frequency of GTCS had increased and myoclonias often preceded the seizures, suggesting a more severe course of the epilepsy. Neuropsychological evaluation revealed nearly total stagnation of development (DQ: 51). The child was very unstable, constantly opposing, and her contact was indifferent. In spite of various treatments, GTCS increased in frequency. However, no episodes of status epilepticus ever occurred. At the age of 5 years 2 months, she was comatose during several days because of severe pulmonary infection, without any seizure. After complete recovery from this event, seizures recurred, but their frequency had decreased to 1 to 2 a month. Concomitantly, myoclonias disappeared. However, the Brunet-Lézine test at the age of 6 years and 2 months, confirmed the lack of progress over the last three years (DQ: 32). Behavior remained quite unchanged with striking affective indifference. During the following months, GTCS rarely occurred, and a few complex partial seizures involving one half of the face were reported. The girl could be retested at the age of 7 years and 2 months. At that stage, developmental progress was noted, especially for language (she could name pictures, which formerly had never been possible) and drawing (she was able to draw a circle whereas she had only scribbled in former tests). At the same time behavior had improved: she was more stable, her emotions were now expressed appropriately and her contact was more differentiated. In fact, these changes were surprising since her behavior in the past could have been considered psychotic, and no dramatic changes were expected. This progress obviously coincided with the improvement of epilepsy.

Two children experienced a different course. One girl (patient 13) was stable at a higher level, over 70 (DQ) from the age of 4 years 9 months, but her development was very heterogeneous. She had had a few episodes of status. The progress generally coincided with a lower seizure activity. Another child (patient 8) had made little progress since the age of 9 months and he presented with autistic features. He had extremely frequent nocturnal convulsive seizures and numerous episodes of status.

2.4.1. Oculomotor Coordination. All patients had poor motor functions. Myoclonias were not the only cause of clumsiness. Patients also had severe visuo-constructive and visuo-spatial deficits. These difficulties were seen even in the child (patient 13) who had shown a relatively good overall performance at 2 years and 6 months and who had been the only one to acquire manual dominance: this patient exhibited major slowing of these learning abilities up to 6 years of age.

2.4.2. Sociability. Interpersonal relationship levels rarely exceeded 2 years and in general, a continuous decrease was observed. Visual contact was not always easy. Prolonged interpersonal contacts were difficult because of hyperactivity. Patients showed little interest for other children, and the relationship was often undifferentiated. However, the children were interested by objects in their environment. All children experienced periods of lesser excitation during which they could better relate with others. From 4 or 5 years on, patients benefited from rehabilitation that had positive effects on behavior and communication. Thus, the presence and degree of psychotic features could vary in a given child.

2.4.3. Language. Only language development showed marked individual differences. Between 2 and 6 years, all children were dysarthric. Three of them could utter only a few isolated phonemes (patients 1, 8, 12). During the 3-year follow-up, most children did not acquire more than 10 words. From the very first stages of speech development, the interest for language seemed to be extremely reduced due to the difficulty to reproduce and to imitate the sounds of language. Patients also suffered from gnostic difficulties related to spoken language: understanding was limited to the recognition of a few linguistic units, often in their familiar context. Six children could associate two words, only 3 (patients 7, 10, 13) could use sentences with the pronoun "I", but the sentences were either stereotyped or without syntax. The girl who showed the best development

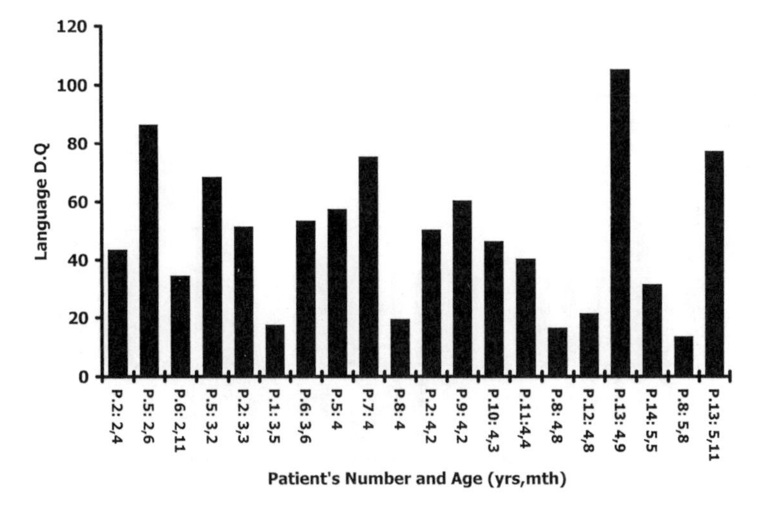

Figure 2. Language developmental quotient evaluated in the second group.

(patient 13) obtained normal results in naming and understanding on the picture assessment task but spontaneous language lacked proper syntax and was most often inappropriate.

2.5. Analysis of the Third Group

Eleven children were evaluated between 6 and 16 years for a total of 28 assessments. The general trend for a decrease in seizure frequency and severity was confirmed in this age range.

In 6 children, this decrease was associated with progress in behavior. Hyperactivity was less marked, contacts were more specific for both people and the environment. However, patients were still under polytherapy, a factor which is have played a role in global slowing.

2.5.1. Analysis of Global Developmental Quotients. Most DQ values were between 20 and 40 which in themselves, do not allow for an accurate description of the individual conditions. For this reason, the following description is based on the analysis of mental age, i.e., performances levels, rather than DQ.

2.5.2. Oculomotor Coordination (OMC). The coordination ability level of most children was between 2 and 3 years. The children could reproduce small buildings with 4 or 5 units. They generally had limited coordination skills and were poorly lateralized. In daily life, they were able to eat alone, but complex gestural sequences like dressing up were very long to perform. Only one girl (patient 13) reached a level of ability that put her between 5 and 6 years. She was also the child who had shown the least amount of deficits in this area when she was tested before age 6. Between 4 and 8 years, her OMC performance had remained stable at the level of 3 years. At 8 1/2 years, she achieved a level of 6 years: she could reproduce complex constructions, geometrical drawings and began to write letters and words. This improvement contrasted with the occurrence of an increase in nocturnal seizures during the previous year.

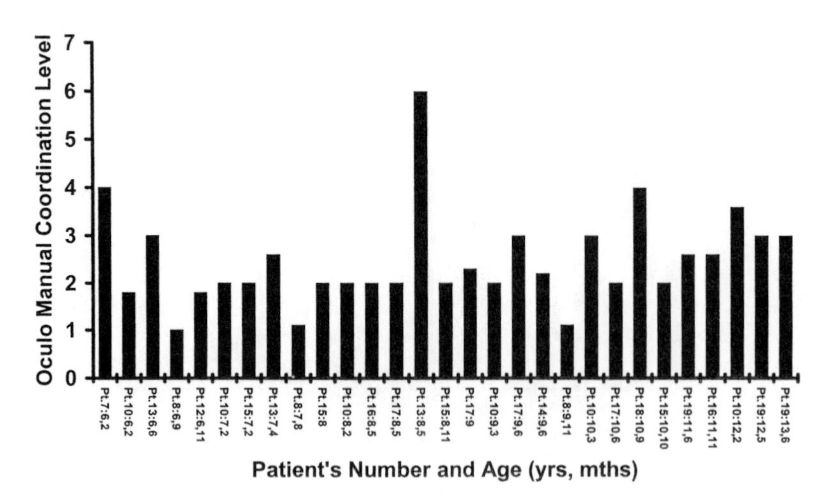

Figure 3. Oculo-Manual Coordination Level measured in Group 3.

2.5.3. Graphism. This activity, which also involves OMC, was either of little interest for these children or it represented a particular investment and could become a stereotyped activity that could not be amenable to change. The reproduction of geometrical drawings of a 3- or 4-year level was possible on request but figurative drawings, such as human beings or animals were very rarely produced.

2.5.4. Spatial Perception. This activity also involves OMC, but it is related to spatial organization. This function is usually studied by evaluating the ability to build puzzles with a model. In half the patients, this activity caught the attention of the children in such a way that they succeeded this test better than all other OMC task. However, the strategy the patients used was completely different from that used by normal children. Rather than trying to reproduce the general representative structure of the object, hey combined the different pieces on the sole basis of their shape. The type of visual analysis was suggestive of a sequential rather than a global process, possibly carried out by the left hemisphere. At this stage, however, we do not have enough information to specify which type of hemispheric specialization may characterize the global performance of these children. As for the other developmental disorders, such as autism and psychosis, this behavior remained unchanged in this age range.

2.5.5. Language. Inter-individual differences could still be observed. Linguistic levels varied between 2 and 6 years and the most deficient patients showed no significant improvements. None of the children displayed a progressive and harmonious language development. Pathological features were found at all ages for both expression and comprehension. Dysarthria often improved spontaneously. The production of single words remained predominant and fluency was generally poor with little spontaneity, except for a few stereotyped topics. Six patients out of 11 (patients 7, 10, 13, 16, 17, 19) were able to complete a naming test. There were many paraphasias (mainly semantic), gender errors and neologisms. This test showed the best improvement, whereas little progress was seen in syntax, fluency and desire to speak. In eight children, the lexical stock was that of an 8-year-old child. The naming test also emphasized features of visual agnosia in 6 children. For example, a radiator could be perceived as a window, a bed as a cube or a fan,

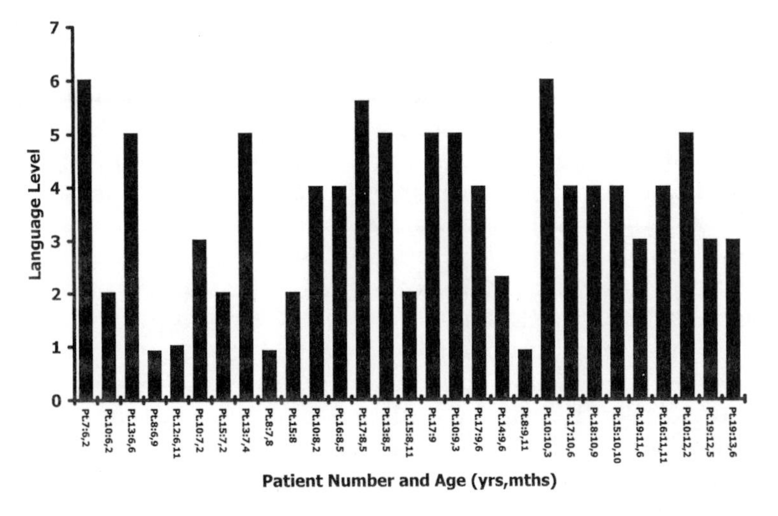

Figure 4. Language level observed in Group 3.

the funnel as a trumpet, the oar as a brush or a shovel. Language comprehension also exhibited agnosic features. Although difficult to test, comprehension was often at the level of about 3 years. Some parents noticed that their child did not seem to react to speech, as if they had trouble decoding the sounds. Four children of this group (patients 7, 13, 16, 17) were able to have a small but adapted conversation. Just as for other cognitive functions, language progressed in a very non-harmonious way compared to normal children. Successive acquisitions did not seem to have any link between them and their sequence lacked precise structure. These limited language abilities were paralleled by poor social competence.

2.5.6. Behavior. In this third group, instability was less marked, which allowed for increased attention towards people and the environment. The inter-individual differences in behavior also appeared to be linked to family relationships and to quality of care. During adolescence, more aggressiveness problems emerged when the family had exhibited rejection reactions over a long period of time.

CONCLUSION

Cognitive and behavioral troubles are homogeneously found in the children affected by SMEI. These problems set in by the end of the first year of life. Neuropsychological deficits concern all skills, indicating a rather diffuse brain involvement. Motor, linguistic and visual abilities are strikingly affected. In the visuo-constructional domain, it appears that the global mode of organizing visual information is more affected than the sequencing or piece-meal approach to construction, possibly reflecting a greater disturbance of right-hemispheric functions. Behavior is marked by hyperactivity, psychotic-type of relationships and sometimes autistic traits.

Three children in this series (patients 10, 13, 17) had a better development and their language was more preserved. This correlated with milder epilepsy (patient 13) or a marked decrease in the number of GTCS from the age of 6 years (patients 10, 17). But in all instances, progress exhibited a dysharmonic mode.

Several factors are statistically linked to cognitive impairment: repeat generalized convulsive seizures and status epilepticus,[1,5,22] prolonged non convulsive status,[10] long lasting interictal EEG paroxysmal activity[25] and antiepileptic drug therapy.[26] Neuropsychological data concerning early-onset epilepsies in childhood are scarce. In West syndrome, Dulac *et al.*[14] found deficits in visuo-motor coordination in children whose SPECT showed parieto-occipital hypoperfusion. Interpersonal contact was affected in many patients with this syndrome, as a consequence of visual agnosia. Dysphasia and hyperkinesia were the prominent features when SPECT disclosed temporal hypoperfusion. Autistic traits were correlated to both frontal and occipital hypoperfusion. Similar psychiatric features were found before the age of 3 years in children with West syndrome by Riikonen & Amnell.[23] Thus early age-related brain dysfunction could be the cause of these symptoms, and there seems to be a correlation between the topography of focal cerebral blood flow abnormalities and the type of neuropsychological disorders. In West syndrome, there also seems to be a correlation between the combination of frequent convulsive seizures with extremely abnormal EEG affecting both hemispheres, thus preventing recovery from the unaffected side, and the rapid clinical deterioration. In contrast, EEG is often normal at the onset of SMEI, and clinical deterioration occurs later and more progressively. Continuous interictal epileptic discharges have not been reported in children with SMEI, but persistent slowing of the background EEG activity may be an indicator for diffuse cortical hyperexcitability.[11] Prognosis in these cases remains poor and dementia is common.[9]

Jambaqué *et al.*[20] found that in tuberous sclerosis, mental retardation is linked to the number and topography of tubers. Ferrie *et al.*[15] studied MRI and functional imaging (PET) in children with SMEI. MRI was normal and no common pattern of hypometabolism could be found. In the present study, all patients had been considered normal before seizure onset. Dravet *et al.*[13] reported similar characteristics. Biological investigations were all normal.

We found a history of convulsions or epilepsy in the families of 50% of the patients, like in others studies.[13,16] SMEI should be classified as a cryptogenic and/or idiopathic syndrome with a strong genetic background. In this context, which mechanisms are likely to account for mental deterioration? Cavazzuti *et al.*[4] studied the influence of epileptic seizures on early psychomotor development in previously healthy children. They observed stagnation of psychomotor development for about 3 months even after a single generalized convulsive seizure, and behavioral disturbances still persisted 12 months later. In the present study, the appearance of neuropsychological disorders seemed to be related to the severity of the epilepsy during the first two years of life. Children with an initially high frequency of GTCS showed earlier slowing of psychomotor development compared to children with a few GTCS in the first two years. Dodrill *et al.*,[8] who studied the neuropsychological outcome of adults suffering from GTCS, believed that an "early onset" (before the age of 5 years) may merely result in a greater total number of seizures. In this series, there were no differences between patients with seizures starting early in the first year compared to the others. Many authors suggest that a great number of GTCS is a risk factor for mental deterioration in epileptic patients.[1,2,24] We found some relation between the total number of GTCS and the degree of mental deterioration since the patients with relatively few GTCS had markedly better test results than the others. The duration of GTCS might play an additional role: long-lasting seizures can lead to temporary stagnation or even regression of psychomotor development, and permanent structural cerebral damage may occur.[1] Paradoxically, the patient in this study who had the best outcome had an increasing number of seizures after 6 years of age, without slowing in her cognitive and behavior acquisitions. The question arises whether seizure impact on development depends of the age of their occurrence, whatever their type and duration.

Polytherapy with two or more drugs is often necessary. We believe that they contribute to the deterioration of cognition and behavior but it is difficult to distinguish the effect of drugs from that of other features. Behavioral disturbances are more frequent in epileptic children with poor cognitive functions,[17] but behavior is influenced by many factors including familial environment. In the present series, psychotic and autistic traits varied over time for a given patient, but were present in a majority of cases, probably as a consequence of severe convulsive seizures in the first year of life.

Stagnation or deterioration was the rule during the first six years of life but an evolution, albeit dysharmonic was seen thereafter. The question arises as to how different functions can develop after the so-called critical period of cortical maturation is over. This is also the case for patients with developmental disorders such as Williams syndrome in which epilepsy is not a major feature. For instance, Udwin *et al.*[27] reported a dysharmonic evolution of language capacities in Williams syndrome. Behavioral disorders observed in this syndrome also consist of hyperkinesia, attention deficit, and some psychotic traits. Children with Williams syndrome have important semantic and pragmatic capacities but low functioning in visuo-spatial skills. The relationship between the evolution of SMEI and other developmental disorders obviously needs to be specified. Further studies are necessary to determine the specific underlying causes of the neuropsychological profile observed in SMEI.

ACKNOWLEDGMENTS

This study was performed with financial support from the Fondation pour la Recherche Médicale, the Fondation Française pour la Recherche sur l'Epilepsie, the Ligue Française contre l'Epilepsie, the Société d'Etudes et de Soins pour les Enfants Paralysées et Polymalformés, the Association pour la Recherche sur les Epilepsies and the Deutsche Forschungsgemeinschaft (DFG) Wo 534/1-1.

REFERENCES

1. Binnie CD, Channon S and Marston D (1990): Learning disabilities in epilepsy: Neuropsychological Aspects. Epilepsia 31(Suppl 4):S2–S8.
2. Bourgeois B, Prensky A, Palkes H, Talent B and Buch S (1983): Intelligence in epilepsy: A prospective study in children. Annals of Neurology 14:438–444.
3. Brunet O and Lézine I (1965): Le Développement Psychologique de la Première Enfance. Paris: Presses universitaires de France.
4. Cavazzuti GB, Ferrari P and Finelli T (1987): Avenir neuropsychologique dans les épilepsies de la première année de vie. In Narbona J and Poch-Olive ML (eds): "Neuropsichologia Infantile". Pamplona: Eurograph, pp 203–205.
5. Chevrie JJ and Aicardi J (1978): Convulsive disorders in the first year of life: Neurological and mental outcome and mortality. Epilepsia 19:67–74.
6. Chevrie-Müller C, Simon A and Decante P (1981): Épreuves pour l'Examen du Langage (Enfants de 4 à 8 ans). Paris: Centre de Psychologie Appliquée.
7. Commission on Classification and Terminology of the ILAE (1989): Proposal for revised classification of epilepsies and epileptic syndromes. Epilepsia 30:389–399.

8. Dodrill CB (1986): Correlates of generalized tonic-clonic seizures with intellectual, neuropsychological, emotional and social function in patients with epilepsy. Epilepsia 27:399–411.
9. Doose HV and Ondorza G (1994): Epileptogene Hirnleistungsstörungen. In Todt H and Heinicke D (eds): "Aktuelle Neuropädiatrie". Ciba-Geigy Verlag. Wehr, pp 129–138.
10. Doose H and Völske E (1979): Petit Mal-status in early childhood and dementia. Neuropediatrics 10:10–14.
11. Doose H and Baier WK (1988): Theta rhythms in the EEG- a genetic trait in childhood epilepsy. Brain Development 10:347–354.
12. Dravet C, Roger J, Bureau M and Dalla Bernardina B (1982): Myoclonic Epilepsies in Childhood. In Akimoto H, Kazamatsuri H, Seino M and Ward A (eds): "Advances in Epileptology: the XIIIth Epilepsy International Symposium". New York: Raven Press, pp 135–140.
13. Dravet C, Bureau M, Guerrini R, Giraud N and Roger J (1992): Severe myoclonic epilepsy. In Roger J, Bureau M, Dravet C, Dreifuss FE, Perret A and Wolf P (eds): "Epileptic Syndromes in Infancy, Childhood and Adolescence". London and Paris: John Libbey Eurotext Ltd, pp 75–88.
14. Dulac O, Jambaqué I and Chiron C (1987): Neuropsychologie des épilepsies de l'enfant. In Narbona J and Poch-Olive ML (eds): "Neuropsychologia Infantile". Pamplona: Eurograph, pp 185–201.
15. Ferrie CD, Maisey M, Cox T, Polkey Ch, Barrington SF, Panayiotopoulos CP and Robinson RO (1996): Focal abnormalities detected by 18 FDG PET in epileptic encephalopathies. Archives of Disease in Childhood 75:102–107.
16. Fujiwara T, Watanabe M, Takahashi T, Yagi K and Seini M (1992): Long-term course of childhood epilepsy with intractable grand mal seizures. Japenese Journal of Psychiatry and Neurology 46:297–302.
17. Gérard CL (1993): Les troubles psycholinguistiques des enfants autistes. Approche Neuropsychologique des Apprentissages chez l'Enfant 5:136–141.
18. Harris A (1947): Harris Test of Lateral Dominance. New York: Psychological Corporation.
19. Herrmann B (1982): Neuropsychological functioning and psychopathology in children with epilepsy. Epilepsia 23:545–554.
20. Jambaqué I, Cusmai R, Curatolo P, Cortesi F, Perrot C and Dulac O (1991): Neuropsychological aspects of tuberous ssclerosis in relation to epilepsy and MRI findings. Developmental Medicine and Child Neurology 33:698–705.
21. McCarthy D (1972): McCarthy Scales of Children's Abilities. New York: Psychological Corporation.
22. O'Leary D, Seidenberg M, Berent S and Boll T (1981): Effects of age of onset of tonic-clonic seizures on neuropsychological performance in children. Epilepsia 22:197–204.
23. Riikonen R and Amnell G (1981): Psychiatric disorders in children with earlier infantile spasms. Developmental Medicine and Child Neurology 23:747–760.
24. Rodin E, Schmaltz S and Twitty G (1986): Intellectual functions of patients with childhood-onset epilepsy. Developmental Medicine and Child Neurology 28:25–33.
25. Tassinari CA, Bureau M, Dravet C, Dalla Bernadina B and Roger J (1992): Epilepsy with continuous spikes and waves during slow sleep-otherwise described as ESES (epilepsy with electrical status epilepticus during slow sleep). In Roger J, Bureau M, Dravet C, Dreifuss FE, Perret A and Wolf P (eds): "Epileptic Syndromes in Infancy, Childhood and Adolescence". London and Paris: John Libbey Eurotext Ltd, pp 245–256.
26. Trimble MR and Thompson PJ (1983): Anticonvulsant drugs, cognitive function and behavior. Epilepsia 24:S55–S63.
27. Udwin O, Davies M and Howlin P (1996): A longitudinal study of cognitive abilities and educational attainment in Williams syndrome. Developmental Medicine and Child Neurology 38:1020–1029.
28. Wechsler D (1981): Échelle d'Intelligence pour Enfants. Forme Révisée (WIC-R). Paris: Centre de Psychologie Appliquée.

15

NEUROPSYCHOLOGICAL STUDIES IN IDIOPATHIC GENERALIZED EPILEPSY

Allan F. Mirsky*,[1], Connie C. Duncan[1,2], and Miriam Levav[1]

[1]Section on Clinical and Experimental Neuropsychology
LBC, National Institute of Mental Health
NIH, 15 North Drive, MSC 2668 Bethesda
MD, 20892-2668, U.S.A.
[2]Department of Psychiatry
Uniformed Services University of the Health Sciences
Bethesda, MD, 20814-4799, U.S.A.

1. DESCRIPTION OF NEUROPSYCHOLOGICAL DEFICITS IN SEIZURE DISORDERS

1.1. The view of Persons with Seizures as Defective in Personality and Intellect

In the descriptions of the psychological characteristics of persons with seizure disorders that were written early in the last century, little attempt was made to distinguish among types of disorders. It was generally assumed that there was a distinctive "epileptic personality," and that persons with seizures would exhibit these characteristics. In 1934, Grinker (cited by Harrower-Erickson[5] in 1941) described the epileptic as "egocentric, selfish, seclusive, [showing] tremendous outbursts of emotions." Other descriptions of persons with seizure disorders are equally or even more pejorative, including such characteristics as malicious, cruel, and tending to develop criminal tendencies.[5] If any objective measures were used to describe persons with seizure disorders, they usually consisted of a measure of IQ and a personality test, such as the Rorschach. The typical conclusion was that such patients were severely retarded intellectually, and that they displayed some or all of the unpleasant traits described above. However, the majority of studies appear to have been conducted on patients with seizures of many years' duration, who

* To whom correspondence should be addressed

Neuropsychology of Childhood Epilepsy, edited by Jambaqué et al.
Kluwer Academic / Plenum Publishers, New York, 2001.

most often had been residents of epileptic "colonies." Harrower-Erickson reviewed research demonstrating that many epileptic persons residing outside of epileptic colonies had normal personalities, and normal or even superior IQs. She concluded that before any generalizations regarding psychological effects of seizures could be drawn, there were many variables that needed to be considered, including the form of the seizures, the duration of illness, and the possible toxic effects of the type and dosage of anticonvulsant medication. Essentially the same conclusions were reached by Folsom[3] in 1953, who recommended that epileptics be subdivided on the basis of cerebral dysfunction and seizure phenomena, and that a "battery of tests, covering a wide range of intellectual processes, including memory, should be used in order to evaluate patterns of cognitive deficit."

1.2. Research Establishing Distinct Neuropsychological Profiles in Idiopathic Generalized Seizures

The studies reviewed in this chapter, beginning in 1960,[13] can be said to have followed the recommendations of Harrower-Erickson[5] and Folsom.[3] We have endeavored to compare and contrast the neuropsychological characteristics of persons with idiopathic generalized seizure disorders with those of persons with focal disorders, and to relate our findings to the known pathophysiological mechanisms underlying idiopathic generalized seizures. Our recent research, which has extended to the family members of patients with seizure disorders, has been designed to evaluate possible familial/genetic influences in idiopathic generalized seizures.

At the outset, we should state that our view of the underlying pathophysiology of idiopathic generalized seizure disorders is that it involves an alteration in the functioning of a central, subcortical integrating system with widespread connections to many parts of the cerebral hemispheres. In a general sense, it can be said that our view is based on the "centrencephalic" theory of Penfield and Jasper,[18] as later modified by Gloor,[4] to the "cortico-reticular" theory. This modification states, essentially, that in idiopathic generalized seizures, there is dysfunction and disturbed relationships among the cortical, thalamic, and reticular components of a major regulatory system within the brain.

It should also be emphasized that most of the information presented in this review is based upon research conducted on patients with typical or atypical absence seizures. There have been relatively few detailed neuropsychological studies on patients with other types of idiopathic generalized seizure disorders. There is reason to believe that there may be some neuropsychological deficits common to most forms of idiopathic generalized seizures. We will present some preliminary information from a study in progress in which we have compared the neuropsychological profiles of three groups of patients: one group with partial (i.e., temporal lobe) epilepsy and two groups with different forms of generalized epilepsy (absence and juvenile myoclonic epilepsy). However, until the neuropsychological similarities and differences among the various forms of idiopathic generalized seizure disorders have been demonstrated, it must be assumed that our results may not be applicable to all idiopathic generalized seizure disorders. We will discuss this issue later in this review.

Our initial investigation employed a large battery of psychological tests, including tests of personality, memory, problem solving and attention.[13] The most salient finding of that investigation was that even when matched for IQ (either by selection of cases or by statistical means), patients with idiopathic generalized seizures performed more poorly on tests of visual sustained attention than patients with focal frontal or temporal lobe foci. Visual sustained attention was measured with a task (the Continuous Performance Test or CPT) that required the subject to maintain a focus of attention on a visual display for an appreciable interval of time, for intervals ranging from 6 to 20 minutes.[19] We used two versions of the CPT, one that required the patient to press a response button whenever the letter "X" appeared in a visual display, and another, more difficult version, that required the patient to press the response key only when the letter "X" followed immediately after the letter "A." This test had been shown to be sensitive to the effects of cerebral dysfunction:[9] in the case of idiopathic generalized seizures, the more difficult "AX" version of the CPT was found to be more sensitive to injury than the "X" task.

The results of subsequent research confirmed the finding of impaired sustained attention in patients with idiopathic generalized seizures as compared with patients with focal seizure disorders; Fig. 1 summarizes the findings obtained with the AX task of the CPT in studies conducted over a 40-year period.

The studies represented in the figure were conducted between 1954 and 1991. Focal controls were usually patients with temporal lobe epilepsy, except for the 1960b focal controls, which includes both frontal and temporal lobe cases. The scores listed at the bottom of the figure are the mean IQ scores, and the numbers of subjects represented in each column. The studies conducted in the years 1969d and 1990f were with children, whose ages ranged from 6 to 18 years. The other studies (1954a, 1960b, 1964c, and 1988e) involved adult subjects. Adapted from Mirsky.[10]

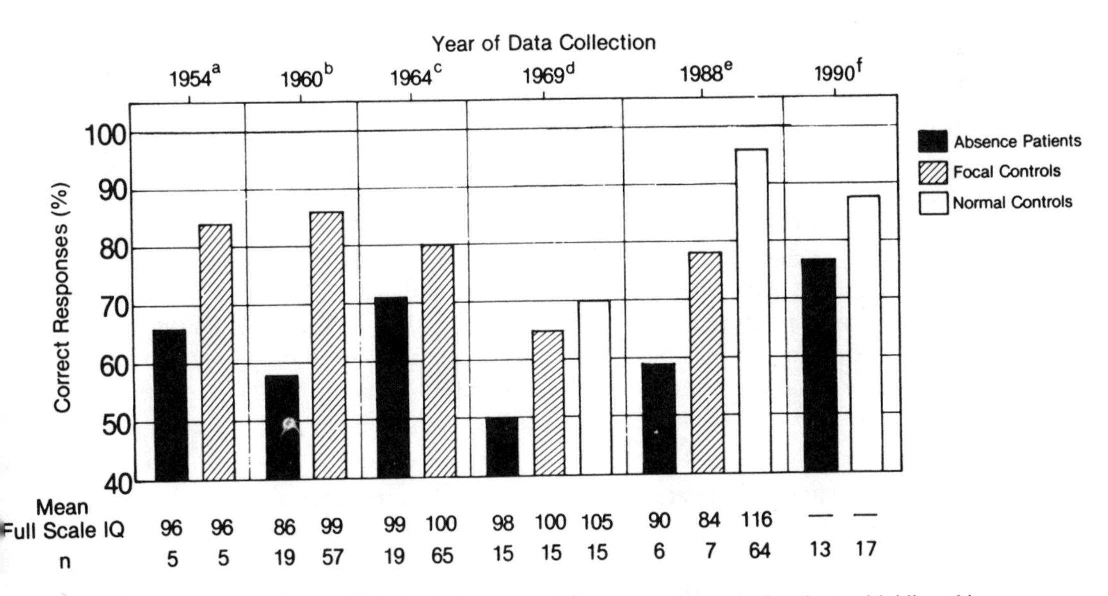

Figure 1. Scores on the visual CPT-AX task (percentage of correct responses) of patients with idiopathic generalized seizures (usually absence epilepsy) compared with scores of patients with focal seizures, and normal controls.

2. SPECIFICATION OF THE NATURE AND EXTENT OF THE ATFENTION DEFICIT IN IDIOPATHIC GENERALIZED EPILEPSY

2.1. Defining the Nature of Cognitive Impairment in Patients with Idiopathic Generalized Seizures

The question arises as to whether patients with idiopathic generalized seizures are impaired generally on cognitive tasks. To some extent, this result would be expected, given that impaired attention should have effects on the performance of any behavioral task. Our research, conducted over the past four-and-a-half decades, indicates that patients with idiopathic generalized seizures, usually of the absence type, do tend to perform more poorly on most cognitive tasks than normal controls matched for age and education. However, their impaired performances are different from those seen in patients with focal seizure disorders. In fact, each patient group has a unique cognitive profile, corresponding to the part of the brain that is impacted by the seizure disorder.[10] In the case of patients with idiopathic generalized seizures who are of normal intelligence, the impairment is more likely to be seen in measures of sustained attention. In cases of focal temporal lobe epilepsy, by contrast, the impairment is more likely to be seen in tests of verbal or visual-spatial memory.[10,22] And it is well known that patients with seizure disorders, generally, tend to be one or more years behind their age mates in academic achievement. Whether this is due to the presence of seizures alone, to the toxic effects of antiepileptic medications, or to some combination of the two is difficult to determine. However, existing research provides some support for all of these explanations.

2.2. Visual and Auditory Attention

In addition to the findings of impaired visual sustained attention in patients with idiopathic generalized seizures, summarized in Fig. 1, it has also been shown that these patients are impaired in auditory attention tasks, and that the impairment may be greater than that seen in visual tasks. Mirsky and Van Buren[14] reported deficits on both visual and auditory CPT tasks in patients with absence epilepsy. Duncan[1] compared the performance of patients with idiopathic generalized seizures and normal controls on both visual and auditory CPT tasks; she employed "X" and "AX" versions of each task. She found that whereas the patients tended to perform more poorly on all sustained attention tasks, auditory tasks were significantly more difficult than visual tasks for the patients. Some of Duncan's findings are reproduced in Fig. 2. These results suggest that the impaired capacity for sustained attention of patients with idiopathic generalized seizures is general in nature, and not confined to either visual or auditory processing. A number of questions arise from these findings, which are discussed below.

2.3. The Relationship Between Attention Test Performance and Paroxysmal EEG Activity

In 1939, Schwab[20] published the results of a study in which he recorded EEG and reaction time simultaneously in a group of patients with absence epilepsy. He reported that responses to auditory or visual stimuli were substantially delayed or did not occur when the stimulus coincided with paroxysmal spike-wave discharges. In a later, detailed

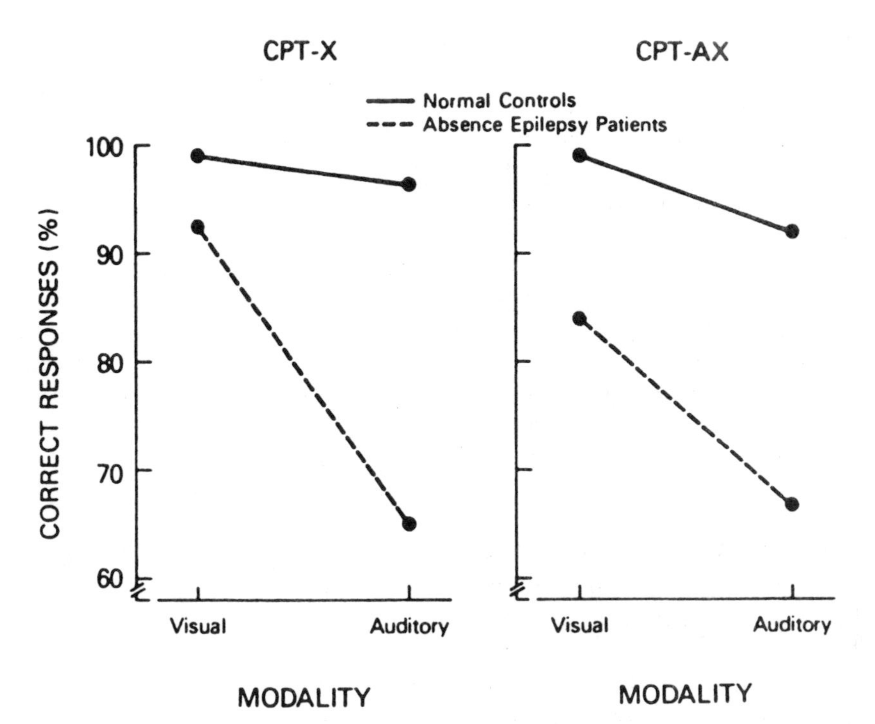

Figure 2. Mean percentages of correct responses for normal control subjects (solid lines) and patients with idiopathic generalized (absence) epilepsy (dashed lines) in the visual and auditory versions of the CPT-X (left panel) and CPT-AX (right panel) tasks. Note the lower scores on auditory than on visual versions of the CPT in the patients. Reproduced from Duncan.[1]

study of the relationship between attention performance and the occurrence of bursts of paroxysmal EEG activity associated with idiopathic generalized seizures, Mirsky and Van Buren[14] confirmed and extended Schwab's results. They found that although there was a strong association between performance and EEG bursts (i.e., the bilaterally synchronous and symmetrical three-per-second spike-and-wave activity), the two phenomena were, in fact, to some degree independent. Thus, some patients were able to attend and respond to task stimuli in the presence of spike-wave bursts, whereas others could not respond at such times. The capacity for response was affected by such factors as degree of left-right symmetry in the EEG paroxysm, burst length, and time of occurrence of the stimulus during the burst (i.e., beginning vs. end). Moreover, Mirsky and Van Buren[14] found that even when those errors in performance associated with bursts were excluded, patients with idiopathic generalized seizures made more errors in the interictal periods than controls.

2.4. Attention Test Performance in The Interictal Period in Patients with Idiopathic Generalized Seizures

The finding that patients with generalized seizures tended to make more errors in the interictal periods than controls led to an intensive study of the EEG of such patients. This work involved not only analysis of the power spectrum (amount of "power" in various EEG frequencies) between seizures in such patients,[21] but also studies of the capacity for information processing, as measured with event-related brain potentials (ERPs).[1] The major finding of those studies is that in persons with idiopathic generalized seizures there is a more-or-less continuous subclinical abnormality in the brain's electrical activity that erupts from time to time into EEG paroxysms. Our assumption is that the tendency for patients with generalized seizures to make errors in sustained attention tasks even in the times between seizures is a reflection of this subclinical abnormal brain activity.

Duncan[1] provided direct evidence of the impaired capacity for mobilizing attention in the interictal period in idiopathic generalized epilepsy; such patients showed reduced amplitudes of ERP components associated with the mobilization of attentional resources, such as the P300 and slow wave components.[1] Such reductions appear more pronounced in the processing of auditory than visual task stimuli. Moreover, patients with focal, temporal lobe seizures did not show the type of reduction in P300 and slow wave amplitude seen in patients with idiopathic generalized seizures.[1,2]

2.5. Is the Impairment in Attention Due Primarily to the Interruption of Sensory Input, to Motor Inability, or to Other Factors (A More Central, Attentional Weakness)?

We have found evidence for reduced sensory input during bursts of paroxysmal EEG activity associated with generalized seizures (spike-and-wave activity). Orren (cited in 9, 16) has documented changes in visual ERPs occurring before and during spike-wave bursts. In addition Mirsky[9] reported reduced amplitude electroretinograms (ERG) during spike-wave bursts, suggesting that compromised visual processing during such events may involve the retina, and is not necessarily confined to more central cerebral structures. With respect to auditory processing, Skoff and Mirsky, (cited in 9), reported evidence of altered brainstem auditory evoked responses (BAERs) during spike-wave bursts. Duncan[2] also found altered BAERs in patients with absence seizures in the interictal period. These findings suggest that part of the behavioral alteration seen in patients with idiopathic generalized seizure disorders, i.e., the impaired ability to respond to visual and auditory stimuli, is related to reduced sensory input, an effect that is exacerbated during paroxysmal burst activity.

Other research has indicated that motor ability may also be altered during burst activity. Thus, it has been shown that in some patients, the ability to execute practiced, repetitive motor responses continues undiminished during spike-wave bursts, although the timing and pace of the responses may change.[14,16] In other patients, motor responsiveness terminates with the appearance of spike-wave bursts, or after the burst has been evident in the EEG for a few seconds.[14]

We have studied the effects of spike-wave bursts on memory, and have found that visual or auditory information presented to patients during, or shortly before, such epochs is recalled poorly, if at all.[14] The paroxysmal burst activity apparently interferes with the consolidation of short-term memory, although it is obvious from the material presented above that effects on attention cannot be ruled out. Regardless, it would be expected that a child suffering from many brief bursts of spike-wave activity would be substantially impaired in the performance of virtually any cognitive activity dependent upon attention or memory.

The research on behavioral impairment associated with spike-wave paroxysms has indicated that there are sensory and motor effects of the EEG bursts. However, the ERP studies of Duncan[1,2] have shown that there are deficits seen in patients with generalized seizures of the absence type that are not attributable to simple sensory or motor effects, but to impairment of a more central, integrative process, the mobilization of attention. We believe that, according to either the centrencephalic or the cortico-reticular theory of absence epilepsy, this attentional effect is attributable to disturbed functioning of a cortico-thalamic-reticular network.

3. IS THERE A UNIFORM PATTERN OF COGNITIVE DEFICIT IN ALL FORMS OF IDIOPATHIC GENERALIZED EPILEPSY?

To address this question adequately would require a much larger treatment than the present chapter would permit. Moreover, there have been relatively few studies that have attempted to compare the neuropsychological test results of patients with differing forms, or different ages of onset, of generalized seizure disorders. It is clear that some forms of generalized seizures, such as cryptogenic Lennox-Gastaut syndrome, are associated with profound mental retardation and generalized cognitive deficits. In contrast, patients with absence seizures may, by comparison, have relatively mild deficits. In order to answer the question posed earlier in this review, as to whether there are common forms of deficit in different types of generalized seizure disorders, we have recently begun a neuropsychological investigation on two similar, although clinically distinct groups of patients: cases with absence epilepsy; and cases with juvenile myoclonic epilepsy. Patients with temporal lobe epilepsy and normal healthy persons serve as control subjects.

Our preliminary findings have indicated that patients with either absence or juvenile myoclonic epilepsy may share certain neuropsychological deficits. Figure 3 illustrates some of our results: the absence and the juvenile myoclonic cases show higher error scores on three CPT tasks than the temporal lobe cases or the healthy controls. The error scores of the two patient groups with idiopathic generalized epilepsy thus tend to cluster together, and to be higher than the scores of the temporal lobe epilepsy and control groups. The three groups with seizure disorders did not differ among themselves on most subtests of the Wechsler Adult Intelligence Scale-Revised; however, the healthy control subjects tended to achieve higher scores on most subtests.

4. DO PERSONS GENETICALLY RELATED TO PROBANDS WITH SEIZURES SHARE THEIR NEUROPSYCHOLOGICAL DEFICITS?

The inheritance of some forms of idiopathic generalized epilepsy was reported nearly 40 years ago.[7,8] There have, however, been few if any neuropsychological studies of the first-degree relatives of probands with these disorders. Thus, if patients with absence epilepsy have characteristic deficits in visual and auditory sustained attention, would relatives who share the genetic makeup of the probands also share the behavioral symptom of the probands? This question has been pursued successfully in the behavior genetic analysis of schizophrenia, using such strategies as comparing the incidence and severity of symptoms in identical vs. fraternal twins (one of whom has the disorder), in adoption studies and in familial incidence studies. Mirsky *et al.* and others have shown

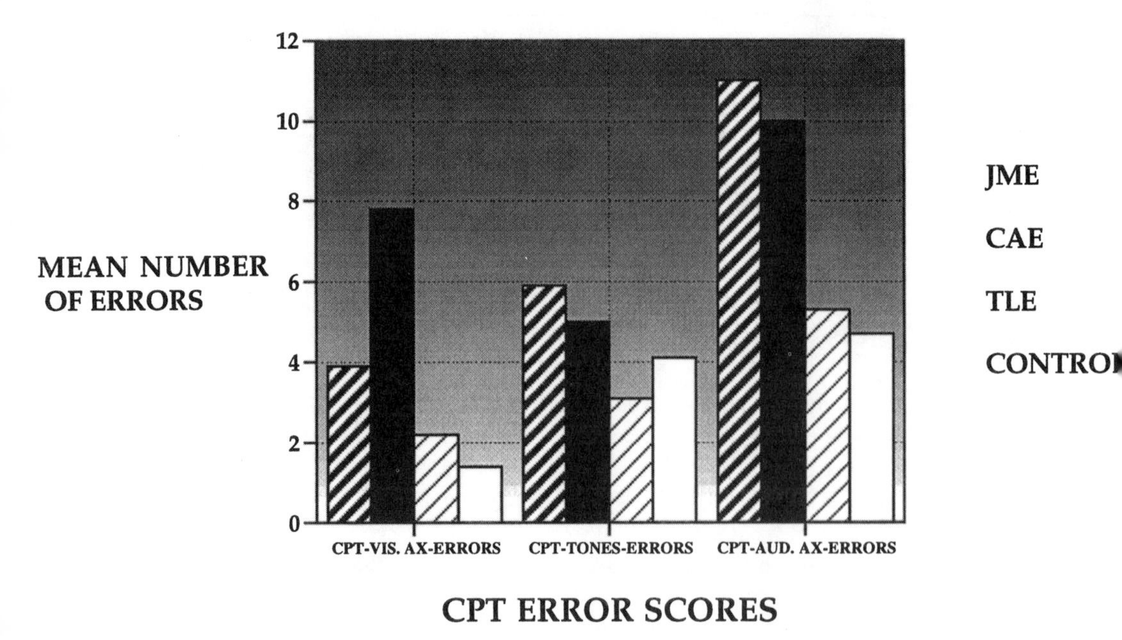

Figure 3. Mean number of errors (responses to wrong or non-target stimuli) for three versions of the CPT task. CPT-VIS.AX requires the subject to respond to the visually-presented letter "X" if it follows the letter "A;" CPT-Tones requires responses to the highest pitch tone of three; CFFAUD.AX requires responses to the spoken letter "O" if it follows the letter "L." JME = mean scores of 11 patients with juvenile myoclonic epilepsy; CAE = mean scores of 10 patients with childhood absence epilepsy; TLE = mean scores of patients with temporal lobe epilepsy with ns varying between 21 and 30; CONTROLS = mean scores of 55 healthy controls. Note that the error scores of the two patient groups with idiopathic generalized epilepsy tend to cluster together, and to be higher than the scores of the temporal lobe epilepsy and healthy control groups. From unpublished observations of Levav, M., Mirsky, A., Herault, J., and Andermann, E.

that the first-degree relatives of patients with schizophrenia may show in milder form some of the attentional deficits seen in the patients themselves.[15]

Levav and Mirsky[6] have been conducting investigations of the familial incidence of cognitive deficits in generalized epilepsy of the absence type, in collaboration with investigators in Israel and Canada. The findings are still preliminary; however, there is evidence of mild impairment in tests of sustained attention, such as the CPT, in the female, but not the male parents of probands with this disorder. Figure 4 shows the scores on some measures of sustained attention, derived from the CPT, of probands and relatives of probands with absence epilepsy. The figure shows that the mothers of the probands were less accurate than the fathers and also that the female probands tended to be slower and less accurate on tests of sustained attention than the male probands. These data are consistent with reports of a maternally transmitted influence on seizure susceptibility.[17]

5. SUMMARY

Patients with idiopathic generalized seizure disorders appear to have neuropsychological deficits that are distinct from those seen in patients with partial seizure disorders. We have studied such patients with a variety of methods, including neuropsychological

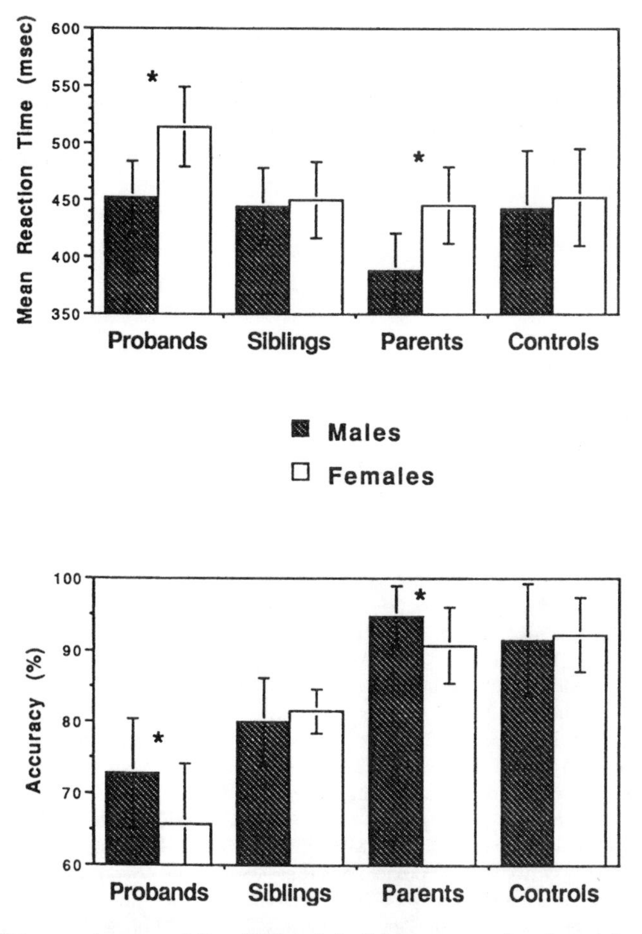

Figure 4. Mean CPT scores (average of four CPT tasks) of four groups of male and female subjects; child probands with absence epilepsy, unaffected siblings of the probands, parents of the probands, and matched adult healthy controls. Top graph, reaction times, bottom graph, accuracy scores. The asterisks at the top of the columns indicate that the male and female subjects within the group differed significantly. The female probands performed significantly worse than the male probands, and the mothers of the probands performed significantly worse than the fathers. From Levav,[6] reprinted from Mirsky *et al.*[12]

tests, sensory and motor tests, and EEG. The underlying deficit is in the capacity for visual and auditory sustained attention (vigilance), although other cognitive capacities that are dependent upon intact vigilance may be impaired, as well. Whereas the impairment is usually linked to the appearance of the characteristic abnormal EEG pattern in the disorder (i.e., bursts of bilaterally synchronous and symmetrical three-per-second spike-and-wave activity), the deficit is also evident in intervals between bursts. This suggests that in some forms of idiopathic generalized seizure disorders, such as absence epilepsy, there is a more-or-less continuous disturbance in the functioning of a brain system that has the responsibility for the maintenance of attention. Analysis of the deficit in vigilance in absence epilepsy indicates that it involves reduced sensory input, altered motor ability, and reductions in central attentional resources, as well. It is not clear whether all types of idiopathic generalized seizure disorders have similar cognitive deficits, although some preliminary work indicates that patients with absence and juvenile myoclonic epilepsy may share problems in attention. We have also found that first-degree relatives of patients with absence epilepsy may show, in milder form, some of the same deficits shown by the patients themselves. Future research should be extended to all types of generalized seizure disorders, and to the first-degree female relatives of patients with these disorders, to clarify the nature of the familial/genetic influences in idiopathic generalized seizures.

REFERENCES

1. Duncan CC (1988): Application of event-related brain potentials to the analysis of interictal attention in absence epilepsy. In Myslobodsky MS and Mirsky AF (eds): "Elements of Petit Mal Epilepsy". New York: Peter Lang, pp 341–264.
2. Duncan CC: Unpublished observations.
3. Folsom A (1953): Psychological testing in epilepsy. I. Cognitive function. Epilepsia 2:15–22.
4. Gloor P (1988): Neurophysiological mechanism of generalized spike-and-wave discharge and its implication for understanding absence seizures. In Myslobodsky MS and Mirsky AF (eds): "Elements of Petit Mal Epilepsy". New York: Peter Lang, pp 159–209.
5. Harrower-Erickson M (1941): Psychological studies of patients with epileptic seizures. In Penfield W and Erickson TC (eds): "Epilepsy and Cerebral Localization". Springfield: Thomas, pp 546–574.
6. Levav M (1991): Attention performance in children with absence epilepsy and their first-degree relatives. Unpublished Ph.D. Dissertation, University of Maryland, College Park, Maryland.
7. Metrakos K and Metrakos J (1961): Genetics of convulsive disorders I. Neurology 10:228–240.
8. Metrakos K and Metrakos J (1961): Genetics of convulsive disorders II. Neurology 11:474–483.
9. Mirsky AF (1988): Behavioral and psychophysiological effects of petit mal epilepsy in the light of a neuropsychologically-based theory of attention. In Myslobodsky M and Mirsky AF (eds): "Elements of Petit Mal Epilepsy". New York: Peter Lang, pp 311–340.
10. Mirsky AF (1992): Neuropsychological assessment of epilepsy. New Issues in Neurosciences IV:25–39.
11. Mirsky AF, Anthony BJ, Duncan CC, Ahearn MB and Kellam SG (1991): Analysis of the elements of attention: A neuropsychological approach. Neuropsychological Review 2:109–145.
12. Mirsky AF, Duncan CC and Levav M (1995): Neuropsychological and psychophysiological aspects of absence epilepsy. In Duncan JS and Panayiotopoulos CP (eds): "Typical Absences and Related Epileptic Syndromes". London: Churchill, pp 112–121.
13. Mirsky AF, Primac DW, Ajmone Marsan C, Rosvold HE and Stevens JA (1960): A comparison of the psychological test performance of patients with focal and nonfocal epilepsy. Experimental Neurology 2:75–89.
14. Mirsky AF and Van Buren JM (1965): On the nature of the "absence" in centrencephalic epilepsy: A study of some behavioral, electroencephalographic and autonomic factors. Electroencephalography and Clinical Neurophysiology 18:334–348.
15. Mirsky AF, Yardley SJ, Jones BP, Walsh D and Kendler KS (1995): Analysis of the attention deficit in schizophrenia—a study of patients and their relatives in Ireland. Journal of Psychiatric Research 29:23–42.
16. Orren MM (1974): Visuomotor behavior and visual evoked potentials during petit mal seizures. Unpublished Ph.D. Dissertation, Boston University, Boston, Massachusetts.
17. Ottman R, Annegers JF, Hauser WA and Kurland LT (1988): Higher risk of seizures in offspring of mothers than of fathers with epilepsy. American Journal of Human Genetics 43:257–264.
18. Penfield W and Jasper HH (1954): Epilepsy and the Functional Anatomy of the Human Brain. Boston: Little, Brown.
19. Rosvold HE, Mirsky AF, Sarason I, Bransome ED Jr and Beck LH (1956): A continuous performance test of brain damage. Journal of Consulting Psychology 20:343–350.
20. Schwab R (1939): A method of measuring consciousness in petit mal epilepsy. Journal of Nervous and Mental Disease 26:690–691.
21. Siegel A, Grady CL and Mirsky AF (1982): Prediction of spike-wave bursts in absence epilepsy by EEG power spectrum signals. Epilepsia 23:47–60.
22. Trimble, M (1992): Cognitive problems of patients with seizure disorders. New Issues in Neurosciences IV:17–24.

16

PAROXYSMAL AND NON-PAROXYSMAL COGNITIVE-BEHAVIORAL EPILEPTIC DYSFUNCTION IN CHILDREN

Thierry Deonna

CHUV, Neuropediatric Unit
Rue du Bugnon 46, 1011 Lausanne
Switzerland

INTRODUCTION

Among all deleterious factors that can influence the cognitive functions and behavior of the epileptic child, the importance of the direct effects of epilepsy is increasingly being recognized. These effects can manifest themselves in various ways which are summarized in Table 1.

The recognition that temporary cognitive and/or behavioral disturbances alone can be epileptic manifestations requires a drastic conceptual change from what clinicians usually consider an epileptic seizure: brief episodes lasting seconds, minutes and more rarely hours with complete recovery involving motor, vegetative and/or simple changes in awareness. Another time dimension must also be introduced, that is the notion that epileptic manifestations in the cognitive domain can be quite durable.

It is now acknowledged that a gradual loss of cognitive functions, arrest or regression in development or a behavioral disorder may be the presenting problem with few or no hints as to its epileptic origin. The clinical disturbances must sometimes be monitored during days or weeks and in special situations, possibly months or years. In addition, one must realize how difficult it is to recognize subtle associated seizures, especially in very young children.

From the electroencephalographic point of view, it is important to remember that large parts of the brain which are important for behavior and cognition, such as the temporal and frontal lobes are far from the brain surface, so that the interictal scalp electroencephalogram (EEG) may be normal and may show spikes, although not always, during clinical seizures only. New data in this domain have been brought about mostly by studying surgical epilepsy patients who have had several prolonged EEG recording both on the scalp and with electrocorticography.

Neuropsychology of Childhood Epilepsy, edited by Jambaqué et al.
Kluwer Academic / Plenum Publishers, New York, 2001.

Table 1. Effects of epilepsy on cognition in children

Effects	Comment	Warning
brief unavailability* of system for cognitive activity	"transient cognitive impairment" "TCI"	variable consequences (possibly none)
prolonged cognitive "epileptic" dysfunction* (EEG parox. status/ postictal)	regression (non development) of given cognitive function	"cognitive epilepsy" not typical epilepsy
pathology induced by epilepsy	epileptic "damage" to developing networks	lessons (and limits!) of experim. epilepsy

*depending on age, system involved, learning period: failure to consolidate new knowledge.

The nature and organization of higher cortical functions, especially in the developing child, also calls for an analysis of epileptic symptoms which is very different from that of observing a temporary motor or vegetative phenomenon and simple alteration of vigilance or "consciousness". The younger the child, the more difficult it is to evaluate these dimensions. Indeed, a temporary loss of performance or an unusual new and changing behavior in a very young child may be easily interpreted as belonging to variations in the dynamics of development and its multiple hazards.

1. PAROXYSMAL PHENOMENA

1.1. Positive and Negative Symptoms

The idea that a temporary loss or distortion of the normal continuity of mental function without evident changes in awareness can be an epileptic manifestation is well accepted in adults and is referred to as a cognitive seizure. It may have either negative or positive features, or both. To suspect a "cognitive seizure", one must interact and not only watch the child to know what aspect of mental function is preserved or altered. In such circumstances, it is difficult to plan a systematic study in advance.[7] The emotional reaction of the child to the seizure may preclude any collaboration during that period.

1.1.1. Positive Symptoms. Positive symptoms refer to phenomena such as hallucinations, memory recollections or acute emotional states (fear, ecstasy). In the adults, they are clearly perceived as abnormal and are integrated within the conscious experience of the individual. They do not disturb his/her capacity to think or interact with the environment beyond the brief period of the seizure. Children are not always able or willing to express these experiences and what one knows can only be an underestimate. It is also possible that they are at times so "concentrated" on the bizarre sensations they have or experience fright that can be mistaken for an "absence" or a panic reaction.

1.1.2. Negative Symptoms. Negative phenomena refer to the temporary loss or alteration of a cognitive/behavioral function.

Temporary cognitive disturbances can be due either to focal epileptic discharges in brain areas which mediate a particular cognitive or behavioral function or to generalized discharges which interfere with more global aspects of mental function (vigilance, execution). The observed deficit corresponds either to the ictal or to the postictal phase.

Depending on the function of the part of cerebral cortex involved in the initial discharge and its propagation, any specific cognitive disturbance can occur such as aphasia, apraxia, executive dysfunction, amnesia, visuospatial disability often with associated behavioral disturbances. These latter problems can actually be the leading or only symptoms recognized. Not surprisingly, there are very few well-documented examples of ictal cognitive deficits in children, especially in very young ones.[4,9]

Selective memory impairment as the sole manifestation of a complex partial seizure has sometimes been described in adults as "epileptic amnesic attacks".[6] The patient is able to perform at a high level of cognitive functioning (reading, counting, writing, etc.) but s/he cannot form new memories. Except for the perplexed anxious attitude and repetitive questioning, there are no signs of seizure activity. These episodes are usually preceded by a brief loss of responsiveness with automatisms (which can be unwitnessed) followed by the amnesic state which can last for a long period. To our knowledge, this has not been precisely reported in children. This type of deficit can easily be missed in children who may not consciously realize the significance of a memory gap in their recollection of current life events. How often can such unrecognized seizures affect learning ability and school achievement in children, in particular transient memory failures, is presently a "*terra incognita*".

2. BRIEF AND PROLONGED COGNITIVE/BEHAVIORAL MANIFESTATIONS

2.1. Transient Cognitive Impairment (TCI)

The cognitive epileptic manifestation which can be considered the "minimal" cognitive seizure is the "transient cognitive impairment" (TCI). TCI are observed during electroencephalographic epileptic discharges on the EEG in children who have otherwise no recognizable clinical sign of seizure activity and correspond to a decrease or loss of efficiency of the performance during the actual discharge.[1] This type of clinical-EEG correlation can be done only in older cooperative children and is limited to the relatively simple level cognitive processes at work during these brief time intervals (reaction time, discrimination tasks, etc.).

Responses to verbal as opposed to nonverbal stimuli are differentially affected by left versus right-sided discharges. From these data, it may be inferred that other specific cognitive functions can be interfered with by focal "subclinical" EEG discharges, and this notion has important implications.[10]

2.2. Recurrent Ictal-Postictal Cognitive Disturbances

Prolonged, sometime fluctuating, cognitive "epileptic" disturbances in children are probably more often due to recurrent seizures and postictal states (with no possibility of recovery in between seizures) rather than to a real status epilepticus. It is not surprising that interference with a brain system supporting a recently acquired complex, not automatized, cognitive activity would be more vulnerable. This activity, in turn, would take longer to recover after an epileptic seizure involving the area responsible for its accomplishment. Interestingly, the EEG in these situations does not show a persistent focal epileptic activity, as one would expect in focal status epilepticus.

2.3. Status Epilepticus

The best known of these purely or mainly cognitive manifestations of epilepsy is the "non-convulsive status epilepticus". It corresponds to a prolonged alteration of the mental state with continuous or sub-continuous generalized spikes or spike-wave discharges on the EEG. The prototype is the "absence status" ("petit mal status", "spike-wave stupor", "minor epileptic status"). In these situations, there is clouding of awareness, but presumably no specific cognitive deficit. Apathy, drowsiness, slowness, inattention, perplexity, strange affect, amnesia, slow speech are the most obvious abnormal behaviors. It is not clear how much higher-level mental functions are truly preserved and what new memories can be formed during these episodes because this has almost never been studied in detail.

According to informal case descriptions, the children are sometime able to perform quite difficult tasks suggesting that high-level functions are spared and that more recollection of events has occurred during that period than what one might believe. More complex disturbances of mental state with specific cognitive or behavioral alterations do probably occur in some of these children but are masked by the lack of vigilance, the apathy or bizarre behavior which are usually the most evident features and cannot be further defined unless detailed testing of several functions is done.

In "complex partial status epilepticus", only limited purposeful organized activity and responses to outside stimuli are possible and there is usually a total amnesia for the period. The epileptic nature of the disorder is recognized by the presence of automatic motor behaviors and vegetative phenomena, although this aspect may be subtle.[12]

The cases in which a prolonged specific cognitive or behavioral deficit constitutes the epileptic manifestation are less well recognized. The recently introduced terms of "cognitive status" or "behavioral status" probably refer to some of these situations, although the physiopathology may be different from the classical status epilepticus.

2.4. Other Focal Inhibitory Mechanisms: Landau-Kleffner Syndrome and Syndrome of Continuous Spike-Waves During Sleep

The syndrome of acquired epileptic aphasia (AEA) or "Landau-Kleffner syndrome") and epilepsy with continuous spike-waves during slow wave sleep (ECSWS) deserves a special place in this discussion of "cognitive" epilepsies.[3,14]

These disorders are often insidious and progressive over the years with devastating effects on cognition (language, thinking) and/or behavior whereas the children have generally few or sometimes no "classical" seizures, i.e., the "visible" seizure disorder is usually "benign" in conventional terms. There is no consistent demonstrable focal brain lesion and the diagnosis rests on the presence of severe persistent (although fluctuant) focal paroxysmal EEG abnormalities.

These very special situations have led to two opposite physiopathological hypotheses, particularly concerning AEA, which has been the most studied. One is the idea that the abnormal electroencephalographic discharges are epiphenomena of a so far unknown underlying encephalopathy. On the opposite, it has been suggested that the pathological EEG simply reflects a secondary focal epilepsy with aphasic symptomatology. Some authors have recently reported children diagnosed as AEA, in whom various brain lesions (tumor, encephalitis) in language areas were found, apparently supporting this point of view. A prolonged aphasia can certainly result from a severe focal lesional epilepsy in relevant brain areas (not always seen on brain imaging, such as a focal dysplasia), but the physiopathology of the epilepsy in most cases of AEA is probably different.

Convergent clinical data and new evidence drawn from electrophysiological studies and functional imaging[11] suggest that the cognitive and behavioral dysfunctions are directly related to the particular epileptic dysfunction affecting specific cortical areas during development. This abnormal epileptic dysfunction would appear to differ, in a manner that not yet well understood from the usual lesional focal epilepsies.

There are many clinical and EEG similarities between AEA (and ECSWS) and some genetically determined epileptic syndromes (mainly the "benign partial epilepsies"). It is now well documented that some cases of typical benign partial epilepsy can develop auditory agnosia or speech problems, usually transiently.[3]

2.5. Consequences of Prolonged Functional Unavailability of the Affected Zone(s) at Critical Developmental Stages or Learning Periods

In children who must continuously acquire new information, failure to encode, store and consolidate newly formed memories during or after seizures is probably an underscored cause of learning problems. This may occur especially during occult nocturnal episodes or with seizures affecting the limbic system.

One must also consider that prolonged functional unavailability of involved zone(s) at critical developmental periods may hinder the children's capacities for new learning or consolidation of prior knowledge, even after the actual seizure activity and postictal period has subsided.

Besides, as a result of a cognitive seizure, decreased alertness, or suboptimal functioning of specific areas involved in the initial stage of selecting and recording information from a sensory channel as well as a more general failure to integrate new data in a meaningful way by making use of available experience and reasoning, can account for some learning disturbances, particularly the failure to store new items in memory during the minutes, hours or possibly days after a seizure.

3. THE ROLE OF EPILEPSY IN SOME DEVELOPMENTAL DISORDERS

A natural extension of the present discussion relates to the possible role of a "hidden" epilepsy starting at a very early age, even at birth. The possible direct causal role of epilepsy in some children who present with developmental disorders and associated clinical seizures or with only paroxysmal EEG discharges is now being increasingly considered (Table 2). Stagnation, fluctuation or regression after early normal development (i.e., autistic or language regression) in direct correlation with the epilepsy or particular patterns of paroxysmal EEG abnormalities can occur but is rarely convincingly demonstrated.[5,13]

Table 2. Possible consequences of persistent (non-paroxysmal) cognitive-behavioral "epileptic" dysfunction in children

Very young children (0–3 y)	Older children (>3–4 y)
Aberrant development	Slow learning
Retarded development	Cognitive arrest
Developmental regression ("pervasive disintegr. dis")	Dementia Specific cognitive Psychiatric dis.(+-ADD)

Experimental data in animals show that focal epileptic discharges early in development can modify the structural development of the brain and this is giving support to the probable negative role of some early-onset epilepsies on cognitive development.

Infantile spasms with hypsarrythmia (West syndrome) are the most striking example of the cognitive and behavioral consequences of very early epilepsy. In this situation, cognitive and or behavioral changes occur usually before the motor manifestations and are sometimes the only sign of the disease.[8]

An early focal epileptic activity involving a part of the developing cerebral cortex which mediates a particular cognitive function at a given age, level of experience and learning, can be expected to preclude the normal development of that function, regardless of the presence or absence of an underlying permanent lesion. It is increasingly acknowledged that the behavioral consequences of a permanent focal brain lesion differ when the lesion has been sustained prenatally or very early after birth from when it occurs later in life.[2] The situation may be even more complex when the formation and consolidation of a network supporting a specific developing function is intermittently hampered by the epileptic activity.

However, clear correlations between clinical symptoms and the location (laterality and site within one hemisphere) of epileptic foci are difficult to find since many variables are involved. Uncertain age of onset of epilepsy, duration, severity of the epilepsy and the level of functional commitment of the area (areas) involved by the epileptic process are all factors that need to be taken into consideration.

Nevertheless, extrapolating from data obtained in older children with "cognitive" epilepsies and knowing that subtle seizures in very young children are difficult to diagnose and that data from surface EEG recordings are limited, one may conclude that the importance of epilepsy in some developmental disorders is grossly underestimated.

COGNITIVE-BEHAVIORAL DISTURBANCES AS EPILEPTIC MANIFESTATIONS: HOW TO SUSPECT IT?

The increasing awareness that prolonged cognitive or behavioral disturbances can be the only or main manifestation of epilepsy raises many difficult practical questions. In the absence of typical seizures or clear-cut paroxysmal EEG abnormalities, how can one suspect that an unexplained cognitive-behavioral disturbance could be an epileptic manifestation?

First, a cognitive dysfunction may be the first manifestation of the epileptic disease which is not yet recognized. Some features of the history can be suggestive.

Frequent mood changes ("good boy" or "bad boy"), fluctuating attention, "forgetfulness", uneven memory skills ("sometimes forgets everything"), variable school results ("can do the best or the worst"), variable speed of performance ("sometimes very fast, sometime endless"), transient failures in specific domains are sometimes reported by parents and teachers of epileptic children and are useful hints. Although none of these complaints is specific, their combination and usually unexplained sudden occurrence and recurrence can be very suggestive of an unrecognized "cognitive" seizure interfering with vigilance, attention, or with more specific functions, such as language or memory. Misinterpretations of the causes for these fluctuations in behavior or performances are easy. One can often find psychological explanations without looking further.

It is not exceptional to see children with a newly diagnosed epilepsy (even a few seizures) who improve very much in terms of behavior, sleep and cognitive function after antiepileptic therapy. In these situations, parents realize in retrospect that epilepsy did indeed affect behavior and cognition. When epilepsy was diagnosed, this aspect had been either overlooked, considered minor or unrelated to the epilepsy. These situations are very important to recognize because they indicate that the positive effects of antiepileptic therapy on cognition clearly outweighs the possible negative psychological impact of the diagnosis of epilepsy itself and the drug side-effects. The parents are usually very aware of this and do not object to the continuation of treatment in these circumstances. It is very important to consider this aspect when one discusses duration of therapy.

Secondly, in children who have had previous documented seizures, even only very few, the parents can sometimes identify that transient unusual behaviors or mental changes are very similar to those they had seen in a postictal state. This can be an important clue that these cognitive or behavioral dysfunctions are probable epileptic manifestations.

Finally, the possibility that "cognitive seizures" or persistent paroxysmal EEG activity play an additional negative role in brain-damaged epileptic children with chronic learning disabilities or behavior disorders related to their basic pathology must be strongly considered. A recent study in children with congenital hemiplegia suggests that epilepsy can be responsible for or aggravate the chronic cognitive disturbances usually attributed to the basic focal lesion. This type of data and the recently demonstrated possible effects of "subclinical" focal EEG discharges on cognition should lead one to pay close attention to the possible negative role of epilepsy or paroxysmal EEG discharges on cognitive functions in these situations.[15] One must admit, however, that is very difficult to show that fluctuating performances or regression in a domain which is already weak can be due to the additional direct effect of epilepsy.

REFERENCES

1. Aarts HP, Binnie CD, Smit AM and Wilkins AJ (1984): Selective cognitive impairment during focal and generalized epileptiform EEG activity. Brain 107:293–308.
2. Bates E, Thal D and Janowsky JJ (1992): Early language development and its neural corrrelates. In Segalowitz SJ, Rapin I (eds): "Handbook of Neuropsychology", Vol 7. Amsterdam: Elsevier, pp 94.
3. Deonna T (1991): Acquired epileptiform aphasia (Landau-Kleffner syndrome). Journal of Clinical Neurophysiology 8:228–298.
4. Deonna T (1993): Annotation: Cognitive and behavioral correlates of epileptic activity in children. Journal of Child Psychology and Psychiatry 34:611–620.
5. Deonna T (1999): Developmental consequences of epilepsies in infancy. In Nehlig A, Motte J (eds): "Childhood Epilepsies and Brain Development". London: John Libbey, pp 113–122.

6. Galassi R, Morreale A, Di Sarro R and Lugaresi E (1993): Epileptic amnesic syndrome. Epilepsia 33 (Suppl 6):S21–S25.
7. Gloor P (1991): Neurobiological substrates of ictal behavioral discharges. Advances in Neurology 55:1–34.
8. Guzzetta F, Crisafulli I and Crine M (1993): Cognitive assessment of infants with West syndrome. How useful in diagnosis and prognosis? Developmental Medicine and Child Neurology 35:379–387.
9. Jambaqué I and Dulac O (1989): Syndrome frontal réversible et épilepsie chez un enfant de 8 ans. Archives Françaises de Pédiatrie 46:525–529.
10. Kasteleijn-Nolst Trénité DGA, Siebelink BM, Berends SGC, van Strien JW and Meinardi H (1990): Lateralized effects of subclinical epileptiform discharges on scholastic performance in children. Epilepsia 31:740–746.
11. Maquet P, Hirsch E and Metz-Lutz MN (1995): Regional cerebral glucose metabolism in children with deterioration of one or more congitive functions and continuous spike-and wave discharges during sleep. Brain 118:1497–1520.
12. McBride MC, Dooling EC and Oppenheimer EH (1981): Complex partial status epilepticus in young children. Annals of Neurology 9:526–530.
13. Rapin I (1995): Autistic regression and disintegrative disorder: How important the role of epilepsy? Seminars in Pediatric Neurology 2:278–285.
14. Roulet-Perez E, Davidoff V, Despland PA and Deonna T (1993): Mental deterioration in children with epilepsy and continuous spike-waves during sleeep: Acquired epileptic frontal syndrome Developmental Medicine and Child Neurology 33:495–511.
15. Varga-Khadem F, Isaacs E, Van Der Werf S, Robb S and Wilson J (1992): Development of intelligence and memory in children with hemiplegic cerebral palsy. The deleterious consequences of early seizures. Brain 115:315–329.

17

TRANSITORY COGNITIVE DISORDERS AND LEARNING IMPAIRMENT

Marie-Noëlle Metz-Lutz*[,1], Anne de Saint Martin[1,2], Rita Massa[1,3], and Edouard Hirsch[1]

[1]INSERM U398, Clinique Neurologique
Hôpitaux Universitaires de Strasbourg
67091 Strasbourg, France
[2]Service de Pédiatrie 1, Hôpital de Hautepierre
Hôpitaux Universitaires de Strasbourg
67098 Strasbourg, France
[3]Istituto di Neurologia
Università degli Studi di Cagliari
Cagliari, Italy

1. COGNITIVE DISORDERS RELATED TO SUB CLINICAL DISCHARGES

In 1936, Gibbs, Lennox and Gibbs[5] observed that spike and wave discharges could occur without concomitant overt clinical manifestations. These EEG events were diversely labeled sub clinical or interictal discharges. A few years later, using a simple reaction time task performed during EEG recording, Schwab[15] demonstrated that such brief generalized spike-wave discharges interfere with perceptual motor performances even in absence of obvious disruption of consciousness. Since this seminal study, many investigators have confirmed that responses to psychological test may be impaired during infra clinical EEG discharges. Aarts *et al.*[1] coined the term of transitory cognitive impairment (TCI) to "designate the functional deficits demonstrable by suitable testing," related to sub clinical spike or spike-wave discharges. In children, the effect of sub-clinical discharges on cognitive performance has been studied mainly in two forms of childhood epilepsy. The first studies were concerned with primary absence epilepsy[8] whereas the more recent ones dealt with benign focal epilepsy, one of the most common forms of epilepsy in childhood.[2,12]

* To whom correspondence should be addressed

Neuropsychology of Childhood Epilepsy, edited by Jambaqué et al.
Kluwer Academic / Plenum Publishers, New York, 2001.

1.1. Detection of Transitory Cognitive Impairments During EEG Recording

To establish the direct relationship between cognitive impairments and the occurrence of abnormal EEG discharges, it is necessary to monitor, by means of v ideo-EEG recording, the EEG and the patient's behavioral changes on line with the subject's performances to test. Indeed, absence of response or delayed response to stimuli may be related to very brief clinical changes that may pass unnoticed to the examiner. TCI can be evidenced only through tests covering a sufficiently long period of EEG recording allowing to record a large number of responses coincident with the occurrence of discharges as well as responses obtained with a normal background EEG. This requires the cooperation of the patients throughout the whole testing which often lasts at least half an hour for a single testing. The task must be attractive enough to keep the child interested throughout the testing period. It does not need to be too easy but its level of difficulty must be adapted to the child's performance. Today computerized tests may be presented to children like a video game with attractive sounds or colored pictures.

1.1.1. Type of Discharges Inducing Cognitive Impairments. The occurrence of subtle cognitive impairments has been first observed in presence of generalized spike-wave discharges. Further investigators demonstrated that the impairments depend on several factors related to the temporal, spatial and morphological features of the epileptiform activity. The few studies that compared the effect on cognitive performances of various kinds of paroxysmal activity showed that the morphology of discharges was an important determinant of cognitive impairments. Symmetrical spike-and-wave activity was shown to be the most strongly associated with disruption of cognitive performances, whereas slow-wave activity and burst of spikes do not appear to disturb significantly the subject's responses to various cognitive tasks.[1,7] The early studies emphasized the greater effect of generalized spike-wave discharges compared to lateralized or focal spike-waves. However, Aarts *et al.*[1] demonstrated that focal and clearly lateralized spike-wave discharges have dissociated effects on spatial and verbal tasks.

From the first study, the variability of spike-wave discharges effects on responses, ranging from delayed reaction times to no responses to the stimulus, were attributed to the duration of generalized paroxysms. The correlation between reaction time to stimulus and duration of generalized spike-wave discharges was clearly demonstrated by Tizard and Margerison[18] who showed that even very brief discharges lasting less than one second induced significant increase of response time. Figure 1 shows the increase of response time as a function of number of generalized spike-wave discharges recorded within the 8 sec preceding the presentation of the critical stimulus to be detected. These reaction times were obtained in a 9-year-old child with benign focal epilepsy of childhood performing a visual detection task. The child was presented with a series of meaningless colored drawings. He was asked to respond as fast as possible when the same drawings were presented successively. Drawing repetition occurred at random with a rate of about 20 percent.

1.1.2. Effect of Test Situations. It is well known that absence seizures are more liable to occur when the patient is drowsy or inactive. The issue of variations in spike-wave activity related to changes in awareness has been examined in a few experimental studies. Tizard and Margerison[18] measured the involvement of patients performing various cognitive tasks using autonomic variables like pulse and respiration rate while the patients executed

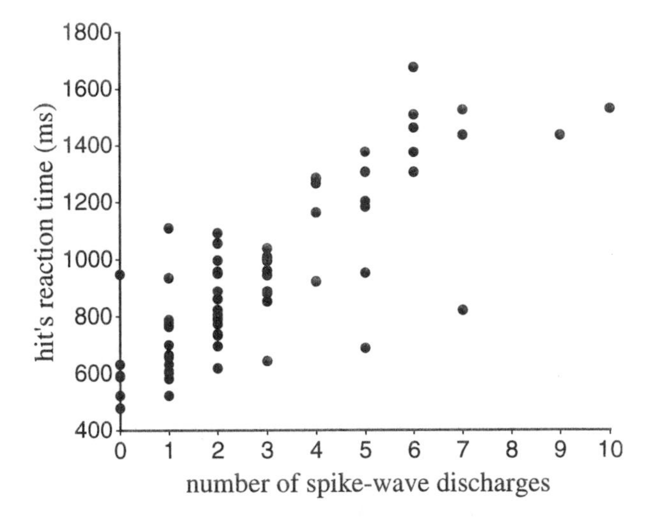

Figure 1. Distribution of reaction times as a function of the number of generalized spike-wave discharges occurring within the 8 seconds preceding the presentation of the target in a detection task

the tasks under EEG monitoring. They found that pulse and respiration rate changes according to the type of test. The rate of pulse and respiration was significantly faster when the patients were performing the task they considered as the less boring, whereas in a simple auditory monitoring task, these variables did not differ significantly from the resting period. They observed parallel to the increase of pulse and respiration rate a decrease of spike-wave activity on the EEG recording. This result shows that it is very important to take into account the effect of test situation on the occurrence of epileptiform discharges when building testing device for the evaluation of transitory cognitive impairments in epileptic patients. However, this study concerned two patients with absence epilepsy. In our experience of childhood epilepsy, the occurrence of focal, bilateral or generalized spike-wave discharges appears to be less sensitive to test situations in benign partial epilepsy of childhood than the 3 Hz generalized spike-and-wave activity of primary absence epilepsy. In another study, the rate of discharges appeared to be higher when the epileptic children were performing tasks of reading and mental arithmetic that was appropriate to their level of ability than for more difficult or easier tasks.[9]

On the other hand, almost all studies concerned with the detection of changes in cognitive performance associated with epileptiform discharges, showed that easy tasks are insensitive to the occurrence of discharges whatever the extent of EEG discharges or the type of epileptic syndrome. In an experimental study of 20 epileptic children with sub clinical paroxysms of spike-wave, Hutt, Newton and Fairweather[8] examined the degree of impairment as a function of the level of task difficulty. They observed that increases in reaction times and rate of error were more pronounced when the tasks required to process more information. The authors introduced the concept of "neural noise" to describe the emergence of a significant interaction of spike-wave discharges with cognitive tasks that require the processing of signal involving greater information content. They suggested that the greater the information content of the signal to be processed the greater the effect of neural noise. In a very refined single case study of a 15-year old epileptic patient with frequent generalized spike-wave discharges, Hutt and colleagues confirmed this assumption.[7] The subject performed a serial performance task whose difficulty changed parametrically according to information content of signal and rate of presentation. Random series of 2, 4 and 8 digits were presented at different rates: one per second, 1 per 2 sec. and 1 per 4 sec. In presence of spike-wave activity, errors were significantly more frequent in the tasks performed at a rate that was close to the one usually chosen by the subject in a self-paced task. At slower or faster rates of presentation, the occurrence of discharges did not change significantly the subject's performances. Moreover, when the series were superior to the subject's digit span and the rate faster, the occurrence of spike-wave discharges did not significantly worsen his performances. Thus, cognitive tasks that are too easy or too difficult for the patient are less disturbed by the occurrence of interictal discharges.

1.1.3. Selectivity of Cognitive Impairments. Several authors addressed the issue of the selectivity of cognitive impairment during focal and generalized sub clinical spike-wave discharges. The performances of 46 patients of more than 8 years of age whose video-EEG recording evidenced more than one epileptiform discharge per 5 minutes, were compared in two short-term memory tasks, one with verbal, the other with spatial material presented on a computer screen like a video-game.[1] In this study, cognitive impairment was demonstrable in 50 percent of the patients overall. Among the patients with focal epileptiform discharges, 58 percent showed TCI. In this group of patients, the effects were specific to the task lateralized to the hemispheric region where the epileptiform discharges were arising. This suggests that TCI result from a disruption of a specific function rather than from a general impairment of attention or overall slowing of mental processing. Also focal epileptiform discharges arising in the visual cortex have been shown to induce transient dysfunction of visual perception in the contralateral visual field.[16] Further studies confirmed the specific effects of focal or lateralized sub clinical discharges.[3,10] As for generalized spike-wave discharges, they appear to disturb more readily spatial tasks than verbal tasks.

1.1.4. Stage of Information Processing Impaired by Sub Clinical Discharges. It has been shown that TCI are more likely to occur in detection task, decision reaction time or working memory tasks. Contrary to simple reaction-time measurements or repetitive motor tasks that have been shown to be less sensitive to discharges, these tasks involve specific stages of information processing. Besides the perceptual stage, they require a recognition stage involving the matching of the result of perceptual processing to an internal representation stored in short-term memory. Following this matching, the detection or the decision response can be prepared. Then, the appropriate motor response can be executed. Therefore, the response depends directly on the result of the stimulus recognition. This would indicate that the interference between discharges and cognitive processing might be localized at the level of stimulus recognition. Indeed, one study investigating the perceptual processing level demonstrated that auditory threshold as well as latency of visual evoked potentials remained unchanged during spike-wave discharges.[6] This study also showed that the decrease of responses in a signal detection paradigm is better explained by change in the decision process due to increased criterion for the acceptability of the signal. However, the level of interference between discharges and stage of processing seems also to depend on the kind of information to be processed. For example, spike-wave paroxysms occurring within a few seconds preceding or during the presentation of a critical stimulus impair both verbal and non verbal processing. As for discharges occurring between the stimulus presentation and response, they interfere significantly only with non-verbal tasks.

1.2. Usefulness of TCI Detection in a Routine Battery

TCI may account for cognitive and behavioral problems encountered, particularly, in children with benign focal epilepsy. In these children who have almost normal IQ, TCI may explain the variability of performances at school or during psychological testing. Siebelink *et al.*[17] investigated the effect of sub-clinical EEG discharges on several subtests of a general intelligence test in 21 epileptic children. They observed that children with frequent discharges during testing did not show significantly lower IQ but showed an abnormal test profile due to poor performances in one test concerned with short-term memory. Another study demonstrated the individual effects of sub clinical discharges on reading performances. Therefore, a brief impairment in cognitive functioning may be considered as a clinical expression of epilepsy. As such, they may justify an antiepileptic treatment, particularly, when they are frequent and consistently disturb the child's academic achievement or social behavior.

For this reason, when frequent EEG discharges are recorded on the waking EEG of an epileptic child, it seems very useful to examine their effects on cognitive performances. It has been suggested that computerized neuropsychological testing should be included in a routine battery for the study of epileptic children particularly those with learning disabilities and psychosocial problems. Computerized tests can be used to evaluate cognitive performances during video-EEG recording. However, the detection of TCI needs the use of rather specific test adapted to the individual level of ability and to obtain performance measurements time locked with the EEG events. One may expect that appropriate testing devices will be suitable for that purpose in the near future. As soon as it will be possible to detect TCI in a routine clinical investigation of epilepsy, they will be taken into account for decision about treatment.

Considering that TCI represent a significant disability that is not restricted to the cognitive domain but they also affect the individual's behavior in daily life, some authors addressed the issue of medication to suppress sub clinical discharges and TCI. In one study, based on a double-blind crossover trial with repeated EEG, neuropsychological and behavioral assessments, the reduction of frequency of discharges in 8 out of 10 children with generalized spike-wave activity was associated to a significant improvement of psycho social function.[12] That the detection of TCI is not yet a usual practice may explain the very few studies on the effect of their treatment on children' behavior, neuropsychological and academic performances.

2. LEARNING IMPAIRMENT IN CHILDHOOD EPILEPSY

Learning usually refers to the process of extracting associations from experience and using them to guide behavior, which necessarily includes memory for storing and retrieving associations. Thus memory as a sub process of learning cannot be separated but learning disorder cannot be reduced to memory impairment. Indeed the ability to learn depends on information processing capacities, attention, working memory and material specific memory. If most learning disabilities are related to material specific memory disorder, learning impairments demonstrated in epileptic children seem to be due to difficulty to maintain attention or to control responses to inhibit interfering stimuli. In the early studies, the prevalence of learning disability in epileptic children was viewed as a consequence of repeated seizures, as a consequence of the underlying structural lesion, or as a side effect of anti epileptic drugs.[11] The demonstration of momentary decrements in cognitive functions coincident with interictal EEG discharges, in particular, in benign forms of childhood epilepsy lead to reexamine the relationship between epilepsy and learning performances.

2.1. Relationship Between Transitory Cognitive Impairments and Learning Difficulties

The process of learning involves various cognitive modules. Some modules are specific memory systems devoted to the processing of particular kinds of information. The others are non-specific systems like working memory, explicit memory and implicit memory. These non-specific memory systems are involved in information storing and association fixing. Sub clinical spike-wave discharges have been shown to interfere mainly with cognitive tasks requiring short-term memory. On the contrary, tasks involving the recall of previously learned material appeared to be less sensitive to the occurrence of epileptiform discharges.[17]

In an attempt to understand the relationship between TCI and learning problems, we analyzed the learning performances in a group of 15 children with benign focal epilepsy of childhood. The children, aged from 6 to 12 years were selected upon strict clinical and electrophysiological criteria. The frequency of overt epileptic seizures was low but their EEG evidenced focal or generalized sub clinical spike-and-wave discharges. Only seven subjects were treated. In these cases, the treatment consisted in valproic acid. In specific continuous performance testing for TCI detection, a decrease of performances associated to discharges was observed, but the impairment was significant only in 8 out of 15 children. They performed the Auditory-Verbal Learning Test,[14] following a procedure standardized for children aged from 5 to 16 years. This test measures both short-term memory and long-term memory retention. It also provides a comparison between retrieval efficiency and learning. The task consists of five presentations with recall of a 15-word list. The words of the list are read at the rate of one per second. After each presentation, the patient is asked to repeat the words in the order in which they are recalled. The first presentation gives a measure of immediate word span recall. The learning curve indicates the rate of learning. After interfering verbal tasks, including a verbal fluency test and two repetition tasks, the child was asked to recall as many words as possible of the list they learned about 20 minutes before. In the recognition testing, the examiner asked the patient to identify as many words as possible in a short story read by the examiner or by the child himself. The learning curve of each child was compared to that of a group of 20 children of the same age.

Except for three children, the performances of epileptic children in the first trial did not show significantly reduced immediate word span in comparison to controls. Among the 15 children included in this study, 11 showed impaired learning performances with either an overall slowing of learning or absence of new recall after three or four repetitions of the list. The children with an impaired learning curve were those who had frequent spike-wave paroxysms. These results are illustrated in Fig. 2 which shows the learning curve of two 10 year-old epileptic children compared to the mean curve of age-matched controls.

The epileptic children also had lower performances in delayed recall, but on recognition their performances did not differ significantly from the control ones. Figure 3 shows the result of the same two epileptic children.

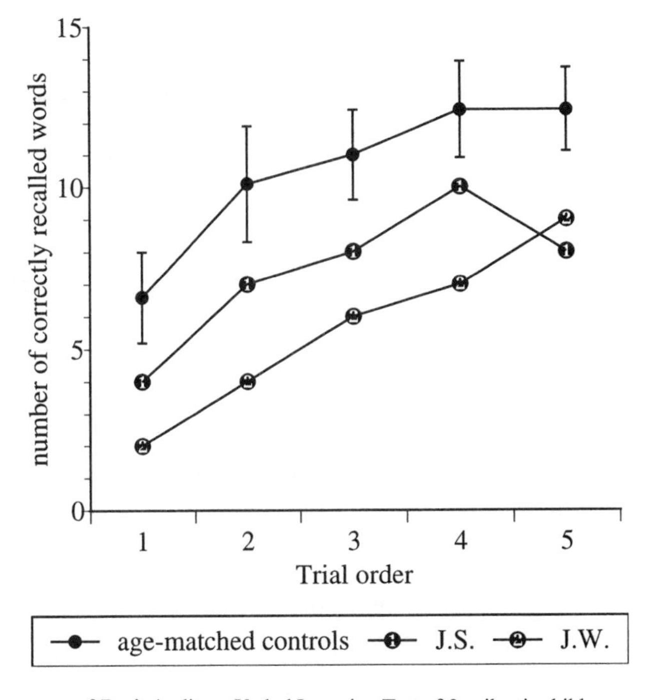

Figure 2. Learning curves of Rey's Auditory Verbal Learning Test of 2 epileptic children compared to the mean curve of a group of 20 controls of the same age

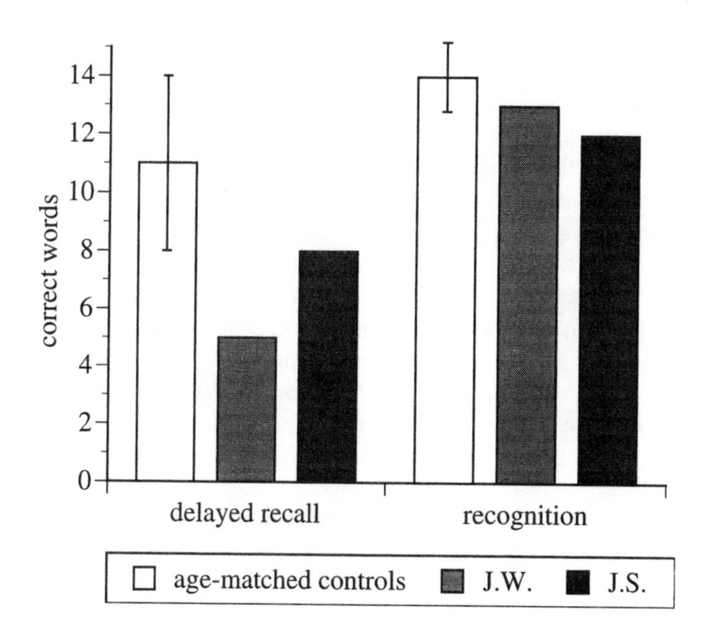

Figure 3. Performances of two epileptic children in delayed recall and recognition in Rey's Auditory Verbal Learning Test compared to the mean results of a group of 20 controls of the same age

From this study, it appears that during learning, TCI or EEG discharges interfere with information storing and fixing. During this particular verbal learning task, the working memory demand is increasing from the first to the fifth trial. So after two or three trials, the slowing of learning may be explained by a higher sensitivity to the occurrence of discharges. The poor performances in the delayed recall of the words also point out a disturbance in the process of information storing. However, the almost normal performance in the recognition test contrasting with a poor retrieval in delayed recall seems to disclose a lack of learning strategy or fixing association. Building a strategy efficient for learning and retrieving learned information needs sustained attention to control the learning process over time.

2.2. Attention Deficit and Learning

The learning process requires memory dependent upon attention and awareness of the stimuli. Learning, as memory, is based on a progression of fundamental processes that can be affected by attention. Without attention to stimuli there is still encoding through more automatic processes. However automatic processes alone cannot accomplish a complete encoding, a long lasting activation and a deliberate conscious retrieval process.[4] Thus learning disabilities may result from a deficit of attention. Although the studies previously reviewed demonstrated that TCI are rather specific deficits, their occurrence during the process of learning may be compared to a disruption in active attentional focus. Clinically, learning failures observed in epileptic children are often perceived as related to attentional problems.

We evaluated the attentional ability of the epileptic children having an impaired learning curve on Rey's Auditory Verbal Learning Test. Attention was assessed in two tasks involving reaction time measurement. The first one was a visual reaction time task. The child responded to the presentation of a small red ball appearing at random with variable inter trial interval and variable position on the screen. Each epileptic child's performances were compared to those of an age-matched control group. This reaction time test was intended to measure along with the processing speed, the attentional capabilities. Faster reaction time depends on the information processing but also on attention and concentration that allow to improve response time as a function of training to respond. Reaction time variability is often considered as a better measure of attention, and particularly of the capacity to focus attention. Variability decreases when attention is better. Comparison of mean reaction times and standard deviation at the beginning and the end of testing provides an evaluation of sustained attention. All the epileptic children with interictal spike-wave discharges showed significantly slower reaction times but also increased variability of reaction time. Contrary to normal controls, who showed faster reaction time with lower standard deviation in the second part of the test, the epileptic children did not show any training effect on performances. Figure 4 shows the pattern of performances of the two 10-year-old epileptic children compared to the mean performances of the age-matched controls.

The second test was the Conner's Continuous Performance Test.[4] The test presents letters and the subjects are required to press the space bar of the computer keyboard when any letter except "X" appears on the screen. The letters are displayed for 250 milliseconds with an inter-stimulus interval ranging from 1 to 4 seconds. A total number of 360 letters is presented in blocks. From one block to the other the inter-stimulus interval changes. The test takes approximately 14 minutes to complete. This test, used for the

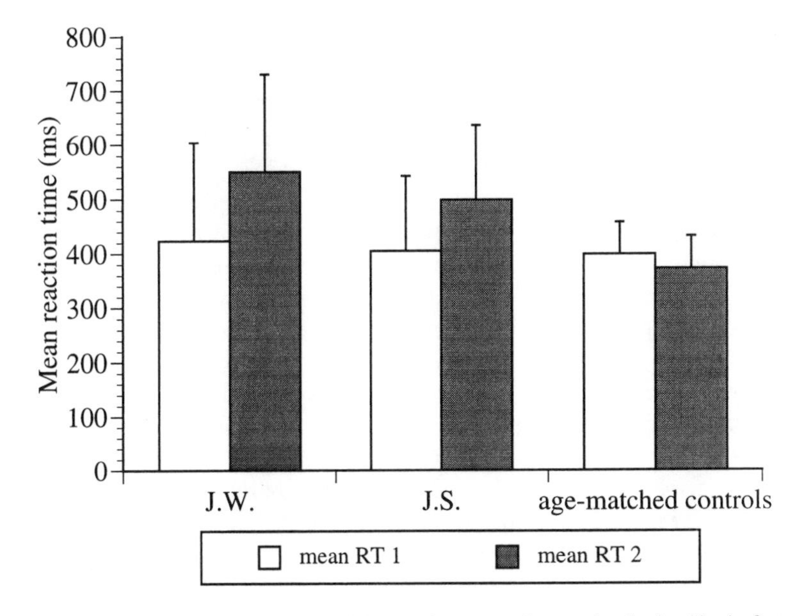

Figure 4. Mean reaction times of two epileptic children and a group of controls, obtained in the first and second parts of a visual detection task

diagnosis of attention deficit disorders with hyperactivity, measures the accuracy of response with a reference to the ability to inhibit responses. Reaction times and variation of reaction time over time and according to changes of the inter-stimulus interval between blocks are computed. For each subject these measures are compared to the norms of the general population.

In the group of children with benign partial epilepsy, all subjects had atypical scores on at least three measures out of the 12 measures of the test. All showed slow mean reaction time and atypical changes of reaction time to inter stimulus changes. Eleven out 15 subjects showed a higher rate of errors and markedly atypical variability of reaction time. Nine children had abnormal score on the measure of how well they discriminated between targets and non-targets. According to the overall index obtained from the weighted sum of all measures of the test, eight children were considered as having attention deficit. In this group study the subject's scores appeared to be correlated to the frequency of generalized discharges recorded at the time of testing, but we did not find any significant effect of the lateralization of the spike-wave discharges.

Piccirilli *et al.*[13] investigated the relationship between the side of the epileptic focus and the attention problems in epileptic children with partial epilepsy with rolandic paroxysms. They observed that children with right-sided or bilateral focus of rolandic paroxysmal discharges have impaired performances in a complex cancellation task. As for the children with left-sided rolandic discharges, they performed as well as the controls. The task involved visual scanning that evaluated attention mechanisms and decision on oriented figures that measure the abilities to process visuospatial information. The authors observed that the errors could not be interpreted as a consequence of visuospatial neglect, but they rather indicate attentional difficulties. They suggested that, besides the direct effect on the function dependent on the cortical area in which the discharges arise, the presence of an abnormal neuronal activity also alters the functional balance between the two cerebral hemispheres.

CONCLUSION

The findings of the studies reviewed in this chapter show that an abnormal neural activity may be responsible for cognitive impairment even when overt clinical seizures are very rare. This abnormal neural activity, only evidenced on EEG, has not only immediate consequences on cognitive performances but also long-term effects. As they occur in childhood at a critical age for the development of most cognitive abilities and the acquisition of knowledge, they may be deleterious for further academic and psychosocial achievement. The demonstration that transitory cognitive impairments may be responsible for attention and learning problems in the so-called benign epilepsy suggests that the frequency of EEG abnormalities particularly spike-wave discharges should be taken into account as a criterion for therapeutic decision in the same way as the frequency of clinical seizures. In the same way, when an epileptic child has behavioral problems and learning difficulties that are not explained by frequent seizures or medication, the presence of TCI should be suspected. Finally, in a child with sub clinical discharges on EEG, even when the global IQ is normal, poor performances on specific subtests may be related to TCI.[19]

REFERENCES

1. Aarts JHP, Binnie CD, Smit AM and Wilkins AJ (1984): Selective cognitive impairment during focal and generalized epileptiform EEG activity. Brain 107:293–308.
2. Binnie CD (1993): Significance and management of transitory cognitive impairment due to subclinical EEG discharges in children. Brain Development 15:23–30.
3. Binnie CD, Kasteleijn-Nolst Trenité DGA, Smit AM and Wilkins AJ (1987): Interactions of epileptiform EEG discharges and cognition. Epilepsy Research 1:239–245.
4. Cowan N (1995): Attention and Memory. An Integrated Framework. Oxford Psychology. Vol. 26. New York: Oxford University Press.
5. Gibbs FA, Lennox WG and Gibbs EL (1936): The electroencephalogram in diagnosis and in localization of epileptic seizures. Archives of Neurology and Psychiatry 36:1225.
6. Hutt SJ, Denner S and Newton J (1976): Auditory thresholds during evoked spike-wave activity in epileptic patients. Cortex 12:249–257.
7. Hutt SJ and Fairweather H (1975): Information processing during two types of EEG activity. Electroencephalography and Clinical Neurophysiology 39:43–51.
8. Hutt SJ, Newton J and Fairweather H (1977): Choice reaction time and EEG activity in children with epilepsy. Neuropsychologia 15:257–267.
9. Kasteleijn-Nolst Trenité DGA, Bakker DJ, Binnie CD, Buerman A and Van Raaij M (1988): Psychological effects of subclinical epileptiform EEG discharges in children. I. Scholastic skills. Epilepsy Research 2:111–116.
10. Kasteleijn-Nolst Trenité DGA, Siebelink BM, Berends SGC, van Strien JW and Meinardi H (1990): Lateralized effects of subclinical epileptiform EEG discharges on scholastic performance in children. Epilepsia 31:740–746.
11. Klein SK (1991): Cognitive factors and learning disabilities in children with epilepsy. In Devinski O, Theodore WH (eds): "Epilepsy and Behavior". New York: Wiley-Liss pp 171–179.
12. Marston D, Besag F, Binnie CD and Fowler M (1993): Effects of transitory cognitive impairment on psychosocial functioning of children with epilepsy: a therapeutic trial. Developmental Medicine and Child Neurology 35:574–581.
13. Piccirilli M, D'Alessandro P, Sciarma T, Cantoni C, Dioguardi MMGAI and Tiacci C (1994): Attention problems in epilepsy: Possible significance of the epileptogenic focus. Epilepsia 35:1091–1096.
14. Rey A (1964): L'examen Clinique en Psychologie. Paris: Presses Universitaires de France.
15. Schwab RS (1939): A method of measuring consciousness in petit mal epilepsy. Journal of Nervous and Mental Disease 89:690–691.

16. Shewmon DA and Erwin RJ (1989): Transient impairment of visual perception induced by single interictal occipital spikes. Journal of Clinical and Experimental Neuropsychology 4:675–691.
17. Siebelink BM, Bakker DJ, Binnie CD and Kasteleijn-Nolst Trenité DGA (1988): Psychological effects of subclinical epileptiform EEG discharges in children. II. General intelligence tests. Epilepsy Research 2:117–121.
18. Tizard B and Margerison JH (1963): The relationship between generalized paroxysmal EEG discharges and various test situations in two epileptic patients. Journal of Neurology, Neurosurgery and Psychiatry 26:308–313.
19. Weglage J, Demsky A, Pietsch M and Kurlemann G (1997): Neuropsychological, intellectual, and behavioral findings in patients with centrotemporal spikes with and without seizures. Developmental Medicine and Child Neurology 39:646–651.

18

ACUTE COGNITIVE AND BEHAVIORAL DISORDERS IN FOCAL EPILEPSIES

Isabelle Jambaqué* and Olivier Dulac

Service de Neuropédiatrie
Hôpital Saint Vincent de Paul
82 Avenue Denfert Rochereau
75674 Paris Cedex 14, France

INTRODUCTION

Partial epilepsy may lead to cognitive and/or behavioral dysfunction during the seizure itself, the post-ictal state and, sometimes, during the interictal period when EEG evidenced marked interictal disturbances The variation in clinical symptoms appear related to the main site, local extension and laterality of the epileptic discharge. Recovery can be observed after control of seizures. Both neuropsychological and neurofunctional imaging studies suggest that acute cognitive and/or behavioral disorders are functional disturbances. A few clinical observations of children with transient, seizure-related selective neuropsychological deficits involving aphasia, apraxia, frontal lobe dysfunction, visuospatial disability or memory deficit have been reported.

1. APHASIA

Partial epilepsy seizures may be associated with ictal verbal behavior disturbances in children as in adults. Furthermore, partial epilepsy may lead to long lasting ictal or postictal language disturbances, particularly when seizures are frequent, sometimes with focal status epilepticus. Deonna *et al.*[4] reported speech and oromotor deficits suggestive of an operculum syndrome in three children suffering from partial benign epilepsy with rolandic spikes. Jambaqué *et al.*[8] reported transient motor aphasia in a 12-year-old right-handed boy with left frontal epilepsy. Speech production was severely reduced to

* To whom correspondence should be addressed

Neuropsychology of Childhood Epilepsy, edited by Jambaqué et al.
Kluwer Academic / Plenum Publishers, New York, 2001.

single-word answers. The main characteristics of the language deficit were the severe loss of fluent speech contrasting with preserved repetition. Auditory word comprehension seemed to be preserved. Reading and writing abilities were impaired to the same degree as spoken language and the child also exhibited severe anomia. In addition to aphasia, he had severe attention and memory deficits with aggressive behavior. There was a clear relationship between progressive loss of speech and seizure worsening. Single photon emission computerized tomography (SPECT) study showed a large area of hypoperfusion in the left hemisphere when the patient had frequent seizures. After seizure control of seizures, the patient recovered from aphasia within 3 months and abnormalities shown on SPECT reversed except in the left inferior frontal gyrus.

2. MEMORY

In adults, "epileptic amnesic attacks" are sometimes the main characteristics of temporal lobe epilepsy. In children, such transitory memory impairment could be very difficult to recognize by the child himself or his surroundings. Differentiation of memory deficits as paroxysmal, episodic or continuous nature is also sometimes very difficult. In all cases, testing of postictal verbal and visual memory in patients with temporal lobe epilepsy appeared to be useful in determining the laterality of focus.[7] The ability of school age children to encode, store and consolidate new information is negatively impacted by complex partial seizures, and this is likely to be a cause of learning failure.

3. PATHOLOGICAL LEFT-HANDEDNESS

For a long time epileptic patients have been considered as one of those disease groups with a high percentage of left handers. The main pathological factors are right-hemiparesis of early onset, cognitive deficit and evidence of left hemisphere disease on clinical examination. Dellatolas *et al.*[3] showed a possible role of the parietal lobe in pathological left handedness in adult epileptics. Chiron *et al.*[2] reported a shift of laterality in a child with parietal epilepsy in whom SPECT evidenced a hypoperfused area corresponding to the epileptogenic zone.

4. VISUO-SPATIAL ABILITIES

Ferraro *et al.*[5] reported the case of a 12-year-old girl with occipital epilepsy who exhibited intermittent difficulty in visual recognition of pictures which can be related to high frequency discharges of spike-slow wave in the left occipital area.

5. EXECUTIVE FUNCTIONS

Bonne *et al.*[1] reported the case of a 13-year-old girl with bilateral frontal foci who exhibited a reversible frontal syndrome characterized by sudden behavioral changes such as sexual desinhibition, loss of concern for personal hygiene and physical and verbal aggression. Neuropsychological testing revealed impaired performance on tasks requiring motor speed, attention, planning ability and response inhibition. Seizure control resulted in general improvement of most functions.

Jambaqué and Dulac[6] reported the case of an 8-year-old boy who developed hyperactivity, marked attention deficits and affective changes in conjunction with a focus in the right fronto-temporal region. Although this child obtained a global IQ score in the high average range, he failed to pass subtests requiring motor speed (coding) and planning ability (mazes). His ability to reproduce a sequence of hand movements was severely impaired and he showed dysgraphia. Although language was preserved, verbal fluency was reduced. In fact, the child suffered from seizures of undetermined frequency. Video EEG recording permitted to demonstrate the occurrence of right frontal seizures. SPECT showed a large hypoperfusion area in the right frontal lobe. Seizures were rapidly controlled by therapy and there was good recovery of cognitive functions with normalization of behavior.

6. BEHAVIOR DISORDERS

Behavior problems have long been reported in children with epilepsy. Although a number of factors are probably relevant in the occurrence of such disturbance, there is evidence that the epileptic process itself can be involved in the behavioral symptomatology. Shewmon *et al.*[9] reported behavioral disorders in three children whose clinical seizures had been well controlled but in whom interictal EEG showed constant epileptiform discharges. Strikingly, they benefited from surgical therapy with elimination of interictal disturbances and cessation of seizures. In some cases, surface EEG only discloses rapid rhythms over the malformation, typical of the type of lesion, but intracranially recorded discharges correspond to pure arrest of activity, which is overlooked by the surrounding. One open question with such patients is whether continuous seizure discharges, overlooked by scalp recording and only recorded on intracranial electrodes, have any impact on behavior and learning abilities, and whether removal of the causative malformation could improve the cognitive potential of the child.

CONCLUSION

Despite recent findings in neuropsychological disorders of children with partial epilepsy, reports including children are still lacking. Like the EEG and functional imaging, neuropsychological tests reflect cerebral functioning and provide indications of selective dysfunction. Cognitive and/or behavioral dysfunction may be the first manifestation of the epileptic disease which is not yet recognized. In some children, with a newly diagnosed of epilepsy, it is not exceptional to see rapid unexpected and marked improvement of behavioral or cognitive problems with the introduction of antiepileptic therapy. This is very particular to children. It may involve patients with no evidence of brain lesion, and seizure control in these cases may be followed by total recovery of cognitive functions and seizure-freedom, even after withdrawal of antiepileptic long treatment. Therefore, early recognition of this condition and appropriate treatment can have lifelong impact on the individual.

REFERENCES

1. Boone KB, Miller BL, Rosenberg L *et al.* (1988): Neuropsychological and behavioral abnormalities in an adolescent with frontal lobe seizures. Neurology 38:583–586.

2. Chiron I, Jambaqué I, Raynaud C *et al.* (1989): Longitudinal study of regional bood flow in partial epilepsy of chidhood after cessation of seizures. Neurology 39:110.
3. Dellatolas G, Luciani S, Castresana A, Remy C, Jallon P, Caplane D and Bancaud J (1993): Pathological left-handedness. Brain 116:1565–1574.
4. Deonna TW, Roulet E, Fontan D and Marcoz JP (1993): Speech and oromotor deficits of epileptic origin in benign partial epilepsy of childhood with rolandic spikes (BPERS). Neuropediatrics 24:83–87.
5. Ferraro C and Fiori Daraio M (1997): Unrevealed cognitive defects in a patient with occipital epilepsy. Epilepsia 38 (Suppl 3):135.
6. Jambaqué I and Dulac O (1989): Syndrome frontal réversible et épilepsie chez un enfant de 8 ans. Archives Françaises de Pédiatrie 46:525–529.
7. Jambaqué I, Dellatolas G, Dulac O, Ponsot G and Signoret JL (1993): Verbal and visual memory impairment in children with epilepsy. Neuropsychologia 31:1321–1337.
8. Jambaqué I, Chiron C, Kaminska A, Plouin P and Dulac O (1998): Transient motor aphasia and recurrent partial seizures in a child: Language recovery upon seizure control. Journal of Child Neurology 13:296–300.
9. Shewmon DA, Warwick J and Peacock J (1997): Epilepsy surgery in children with medically controlled seizures and continuous focal interictal discharges. Epilepsia 38 (Suppl 3): 169.

19

NEUROPSYCHOLOGICAL OUTCOME IN CHILDREN WITH WEST SYNDROME

A "Human Model" for Autism

Isabelle Jambaqué*[,1], Laurent Mottron[2], and Catherine Chiron[3]

[1]Service de Neuropédiatrie
Hôpital Saint Vincent de Paul
82 Avenue Denfert Rochereau
75674 Paris Cedex 14, France
[2]Clinique Spécialisée de l'Autisme
Hôpital Rivière des Prairies
7070 Boulevard Perras
Montréal (Québec), H1E 1A4, Canada
[3]INSERM U29, Hôpital Saint Vincent de Paul
82 Avenue Denfert Rochereau
75674 Paris Cedex 14
France et Service Hospitalier Fréderic Joliot
Département de Recherche Médicale CEA
Orsay, France

INTRODUCTION

West syndrome is an epileptic syndrome of the first year of the life which includes infantile spasms with hypsarrythmia. Infantile spasms are characterized by brief flexion, extension or mixed flexion-extension jerks of the muscles of the neck, trunk or extremities. Hypsarrythmia consists of asynchronous diffuse paroxysmal discharges reflecting hyperexcitability of the immature brain. Control of spasms may be obtained with several medications including steroids, valproate, benzodiazepines and vigabatrin. However, cognitive disorders often persist, even in patients who become seizure free by treatment and whose EEG becomes normal. Strikingly, psychiatric disorders, mainly autistic features, are overrepresented in this population.

* To whom correspondence should be addressed

Neuropsychology of Childhood Epilepsy, edited by Jambaqué et al.
Kluwer Academic / Plenum Publishers, New York, 2001.

1. BEHAVIORAL REGRESSION

Mental deterioration is classically described as one major component of West syndrome.[20] It often occurs at the onset of the first spasms, and occasionally follows or even precedes them. The behavioral regression is characterized by loss of psychomotor abilities and a reduction in social interaction.[16] The infants exhibit hypotonia and loss of reaching for objects. They fail to smile, and they show passivity in front of the mirror and lack of eye contact. Often these infants no longer track visually and they eventually appear to be blind. Regression of babbling and poor reaction to voice or nonverbal auditory stimuli are often present. When mental retardation is present before the onset of spasms, mental regression is more difficult to assess. The severity of mental delay and deterioration is undoubtedly related to the nature of the underlying neurological disorder. During steroid therapy, there is often a marked worsening of mental regression.[11] Nevertheless, impairment of sensory input can be said to result from this devastating epileptogenic encephalopathy of infancy through the combination of epileptic spasms and paroxysmal EEG activity. Interestingly, the neuropsychological status observed in patients at the onset of the West syndrome is a useful predicting factor. The best prognostic indexes are visual tracking and reaction to sensory stimuli, rather than less specific features such as muscle tone. Furthermore, there is a significant correlation between developmental quotient at onset of spasms and at 3 years of age.[11]

2. COGNITIVE OUTCOME

West syndrome is one of the most devastating epileptic encephalopathy of early childhood. Patients with West syndrome generally have a poor long-term neurological prognosis. Mental retardation is observed in 80% of patients, and is severe in more than half of them.[27] A major correlate of cognitive deficit in childhood epilepsy relates to age of onset, an important factor because West syndrome appears mainly in infancy and there is evidence of a deleterious effect of seizures on brain development. It is clear that behavioral regression, and not just arrest of development, classically occurs during West syndrome. The prognosis for recovery to premorbid level is poor even if treatment is successful in stopping seizure activity and normalizing the EEG. There is, however, a great variability in severity of cognitive impairment which reflects the heterogeneity of the etiologies and therefore the diversity of brain abnormalities.

A small number of patients (5%) with idiopathic West syndrome recover completely.[11,36] In our series of 12 cases followed to the age of 7–11 years, none showed a relapse to seizures and all had normal IQ with normal academic achievement. The main characteristics of this subgroup were normal development prior to the first spasms and no major deterioration at the onset of the disease particularly no loss of visual tracking. On the other hand, mental retardation is clearly present before the first seizures in a number of patients who have multiple brain lesions and for whom the occurrence of generalized epilepsy is an additional factor in their mental regression. Certain etiologies appear be associated with especially poor outcomes; children with suspected genetic disease, with remote infectious diseases and children with tuberous sclerosis.[28,17] A significant proportion of patients with symptomatic infantile spasms who exhibit severe mental retardation continues to experience intractable generalized epilepsy. When on the contrary seizures are rapidly controlled by therapy, some patients with tuberous sclerosis, for example, can have a more favorable mental outcome. Riikonen[28] reported the only study in which children have been followed to adulthood. Strikingly, 24% of patients have apparently normal or subnormal intelligence and socioeconomic status, including adults whose infantile spasms were either symptomatic or associated with focal EEG findings. Not only early cessation of spasms and disappearance of hypsarrhytmia are important factor for a good outcome but also the location of brain abnormality. The persistence of secondary generalized seizures is, in fact, indicative of worse mental outcome regardless of etiology. For example, a number of patients with West syndrome in early infancy later develop Lennox-Gastaut syndrome which is classically associated with defective cognitive abilities. For medically intractable infantile spasms, surgical removal of the seizure focus may favorably alter the developmental outcome, presumably by protecting the rest of the brain from spreading seizure activity.[7]

3. CENTRAL VISUAL IMPAIRMENT

Long-term outcome of West syndrome showed that many patients have global cognitive impairment but others may have more specific cognitive deficits. Although speech delay is a classic picture, recent studies reported the frequency of non-verbal communication troubles. In 1993, we reported a neuropsychological and neurofunctional imaging study in a subgroup of 8 children who suffered from cryptogenic infantile spasms.[18] During the acute phase, all infants lacked visual tracking and did not reach for objects or smile. They also showed secondary stereotypes like hand flapping and head banging. Control of spasms and disappearance of hypsarrhytmia were obtained within one month of steroid treatment. However, two children remained with residual partial epilepsy. Follow-up until the age of 4–7 years showed a persistent, but specific, central visual impairment in three cases. Two children also exhibited marked speech disorders and the three most severely affected patients remained with autistic behavior. SPECT showed perfusion defects involving the parieto-occipital regions in 6 of 8 patients. Recently, we studied the mental outcome of 30 patients with West syndrome. Selected patients had cryptogenic spasms or focal cortical lesion. Spasms had been controlled before one year with no persisting generalized epilepsy, and a neuropsychological follow-up was conducted for 5 years and more. In this group of 30 patients, neuropsychological outcome showed that:

- 5 patients, aged 5 to 14 years, had predominant language impairment. Four of them were left-handed and mentally retarded. One patient developed language only at the age of 9 years. In this case, SPECT showed, at 5 years, marked hypoperfusions located in the temporal lobes and, at 14 years, hypoperfused areas located in the left temporal lobe.
- 13 children, aged between 5 and 16 years, had predominant visual-spatial deficits. All these children, except one, were right-handed. On the Wechsler Scale, all these patients showed a significant discrepancy between verbal and performance IQs indicating non verbal learning dysfunction (mean verbal IQ of 85, range: 62–107; mean performance IQ of 64, range: 45–89). These children showed deficits of visuo-motor coordination, dyspraxia, piece-meal vision, space orientation and object recognition impairment. Most of them also had difficulties in social interaction. In the 9 patients who underwent neuroimaging, structural lesions and/or hypoperfusion abnormalities involved the parieto-temporo-occipital cortex (Fig. 1).

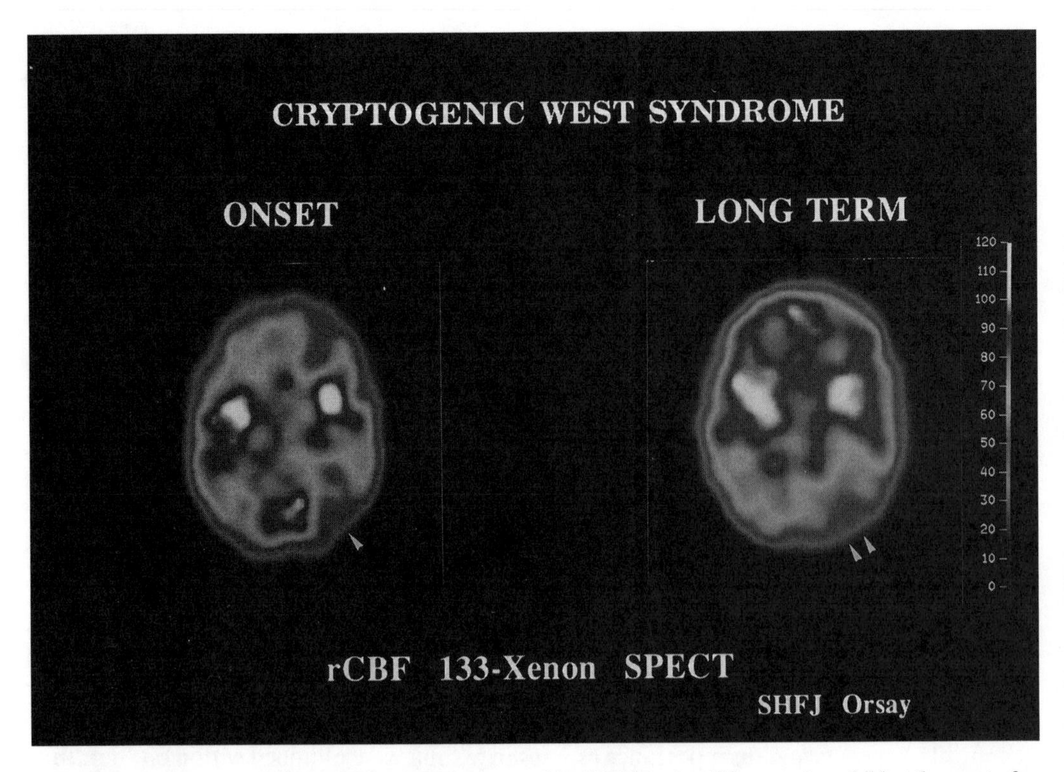

Figure 1. Regional cerebral blood flow (rCBF) measured by SPECT using 133-xenon in a child at the onset of West syndrome (9 months) on the left, and 2 years after cessation of spasms (2 years 10 months), on the right. rCBF is expressed in ml/mn/100 g according to the color scale. There is an area of hypoperfusion in the right temporo-parieto-occipital region which was not modified between the two examinations.

- 12 patients, aged between 5 and 11 years, were severely retarded with a complete autistic picture. Ten of them were left or mixed handed and 8 had no development of language at all. MRI and/or functional imaging by SPECT were performed in 9 patients. Both posterior and frontal lesions or hypoperfusions were identified in this group.

In conclusion, visual unresponsiveness characterizes behavioral regression of West syndrome and neurovisual disorders are quasi-constant sequelae in a history of infantile spasms. Although West syndrome is usually associated with generalized EEG disturbances (hypsarrhytmia), more focal abnormalities are sometimes recorded, particularly in the occipital regions. In addition, functional brain imaging studies (PET or SPECT) have shown the frequent involvement of the cerebral cortex at the junction of parietal-occipital- and temporal lobes.[4,7] There is therefore a strong relationship between neuropsychological sequelae and the neurofunctional imaging data. Furthermore, one important characteristic of West syndrome might be the relation of age and the onset of the disease. Since the syndrome appears mainly in infancy, most often between 3 and 8 months, maturational changes could account for its age relationship. In particular, recent neurofunctional imaging studies have shown changes of metabolism and cerebral blood flow in the visual associative cortex during the first six months of life[3,6] and that the right hemisphere, which sustains visuospatial and emotional abilities, develops its functions earlier than the left.[5]

4. ASSOCIATION WITH AUTISM

4.1. West Syndrome and Autism

Children with infantile spasms show a high frequency of psychiatric disturbances. In particular, the co-occurrence of infantile spasms with autism is well established.[27,33] Several authors have stressed the association between early signs of autism, behavioral regression and West syndrome, the latter condition representing an acquired autistic state. Behavioral regression affects both social interaction and interest in the environment. Hence, infants lack eye contact, fail to acknowledge other people, and do not notice or recognize familiar faces. In addition, they exhibit stereotypes such as hand flapping and head banging. On first thought, behavioral regression can be attributed to the neurosensory and emotional stimulus deprivation produced by epilepsy, causing apparent indifference as a characteristic behavior. In particular, visual unresponsiveness may be an important factor that contributes to the initial disorganization of interpersonal contact in patients with West syndrome.

The association between West syndrome and autism constitutes a "human model" of autism in the same manner that the "animal models" of known conditions are used. West syndrome allows for an examination of the way in which the combined effect of a pathological process (epilepsy) and/or brain lesion (focal hypometabolism) approximates a known clinical entity (autism). West syndrome is also useful for understanding autism due to the reversibility of some autistic states in this context, although a number of patients with West syndrome remain autistic, from 13% in cryptogenic cases to 57% in tuberous sclerosis.[15]

4.2. Epilepsy, West Syndrome and Autism

A ternary combination between early epilepsy, autism, and visual impairment is observed with a high frequency in West syndrome. The rarity of both autism and West syndrome suggests that a fortuitous association between the two syndromes is unlikely, and instead orients toward some type of causal relationship between the two conditions. Nevertheless, the heuristic value of this comorbidity must be interpreted in the more general context of possible relations between autism and epilepsy.

- Lack of specificity of the relation between autism and epilepsy, autism and brain lesions, brain lesion and epilepsy

Autism can occur in association with any clinical type of epilepsy[12] or EEG abnormalities.[34] In addition, autism is found in conjunction with a large and heterogeneous number of medical conditions (e.g., genetic syndromes, perinatal accidents, cortical dysplasia). Even if the exact incidence of medical conditions related to autism is a matter of debate, there is consensus regarding their heterogeneity[1,14] as no common mechanism has been reliably found between these medical conditions. On the other hand, epilepsy is associated with a number of heterogeneous, pathological brain damages. If autism and epilepsy are both associated to a large class of brain abnormalities, then perhaps these two conditions are also associated with a large, heterogeneous, number of medical conditions involving brain injury. Instead of causal relation, there may be only a relation of co-occurrence among favoring factors. An extreme proponent of this position is Minshew,[23] who believes that general statements about autism cannot be drawn from a clinical condition in which a brain lesion is associated to autism and epilepsy (for example, tuberous sclerosis).

- The lower the IQ, the more frequent the association between epilepsy and autism

Recent recognition of a population with a clear picture of autism but without mental deficiency (e.g., high-functioning individuals with autism and Asperger's syndrome), reveals that a low IQ is not a mandatory component of pervasive developmental disorder.[13,30,31,37] Consequently, the increase of epilepsy prevalence among persons with autism, when IQ decreases, suggests that epilepsy is associated not with autism itself, but with the overall magnitude of brain injury. However, even in subjects with a low IQ, this relationship is rather weak. Autism is found only in 38% of subjects with mental deficiency and autism[32]

- Chronological relationship between autism and epilepsy

Three possible patterns of onset between the two conditions can be found. First, in numerous cases, epilepsy appears during the teens years in an individual diagnosed with autism in early childhood. However, late onset of epilepsy is not particularly frequent. For instance, Elia *et al.*[12] studied a group of 28 subjects with autism, epilepsy and mental deficiency, and found a mean age of onset of 4.36 years for epilepsy. Second, as illustrated by West syndrome, epilepsy is the first symptom to occur, and autism is its outcome, but this is unlikely to occur in reverse order, given the early age of onset of West syndrome. However, autism may also develop when epilepsy is controlled, particularly in cryptogenic infantile spasms. Third, a single cause may determine both epilepsy and autistic symptoms.[19]

A mention must be made of cases in which there is behavioral regression following epilepsy. This finding is frequently used as an argument for a causal relation between the two disorders. Several single case studies have shown that autistic symptoms may appear in the year following the onset of seizures.[21,22] The notion of a causal relationship between the two events is reinforced by the improvement of autistic symptoms that is sometimes observed after anti-epileptic treatment.[9,14,26] However, a closer examination of the relationship between autism, regression and epilepsy revealed that epilepsy was only present in one third of autistic children exhibiting a behavioral regression.[35]

In sum, two alternative causal models are represented in the literature. These models are represented in opposition for purposes of clarity, although intermediate models are plausible:

X ⇨ autism
⇨Low IQ
⇨epilepsia
vs.
X ⇨ epilepsia ⇨autism
⇨Low IQ

A, third alternative, hypothesis is that the underlying brain dysfunction itself accounts for autistic symptoms. According to this hypothesis, it is not epilepsy which is relevant for autism, but the location of the processes causing or resulting from epilepsy (i.e., temporo-occipital lesion). The notion that numerous symptoms of autism might be explained by an early perceptual visual impairment that prevents face and emotion perception, as well as by a compensatory orientation toward certain dimensions (movement, contrasts, outlines) in visual displays, has been debated.[25] This hypothesis is supported by the observation that, among children with West syndrome, those with early visual impairment presented with an autistic outcome more frequently.[11,18] It is possible that a dysfunction of temporo-occipital region produces a functional visual agnosia which is responsible for some aspects of autistic syndrome.[24] This could account for the observation that, in multifocal diseases like tuberous sclerosis, the temporo-occipital locations of tubers appears to be associated with an autistic outcome.[1] We reported a SPECT study showing hypoperfusion of frontal lobes associated with a posterior defect.[19] More recently, a positron emission-tomography revealed that children with infantile spasms who have perfusion abnormalities in the temporal lobes are much more likely to develop autism.[8] The higher susceptibility of temporal regions to various types of pathological agents such anoxia, also suggests that epilepsy, whatever its cause, might alter these brain regions when it occurs at a critical period for brain development.

This model may be summarized in the following manner:

X (any factor, e.g., epilepsia) ➪ temporo-occipital lesion ➪ autism
↳ non specific brain lesion ➪ low IQ

In sum, it remains unclear whether autism and epilepsy result from the same group of brain injuries, or if epilepsy as such might be responsible for autistic syndrome (i.e., had the injury not produced epilepsy, it would not have been followed by an autistic regression). Convincing accounts of both situations have been reported. Considering the heterogeneous nature of autistic syndrome, it is risky to select only one medical condition associated with autism to build a general model for autism. Nevertheless, we believe it is worth building etiologic models for particular clinical profiles. Accurate (though partial) etiologic explanations are preferable to false (but general) explanations.

CONCLUSION

There is some evidence that posterior visual regions are frequently involved in West syndrome. Central visual impairment results in long term perceptual deficits and there is a strong relationship between functional neuroimaging and neuropsychological findings. Furthermore, this neuropsychological pattern may have implications for social communication. It is tempting to link lack of communication with the impairment affecting development of face and object recognition. Similarly, impairment in visual motion might result in impairment in perception and production of social gestures. In our experience, the most severely affected patients remain with autistic behavior or other pervasive developmental disorders. Nevertheless, the early identification of visual impairment may guide neuropsychological rehabilitation.

REFERENCES

1. Bailey A, Philipps W and Rutter M (1996): Autism: towards an integration of clinical, genetic, neuropsychological, and neurobiological perspectives. Journal of Child Psychology and Psychiatry 37:89–126.
2. Bolton PF and Griffiths PD (1977): Association of tuberous sclerosis of temporal lobes with autism and atypical autism. Lancet 349:392–395.

3. Chiron C, Raynau C, Maziere B, Zilbovicius M, Laflamme L, Masure MC *et al.* (1992): Changes in regional cerebral blood flow during brain maturation in children and adolescents. Journal of Nuclear Medicine 33:696–703.
4. Chiron C, Dulac O, Bulteau C *et al.* (1993): Study of regional cerebral blood flow in West syndrome. Epilepsia 34:707–715.
5. Chiron C, Jambaqué I, Nabbout R, Lounes R, Syrota A and Dulac O (1997): The right brain hemisphere is dominant in human infants. Brain 120:1057–1065.
6. Chugani HT, Phelps ME and Maziotta JC (1987): Positron emission tomography study of human brain functional development. Annals of Neurology 22:487–497.
7. Chugani HT, Shields WD, Shewman DA *et al.* (1990): Infantile spasms I: PET identifies focal cortical dysgenesis in cryptogenic cases for surgical treatment. Annals of Neurology 27:406–413.
8. Chugani HT, Da Silvea E and Chugani DC (1996): Infantile spasms: III prognostic implications of bitemporal hypometabolism in positron emission tomography. Annals of Neurology 39:643–649.
9. Deonna T, Ziegler AL, Moura-Serra J and Innocenti G (1993): Autistic regression in relation to limbic patholgoy and epilepsy: report of two cases. Developmental Medicine and Child Neurology 35:166–176.
10. Deykin EY and MacMahon B (1979): The incidence of seizures among children with autistic symptoms. American Journal of Psychiatry 136:1310–1312.
11. Dulac O, Plouin P and Jambaqué I (1993): Prediciting favorable outcome in idiopathic West syndrome. Epilepsia 34:747–756.
12. Elia M, Musumeci SA, Ferri R and Bergonzi P (1995): Clinical and neurophysiological aspects of epilepsy in subjects with autism and mental retardation. American Journal of Mental Retardation 100:6–16.
13. Gilberg C and Steffenburg S (1987): Outcome and prognosis factors in infantile autism and similar conditions: a population-based study of 46 cases followed through puberty. Journal of Autism and Developmental Disorder 17:273–287.
14. Gilberg C, Uvebrant P, Carlsson G, Hedstrom A and Silfvenius H (1996): Autism and epilepsy in two pre-adolescent boys: neuropsychiatric aspects before and after epilepsy surgery. Journal of Intellectual Disabilities Research 40:75–81.
15. Hunt A and Dennis J (1987): Psychiatric disorders among children with tuberous sclerosis. Developmental Medicine and Child Neurology 29:190–198.
16. Illingworth RS (1955): Sudden mental deterioration with convulsions in infancy. Archives of Disabled Childhood 30:529–537.
17. Jambaqué I, Cusmai R, Curatolo P, Cortesi F, Perrot C and Dulac, O (1991): Neuropsychological aspects of tuberous sclerosis in relation to epilepsy and MRI findings. Developmental Medicine and Child Neurology 33:698–705.
18. Jambaqué I, Chiron C, Dulac O, Raynaud C and Syrota A (1993): Visual inattention in West syndrome: a neuropsychological and neurofunctional imaging study. Epilepsia 34:692–700.
19. Jambaqué I, Mottron L, Ponsot G, and Chiron C (1998): Autism and visual agnosia in a child with right occipital lobectomy. Journal of Neurology, Neurosurgery and Psychiatry 6:555–560.
20. Jeavons PM and Bower BD (1974): Infantile spasms. In Vinken PJ and Bruyn GW (eds): "Handbook of Clinical Neurology", Vol 15. Amsterdam: Elsevier North Holland, pp 219–234.
21. Kimura K and Nomura Y (1994): Age dependent epileptic encephalopathy (ADEE) associated with autism. The Japanese Journal of Psychiatry and Neurology 4:2–11.
22. Kyllerman M, Nyden A, Praquin N, Rasmussen P, Wetterquist AK and Hedstrom A (1996): Transient psychosis in a girl with epilepsy and continuous spikes and waves during slow sleep (CSWS). European Child Adolesecent Psychiatry 5:216–221.
23. Minshew NJ, Sweeney JA and Bauman ML (1997): Neurological aspects of autism. In Cohen DJ and Volkman FR (eds): "Handbook of Autism and Pervasive Developmental Disorders". New-York: Wiley, pp 344–356.
24. Mottron L, Mineau S, Decarie JC, Jambaqué I, Labrecque R, Pepin JP and Aroichane M (1997): Visual agnosia with bilateral temporo-occipital brain lesions in a child with autistic disorder: a case study. Developmental Medicine and Child Neurology 39:699–705.
25. Mottron L and Belleville S (1998): L'hypothèse perceptive dans l'autisme. Psychologie Française 43:135–145.
26. Perry R, Cohen I and DeCarlo R (1995): Case study: deterioration, autism, and recovery in two siblings. Journal of American Academy of Child and Adolescent Psychiatry 34:232–237.
27. Riikonen R and Amnell G (1981): Psychiatric disorders in children with earlier infantile spasms. Developmental Medicine and Child Neurology 23:747–760.

28. Riikonen R (1996): Long-term outcome of West syndrome: A study of Adults with a history of infantile spasms. Epilepsia 37:367–372.
29. Rossi PG, Parmeggiani A, Bach V, Santucci M and Visconti P (1995): EEG features and epilepsy in patients with autism. Brain Development 17:169–174.
30. Steffenburg U, Hagberg G, Viggedal G and Kyllerman M (1995a): Active epilepsy in mentally retarded children. I. Prevalence and additional neuro-impairments. Acta Paediatrica 84:1147–1152.
31. Steffenburg U, Hagberg G and Kyllerman M (1995b): Active epilepsy in mentally retarded children. II. Etiology and reduced pre- and perinatal optimality. Acta Paediatrica 84:1153–1159.
32. Steffenburg U, Gilberg C and Steffenburg U (1996): Psychiatric disorders in children and adolescents with mental retardation and active epilepsy. Archives of Neurology 53:904–912.
33. Taft LT and Cohen JH (1971): Hypsarrythmia and childhood autism: a clinical report. Journal of Autism and Child Schizophrenia 1:327.
34. Tsai LY, Tsai MC and August GJ (1995): Brief report: implications of EEG diagnoses in the subclassification of infantile autism. Journal of Autism and Developmental Disorders 15:339–344.
35. Tuchman RF and Rapin I (1997): Regression in pervasive developmental disorders: seizures and epileptiform electroencephalogram correlates. Pediatrics 99:560–566.
36. Vigevano F, Fusco L, Cusmai R, Claps D, Ricci S and Milani L (1993): The idiopathic form of West syndrome. Epilepsia 34:743–746.
37. Volkmar FR and Nelson DS (1990): Seizure disorders in autism. Journal of American Academy of Child and Adolescent Psychiatry 29:127–129.

COGNITIVE DETERIORATION IN LENNOX-GASTAUT SYNDROME AND DOOSE EPILEPSY

Virginie Kieffer-Renaux*, Anna Kaminska, and Olivier Dulac

Service de Neuropédiatrie
Hôpital Saint Vincent de Paul
82 Avenue Denfert Rochereau
75674 Paris Cedex 14, France

INTRODUCTION

Epilepsy in children comprises not only seizures, but also a risk for cognitive functions and/or behavior that interfere with social integration and schooling. For no other type of epilepsy is it more severe than for intractable generalized epilepsy, such as Lennox-Gastaut syndrome (LGS) and Doose epilepsy. The incidence of these disorders is low, in the range of 1%.[20] However, since the epilepsy remains intractable for a number of affected patients, the prevalence is in the range of 10%. Patients suffer from daily seizures and major interictal paroxysmal activity. Several factors, including frequency and severity of the seizures[4] and major interictal EEG abnormalities seem to interfere with the development of cognitive functions.

The distinction between these two conditions remains difficult, if not disputable. Some authors consider that there is no clear boundary between them and that they represent a continuum[2] whereas others consider that both etiology and outcome are clearly different.[10] Neuropsychological characteristics are different, however, and this approach may contribute to distinguish both conditions and their etiologies.

* To whom correspondence should be addressed

Neuropsychology of Childhood Epilepsy, edited by Jambaqué et al.
Kluwer Academic / Plenum Publishers, New York, 2001.

1. DEFINITIONS

1.1. Lennox-Gastaut Syndrome

LGS was first identified based on the combination of several types of generalized seizures, generalized slow spike-wave activity and deterioration of cognitive abilities,[7] thus constituting an epilepsy syndrome. The slow spike-wave (SSW) pattern was initially considered as a variant of "petit mal" and called "petit mal variant". However, the SSW pattern was clearly distinct from the 3 Hz tracing of absence seizures, the so-called "petit mal". It differs from the "petit mal" pattern since the spike-wave activity is slow, and onset and end are not clear. Thus, it is often difficult to determine whether the patient is in an interictal or an ictal condition, even with video-EEG recording. SSWs predominate over the frontal lobes.[2,13] This pattern is recorded in patients with a severe condition, including drop attacks, tonic seizures and atypical absences during which there is progressive loss of antigravidic tone, and either the head or the whole body drops. Myoclonic seizures are rare, if they occur at all, but focal seizures do occur.

LGS begins between 1 and 9 years, later in cryptogenic cases than in symptomatic ones, particularly when it follows West syndrome.[2,4,11,15,23] The disease is chronic with frequent episodes of status epilepticus that do not seem to alter the outcome. It affects boys more frequently than girls in a proportion of 1.5 to 1. In most cases, it is possible to identify some kind of non-progressive pre-existing brain damage. Forty percent of patients previously had West syndrome, others partial epilepsy of various causes or either mental retardation or behavior problems without epilepsy.[3,9] It may be difficult to determine in patients who display mental or behavior problems as the first sign whether these result from prior brain dysfunction or represent the first manifestations of the epilepsy.[25] Thus, LGS is usually considered as a symptomatic condition. In the cases without any evidence of brain damage, the condition is qualified as "cryptogenic", which means that the cause, likely brain damage, is "hidden".

1.2. Doose Epilepsy

Doose was also faced with patients who suffered from drop attacks, several types of generalized seizures and generalized spike waves, but in contrast to LGS cases, these patients with high familial incidence of epilepsy exhibited myoclonus and 3 Hz SWs, which he considered as signs of genetic predisposition.[10] However, the course of the disease in these patients was most variable, some of them being very similar to the groups reported by Dravet as "benign" or "severe" myoclonic epilepsies of infancy,[12] and others having a later onset, between 2 and 5 years.[7,13,22] Thus, this condition was not a syndrome but an etiologic concept which Doose called "myoclonic-astatic epilepsy", and which is considered to be genetically determined, or cryptogenic but never due to brain damage.

It is the tableau of a group of patients starting epilepsy after the age of one that will be considered here and that, for practical purposes, we will call Doose syndrome (DS) in the following sections.

1.3. Doose Syndrome

At the onset, patients have generalized tonic-clonic seizures which are soon followed by myoclonic drop attacks whereas absences and tonic seizures are less frequent. DS has a high incidence of familial epilepsy, myoclonus and 3 Hz SW activity. The EEG shows bursts of generalized SWs or polySWs. The outcome is variable. Some patients recover after one to three years.[14] Other patients have poor outcome because of the occurrence of episodes of myoclonic status epilepticus during which the patient is drowsy and exhibits erratic myoclonus of the face and upper limbs, lasting one to several weeks, intermingled with vibratory tonic seizures at the end of night sleep. At this late stage, SSW activity is frequent.[17] Thus, it is only in the course of the disease that SSWs appear in cases with unfavorable outcome. Mental deterioration seems to correlate with the occurrence of myoclonic status and tonic seizures. This group with unfavorable outcome has long been considered as the "myoclonic variant of LGS".[1] However, multifactorial analysis has shown that it is distinct from LGS from the very onset of the disease.[17] While there is growing evidence that LGS is affecting patients with previous brain damage, DS appears to be genetically determined. Unfavorable cases of DS exhibit features of LGS only later in the course of the disease. LGS is likely to combine neuropsychological features related to the topography of the focus and to the frontally predominating SSW activity. In contrast, DS produces myoclonus, and therefore mainly affects the rolandic area. Later in the course of cases with unfavorable outcome, features due to rolandic involvement are combined with features of frontal lobe involvement corresponding to the SSWs occurring at that stage. Both conditions will now be compared.

2. INTELLECTUAL ABILITY

Intellectual ability deteriorates progressively. In a follow-up study of 7 years for LGS and 5 years for DS, conducted by our team,[19] a significant decrease in IQ was observed in both conditions. The significance of this deterioration in IQ has been discussed. Most authors consider that a decrease in mental performance in children does not reflect loss of abilities but merely an arrest of development compared to normal children who continue to make acquisitions. In our series, it was possible to distinguish both conditions: LGS patients clearly showed deterioration whereas DS patients displayed stagnation. In a series of 37 patients, we found that although the mean age of onset was earlier in DS (3 years) than in LGS (6 years), patients with DS exhibited deterioration later in the course of the illness (between 8 and 10 years of age) than those with LGS (between 6 and 8).

In LGS, intellectual deterioration has been found to be frequent[12] and to occur rapidly.[3,16] It is more severe for symptomatic than for cryptogenic patients and for patients with previous West syndrome than for those with later onset. Whether there is ongoing deterioration is still a matter of debate. Ninety-five percent of the patients suffer from mental retardation[23] and 80% have an IQ under 50.[6] In the series reported by Hirosaku *et al.*,[16] 73% (53/72) of the patients demonstrated variable degrees of mental retardation with an IQ under 70. In most instances, the average IQ of the symptomatic group was significantly lower than that of the cryptogenic group (43% of cryptogenic patients had an IQ below 35, compared to 76% of symptomatic patients): An IQ below 55 was observed in 86% of cryptogenic and in 96% of the symptomatic cases.[16]

The outcome is more severe for cases following West syndrome than for those without previous epilepsy, and therefore with a later onset.[23] Thus, the outcome depends on the age of onset of the seizure disorder, with the worst outcome for those with earlier onset.[6] A single series has shown similar outcome for cryptogenic and symptomatic cases.[4]

In our series, we could not evaluate the difference between symptomatic and cryptogenic cases since we restricted our analysis to cryptogenic cases in order to concentrate on the effect of epilepsy as opposed to the effect of pre-existing damage. In the previously mentioned series, we found a median IQ value of 61 from the first year of the seizure disorder. For seventy of the patients, the IQ was below 75 three years after onset,[6,23] and below 45 seven years[19] and 10 year[16] after onset, respectively. In our series of twenty-one cryptogenic cases, the children exhibited progressive intellectual slowing.[4] Associated behavioral disturbances were present in all cases.

In DS, the outcome of intellectual abilities is more variable.[10,19] Following cessation of seizures, it may be favorable, with recovery of intellectual efficiency. Only a few seizure-free patients are left with an IQ below 50.[1,14] In a group of 14, we found a median IQ of 82 for patients who stopped having seizures one to three years following seizure onset.[19] For patients with persisting and intractable seizures, the outcome is very poor. Some become demented.[10] In our series, the median IQ was 52, thus in the same range as that of LGS patients.[19] When following patients with unfavorable myoclonic-astatic epilepsy, it appears that there is a time lag between the onset of epilepsy and mental deterioration, as opposed to LGS. In fact, the age of occurrence of deterioration, between 8 and 10 years of age, corresponds to the age of occurrence of repeated and long lasting episodes of myoclonic status, thus two years later than for LGS patients.

3. SPECIFIC NEUROPSYCHOLOGICAL PATTERNS AT ONSET OF THE DISORDER

During the first year of the seizure disorder, behavior problems are frequent, both in LGS[2,8] and DS.[19] They consist of hyperkinesia[3,8,13] with inability to pursue a given activity during more than a few minutes.[13] Some patients exhibit also autistic or psychotic traits. Patients of both groups have apraxia. They are unable to execute skillful tasks such as puzzles and/or drawings on the Brunet-Lézine, Termann-Merrill or Wechsler scales.[19]

Only LGS patients show major slowing of intellectual functioning. They rapidly develop impairment of motor speed, apathy, slowness and slow expression,[3] which are observable during clinical observation. These children need to be stimulated in order for responses to be obtained.[19] We found that they exhibited inattention, perseveration, instability and memory problems. Two children in our series had spatial disorientation and were unable to perform everyday activities.[19] In contrast, patients with DS suffer more commonly than those with LGS from dysarthria that may result in mutism.[14]

4. LONG-TERM NEUROPSYCHOLOGICAL PROBLEMS

Long-term follow-up studies have revealed differences between favorable and unfavorable cases of DS: Dysarthria was mainly observed in unfavorable cases with DS.[19] The latter and those with LGS showed important ideic slowing and praxic disorders. In the series reported by Boniver *et al.*,[4] patients with LGS were left with hyperkinetic behavior on average 7 years after onset and with slowing of intellectual functioning.[4] Perseverative behavior was observed in all three groups; however, we found that some patients with either LGS or unfavorable DS were so slow and apathetic that any tendency to perseverate would have been masked by the magnitude of the slowness.[19]

5. CHARACTERISTICS

Only patients with unfavorable DS have dysarthria and ataxia.[10] Both patients with LGS and unfavorable DS exhibit mental deficiency, signs of frontal lobe involvement and perseveration. Although slowness and decrease of IQ may result from intellectual deterioration, the occurrence of a frontal lobe syndrome, evolving to dementia, also seems to contribute to deterioration.

6. EFFECTS OF TREATMENT

Very few data are available regarding the effect of callosotomy or steroid treatment on cognitive functioning in these cases. Some authors consider the effect of callosotomy in LGS disputable.[18] Others observed improvement of mental abilities, behavior[19,26] and memory[21,24] when seizure frequency decreased.[16,24]

7. COMMENTS

DS differs from LGS with regard to the occurrence of articulation difficulties, suggesting pre-rolandic involvement. Thus, neuropsychological findings are consistent with the topography of the maximal spiking activity. In LGS, this activity affects most of the brain from the beginning of the disorder, although it predominates over pre-frontal areas. Global mental deterioration is indeed an early manifestation of this syndrome.

At the onset of DS, spike activity predominates over the rolandic areas. In fact, the ictal manifestations mainly consist of myoclonus involving the motor strip. Dysarthria and apraxia, two relatively specific symptoms in DS, involve areas of the brain that are near the motor strip. In this disorder, SSWs occur later, if ever, and severe deterioration is a relatively late manifestation of DS in cases with persistent intractable seizures.

DS also differs from LGS in the course of the disorder and the neuropsychological pattern. DS has variable outcome with either a short or a long course of the disease. Only the subgroup with persistent pharmacoresistant epilepsy is comparable to LGS, and from the nosological point of view it has long been considered as indistinguishable from LGS, therefore the name “the myoclonic variant of LGS”. However, this review demonstrates that the neuropsychological profile is distinct in these two conditions and therefore is likely to contribute to a differential diagnosis. Favorable cases of DS are characterized by an IQ in the lower normal range, apraxia and perseveration. Although the course is relatively short, a number of patients are affected in their cognitive abilities and have major problems. The great number of convulsive seizures that occur in the short course of the disease is likely to contribute to this deterioration. The persistent hyperkinesia requires specific rehabilitation. The effect of drug treatment with methylphenydate has not been evaluated to our knowledge.

Patients with unfavorable DS exhibit an early course similar to those with favorable outcome. It is only after several months to a few years that cognitive functions are affected. Within a few months to one or two years, deterioration is observed as the patient exhibits long episodes of frequent seizures or even status epilepticus. The cognitive profile is specific with apraxia and dysphasia as predominant features, although most functions are affected with deterioration of the IQ. Deterioration of verbal functions is unusual in the course of epilepsy.

REFERENCES

1. Aicardi J and Chevrie JJ (1971): Myoclonic epilepsies of childhood. Neuropédiatrie 3:177–184.
2. Aicardi J (1973): The problem of the Lennox-Gastaut syndrome. Developmental Medicine and Child Neurology 15:77–81.
3. Beaumanoir A and Dravet C (1992): Le syndrome de Lennox-Gastaut. In Roger J (ed): "Les Syndromes Épileptiques de l'Enfant et de l'Adolescent". London: John Libbey, pp 115–132.
4. Boniver C, Dravet C, Bureau M and Roger J (1987): Idiopathic Lennox-Gastaut syndrome. Advances in Epileptology 16:195–200.
5. Bulteau C, Jambaqué I, Kieffer V and Dulac O (1997): Aspects neuropsychologiques du syndrome de pointes ondes continues pendant le sommeil lent. Approche Neuropsychologique des Apprentissages chez l'Enfant 41:5–9.
6. Chevrie JJ and Aicardi J (1972): Chilhood epileptic encephalopathy with slow spike-wave. A statistical study of 80 cases. Epilepsia 13:259–271.
7. Commission on Classification and Terminology of the International League against Epilepsy (1989): Proposal for revised classification of epilepsies and epileptic syndromes. Epilepsia 30:389–399.
8. Costa P and Beaumanoir A (1992): Modalités de début du syndrome de Lennox-Gastaut: à propos d'un cas. Epilepsies 4:221–228.
9. Donat JF and Wright FS (1991): Seizures in series: similarities between seizures of the West and Lennox-Gastaut syndromes. Epilepsia 32:504–509.
10. Doose H (1992): L'épilepsie myoclono-astatique du jeune enfant. In Roger J (ed): "Les Syndromes Épileptiques de l'Enfant et de l'Adolescent". London: John Libbey, 103–114.
11. Dravet C, Natale O, Magaudda A *et al.* (1985): Les états de mal dans le syndrome de Lennox-Gastaut. Revue EEG Neurophysiologie Clinique 15:361–368.
12. Dravet C (1996): Le syndrome de Lennox-Gastaut et ses frontières. Epilepsies 8:77–88.
13. Dulac O and Chiron C (1996): Malignant epileptic encephalopathies in children. Baillères Clinical Neurology 5:765–781.
14. Dulac O, Plouin P and Chiron C (1990): Forme "bénigne" d'épilepsie myoclonique chez l'enfant. Neurophysiologie Clinique 20:115–129.
15. Gastaut H, Roger J *et al.* (1966): Childhood epileptic encephalopathy with diffuse slow spike-waves (otherwise known as "Petit mal variant"). Epilepsia 7:139–179.
16. Hirokazu O, Kitami H and Makiko O (1996): Long term prognosis of Lennox-Gastaut syndrome. Epilepsia 37:44–47.
17. Kaminska A, Ickowicz A, Kieffer V, Bry MF, Plouin P and Dulac O (1997): Statistical analysis of nosologic differences between Lennox-Gastaut syndrome and myoclonic-astatic epilepsy. 22nd international epilepsy congress, Dublin-Ireland, 29th June–4th July.
18. Kazuichi Y (1996): Evolution of Lennox-Gastaut syndrome: A long term longitudinal study. Epilepsia 37:48–51.
19. Kieffer-Renaux V, Jambaqué I, Kaminska A and Dulac O (1997): Évolution neuropsychologique des enfants avec syndromes de Doose et de Lennox-Gastaut. Approche Neuropsychologique des Apprentissages chez l'Enfant 42:84–88.
20. Kramer U, Nevo Y, Neufeld MY, Fatal A, Leitner Y and Harel S (1998): Epidemiology of epilepsy in childhood-a cohort of 440 consecutive patients. Pediatric Neurology 18:46–50.
21. Lassonde M, Sauerwein HC and Geoffroy G (1990): Long term neuropsychological effects of corpus callosotomy in children. Journal of Epilepsy 3:279–286.
22. Minasssian BA, Sainz J and Delgado-Escueta AV (1996): Genetics of myoclonic and myoclonus epilepsies. Clinical Neuroscience 3:223–235.
23. Ohtahara S (1978): Clinico-electrical delineation of epileptic encephalopathies in childhood. Asian Medical Journal 21:499–509.
24. Reutens DC, Bye AM, Hopkins IJ *et al.* (1993): Corpus callosotomy for intractable epilepsy: Seizure outcome and pronostic factors. Epilepsia 34:904–909.
25. Roberto FT (1994): Epilepsies, language and behavior: Clinical models in childhood. Journal of Child Neurology 9:95–102.
26. Sauerwein HC, Lassonde M, Geoffroy G and Mercier C (1996): L'intervention chirurgicale chez l'enfant épileptique. Approche Neuropsychologique des Apprentissages chez l'Enfant, Hors série: 37–42.

21

APHASIA AND AUDITORY AGNOSIA IN CHILDREN WITH LANDAU-KLEFFNER SYNDROME

Anne Van Hout

Catholic University of Louvain
Pediatric Neurology Service
Saint Luc University Clinics
10 Hippocrate Avenue, 1200 Brussels
Belgium

1. DELINEATION OF THE SYNDROME

Initially described by Landau and Kleffner[12] as an "acquired aphasia with convulsive disorders", this rare form of childhood aphasia occurs in the age range of three to eight years old. Seizures are infrequent but may precede, coincide with, or follow the language symptoms. Often of a generalized motor type, they may even not occur at all.[19] EEG is abnormal, with uni- or bilateral 1 to 3 Hz spikes and waves predominating in the temporal regions. The more prominent and early manifestation of the language disturbances is an acquired word deafness (loss of comprehension for words). Receptive disorders may become so pronounced as to evoke peripheral deafness and may develop into auditory agnosia (loss of comprehension even for the environmental sounds). Expressive language deteriorates usually later on. These language manifestations fluctuate and are aggravated in the early stages at least, by the occurrence of seizures. Although they are considered, even in the princeps description, as a direct consequence of the EEG disorders, a clear relationship has not been demonstrated between the course of the seizures (which often subside easily with anticonvulsive therapy), the course of the standard EEG, and the course of the language deficit. Neurologic and ENT examinations are usually normal and no causal etiology has ever been clearly demonstrated either by CT scans, MRI or biopsy. Although a rare disorder, more papers have been devoted to its description than to acquired childhood aphasias of other etiologies together. Major clues for the explanation of this puzzling entity have appeared within the last few years from findings on sleep EEG that link the condition to ESES (epileptic status epilepticus induced by sleep),[17] and from the demonstration of a specific cortical activation pattern on PET scan.[13]

Neuropsychology of Childhood Epilepsy, edited by Jambaqué et al.
Kluwer Academic / Plenum Publishers, New York, 2001.

2. SYMPTOMS

As stated by De Wyngaert and Gommers[2] "little attention has been given to the specific nature of the language disorders" in this syndrome, which is very strange, particularly in view of the recent discoveries in acquired childhood aphasia.[23] The variability and fluctuations of symptoms might explain the relative paucity of linguistic data for this entity. Developmental milestones are usually normal in those children, in particular for language, although rare reports have indicated the contrary.[18] In most cases at onset, the children become unresponsive to calls ("word deafness"): they develop a progressive lack of attention to language, with an ineffectiveness of voice raising (owing to the fluctuations of the symptoms, the troubles may be confused with attention deficit disorder as the "startle" response to loud noises is also usually suppressed). Soon after, the child frequently fails to react even to environmental sounds, thus disclosing "auditory agnosia". Incipient peripheral deafness is evoked but audiometry remains in the normal limits. This striking initial receptive deficit has been particularly emphasized by authors who consider these severe comprehension disorders as the basis of the Landau-Kleffner syndrome and suggest that it be called" verbal auditory agnosia in children".[18] In some children, the deterioration of language affects reception only during some episodes, and, in rare instances, expressive impairment precedes the auditory disorders.[20] Expressive symptoms consist of a difficulty in finding the name of well-known objects with auto-corrections resembling stuttering[12] and a progressive loss of vocabulary. Phonemic or semantic verbal substitutions (paraphasias) are observed and even neologistic jargon (use of words of which at least half the phonemes are substituted so that they become meaningless).[18,19] Sentences are short often in a "telegrammatic style" (use of non-inflected verbs) or with a reduction of "mean length utterance" (mean number of words pronounced in a given unit of time). Within a few days or weeks, the child may become mute, losing both his/her receptive and his/her expressive language abilities. Even in cases with initial jargon, in opposition to adult receptive aphasia where fluency usually does increase, expression soon becomes limited with phonological disturbances. However, the brevity of reports on those phonological disturbances does not always allow to decide if the authors refer to phonemic substitutions (the sounds, although incorrectly located, are well articulated) or to a disorder of articulation *per se* with distortion of sounds pronunciation. In some cases, both conditions may co-occur or appear in succession. There can be a "return" to babbling with perseverative utterances. As with acquired deafness in very young children, this limitation in expressive speech in auditory agnosia has been interpreted in function of Geschwind's hypothesis: "the child's Broca area has not adequate practice to run as freely as that of adults with auditory agnosia".[1] An alternative explanation has been proposed by Van de Sandt-Koenderman *et al.*,[21] who observed that, during acute speech breakdowns, precisely when more neologisms were produced, the children were less "inclined" to speak and had a reduced mean length of utterances, together with periods of production of grammatically complex sentences. In these cases, the reduction in speech utterances seems to be part of an overall hypospontaneity of speech, as often occurs in childhood aphasia.[23] However, symptoms that are usually associated with motor aphasias have been also described in Landau-Kleffner syndrome (dropping of words endings, consonants substitutions, dysprosodia)[18] and purely expressive symptoms with mutism or syntactic deficits have occasionally been reported; but only in one case in ten are there no apparent receptive disorders.[22] Children with late-onset Landau-Kleffner syndrome have been described by Gérard *et al.*:[8] as a rule, there is a mixture of paraphasias and syntactic disorders. For these children, aged 9 and older, the receptive deficits are not prominent in a conversational setting and are discovered by means of tests like the Token test (the child is given instructions of increasing length and syntactic complexity to move tokens varying in color, dimension, and form). However, the expressive symptoms seem rather similar to those occurring in younger children and, contrasting with most recent reports on acquired aphasia in children older than eight years,[23] there is hypospontaneity of speech and non-fluency persisting over a long time. Even for young children, however, there are variations in the individual tempo with which these Landau-Kleffner children lose their language skills, and some authors, like Deonna,[3,4] use the mode of onset as one of the criteria for subclassifications of the syndrome.

3. NATURE OF THE RECEPTIVE DISORDER

The origins of this syndrome are probably as varied as its clinical expressions. Owing to the variability and episodic recurrences of the language deterioration, it is difficult to compare individual cases because assessments were probably made at different stages of the disorder. Among the more widely recognized symptoms is a receptive deficit, often interpreted as a deficit in phonologic decoding. Phonemic discrimination deficit has been suspected as the basic disorder, probably because of the temporal localization of the spikes and waves. However, we do not know for sure if the temporal paroxysmal abnormalities are uni- or bilateral. A bilateral focus would corroborate both the frequent occurrence of auditory agnosia at some stages (even in adults, bilateral temporal involvement is mandatory in this condition) and the limitations of language recovery, as it has been admitted that, for posterior lesions at least, a take-over of functions is mediated by the non-lesioned hemisphere.[23] According to the standard EEG records, the spike and wave concentration vary during the evolution of the disorder, being slightly more often unilateral (and with a left lateralization) in the early course of the syndrome, and, later on, being bilateral, with independent spikes and waves more marked in the posterior temporal regions. In the acute stages, the comprehension of oral language is abolished and later on, receptive abilities (as measured for instance with the syntactic part of the Token test) are in the low range. Repetition is usually disrupted, even for vowels in the acute stages and, later on, particularly for sentences, with the production of phonemic and semantic paraphasias, or of perseverations, although occasional episodes of echolalic speech have also been reported.[19] Rapin[18] found that children with Landau-Kleffner-like conditions become able to discriminate environmental sounds sooner than oral language, so that their difficulty is less that of a cortical deafness than of a specific discrimination for speech sounds (verbal auditory agnosia). As for the nature of this difficulty, she describes[18] a child for whom phonemic discrimination difficulties preceded clearly his problems for sentences comprehension. This temporal succession can evoke a causal mechanism. Korkman[11] compared auditory perception for sentences, words, and phonemes and for non-verbal sounds in six children with Landau-Kleffner syndrome and found a clear superiority for the perception of non-verbal sounds while the children remained unable to identify any verbal material. She argues that the disorder should refer to a specific impairment of phoneme discrimination rather than to auditory agnosia in general. In a longitudinal study of a four-year-old child, Paeteau *et al.*[15] used magnetoencephalography to analyze the generator-brain areas for auditory evoked responses for syllables and for tones. A narrow cortical zone in the superior bank of the left auditory cortex was shown to concentrate nearly all the spikes over the temporal region. This is very close to the area that responds to auditory stimuli in normal children, which suggests a disruption by the spikes of the left-hemisphere responses to verbal stimuli while the right hemispheric response was lacking. They hypothesize that the callosal interhemispheric transfer of verbal information from the left side is impeded by the epileptic spikes, which causes bilateral dysfunction. Confirming those findings, left temporal hypoperfusion has been described on SPECT[15] in the latero-superior part of the lobe near the sylvian fissure in four patients, independently of the variability of their symptoms and of the different localizations of their EEG spikes and waves paroxysms. The recent researches on ESES have suggested that unilateral spikes are observed more frequently during wakefulness and that bilateral temporal discharges occur as a rule during most of the slow-waves stage of sleep. But one has to distinguish between a unilateral focus with its mirror counterpart (by kindling) and two independent foci. Morrell[14] notes that there is a subgroup of Landau-Kleffner children whose EEG focus can be demonstrated to be unilateral, and who differs, for instance, by their sensitivity to subpial disconnection (see lower) from the cases where both foci are independent. In a surgical setting, injecting a drug like "methohexital", which blocks the synaptic excitatory transmission, discloses the difference of reactivity of the temporal foci. The independent homologous focus is very resistant to this injection. In the children for whom the EEG sleep temporal continuous paroxysms were primarily unilateral, this procedure causes the "mirror focus" disappearance. In their PET study, Maquet *et al.*[13] compared the results of dichotic listening to the localization of the abnormal temporal lobe metabolism (dichotic listening consists of a presentation of different verbal material synchronously to each ear: as the temporal bundles carrying verbal information are crossed, there is a suppression of the ipsilateral pathway for synchronous presentations, and in unilateral lesions, this effect is suppressed). As all repetition abilities may be abolished during the acute stages, dichotic procedure can usually only be used during the recovery periods. Extinction (lack of report of the verbal material opposite to the lesion) was observed during the recovery stages for the five cases of this ESES series that fit the Landau-Kleffner profile better. In each case, the extinction was opposite to the temporal lobe disclosing the most metabolic disturbances. This finding favors a unilateral hemisphere dysfunction and supports the suggestion of Morrell on the existence of a subgroup whose temporal focus must be primarily unilateral.

4. ASSOCIATED SYMPTOMS

Non-linguistic cognitive functions and emotional behavior usually remain normal in the "Landau-Kleffner" syndrome, but there are exceptions, with some children having problems with gnosias and praxias (manual dyspraxia or bucco-lingual apraxia, sometimes with associated drooling). Hyperkinesia with attention deficit is not infrequent. Emotional symptoms (depression, aggressiveness, even psychotic conditions) have been interpreted as secondary to the language difficulties, but in some children they preceded the onset of aphasia. Many children who develop the syndrome have not reached school age so they have not yet learned to read, write or calculate. In many, during the recovery periods, secondary learning of written language is possible so it can be used as a means of communication. There are some descriptions of oral reading level superior to comprehension skills, evocative of "hyperlexia". For those who have already learned those skills, deterioration may occur with paralexias and comprehension problems for long or syntactically complex sentences; calculation disorders, although rarely described, may also occur.[6]

5. EVOLUTION

The aphasic symptoms fluctuate and there are no clear relationships of the deterioration or improvement with those of seizures and standard EEG. Some cases have early recovery, within weeks or months, and, providing there is no recurrence, these children may show spontaneous remission.[6] Dugas *et al.*[6] reported a positive relationship between scholastic and language outcome and the severity of the initial symptoms. According to Deonna *et al.*,[4] there are at least three different clinical forms of this syndrome, and recovery occurs in no more than 30% of the patients in long-term follow-up. A striking variability is observed in the severity, outcome, and type of language deficit. An inverse correlation between recovery and age is now acknowledged, and, in opposition with current theories on aphasia recovery in children, the prognosis is better in younger children: the earlier the age of onset, the worse the prospects of recovery.[1] Besides this relationship, no clinical prognostic features appear to emerge, and it has been recommended that substitutive language be used as soon as possible because of this unpredictability.[4] However, the recent discovery of ESES[16] has shown their importance for language recovery: according to Dulac,[7] ESES disappearance must always precede any stable language improvement.

6. ETIOLOGY

Similarities between the Landau-Kleffner electroclinical pattern and the one which occurs during "benign partial epilepsy in childhood" (a benign form of childhood epilepsy, with bucco-facial symptoms and speech arrests) have been underlined. In that condition, seizures are benign but the electroclinical pattern is enhanced during sleep and the children do frequently present with associated language or learning problems. Whatever the origin of the disorder in the Landau-Kleffner syndrome, it appears able to affect differentially but selectively the different subsystems mediating language. This "modularity" of the disorder has always led to intense scientific curiosity. This might explain why there have always been more publications on this rare condition than on the other aphasic conditions in children. This modularity bears even on the subcomponents of language if one accepts the suggestion that expressive disorders may be secondary to the major receptive deficit. Personal and familial history is usually irrelevant. Hypothesis for an infectious (slow viral chronic encephalitis in particular), parasitic, immune or even auto-immune origin, have not been clearly demonstrated and brain biopsies have usually remained normal. Suspected lesions, vascular or tumoral, have proven to be artifactual or purely coincidental. Landau and Kleffner[12] suggested that the spikes observed on standard EEG were responsible for a "lesional" effect on the language areas where they predominated, giving rise to their "functional ablation". In 1991, Landau himself remarked that, since his initial suggestion, no progress had been made in the understanding of the cause of "his" eponymic syndrome! On the contrary, for Holmes,[9] the language disorders and EEG paroxysms might be independent phenomena, reflecting an underlying common lesional disorder or dysfunction in the speech areas. The discovery of ESES or "electrical status epilepticus induced by sleep" by Patry *et al.*[17] shed new light on the understanding of the determinism of the syndrome. The six patients described presented a language deficit rather similar to the Landau-Kleffner syndrome and repeated sleep EEG disclosed an increase of paroxysms who took on a "petit mal" status pattern. The criteria for ESES were defined by Patry as continuous paroxysms occurring in at least 85% of slow sleep records; they also had to be recorded on three occasions at least during a one-month period. This discovery adds arguments to the original hypothesis of Landau and Kleffner: the electrical "bombing" of the speech areas was even more long lasting in sleep than when awakening. However, one may wonder why, in the Landau-Kleffner syndrome, the deficit remains limited to the areas of language although the EEG paroxysms were diffuse. As a matter of fact, the children described by Patry had more extensive cognitive problems than those described in the Landau-Kleffner syndrome, and their intellectual quotients were in the defective range. Cases of ESES were later reported in which the language problems were not prominent in relation to other cognitive deficits and in which, when occurring, they differed from the classic features of the Landau-Kleffner syndrome. Tassinari[20] also emphasized a different age range for children with ESES and with Landau-Kleffner syndrome: between 4–5 and 14 years old in the former and less than 8 years old in the latter. However, the pattern of ESES could be found in children with the Landau-Kleffner syndrome: Dulac,[7] for instance, found ESES during slow sleep in 7 out of his personal series of 10 children and emphasized the noxious influence it could have on speech areas. However, there is a major difference in the sleep distribution of the abnormal sleep spikes and waves in the two conditions: in the Landau-Kleffner syndrome, they also affect the REM phase (the dreamy, rapid-eyes-movements stage of sleep) and, in pure ESES, they are strictly limited to the slow stages of sleep. It was later demonstrated that language in Landau-Kleffner syndrome could only recover after the disappearance of ESES (sometimes many years after onset). The findings by Maquet *et al.*[13] of a specific PET (positron emission tomography) profile in ESES affords a tentative explanation of those syndromes. They showed cortical hypermetabolism during sleep with ESES, when paroxysms were at their maximum. Some regions were more affected according to the case and for the children fitting better the diagnosis of Landau-Kleffner, the abnormalities were more marked in the temporal regions. The topography of the maximal metabolic disturbances was congruent with the neuropsychological examination: more marked temporal localization in cases with severe comprehension deficit and more marked parietal extension in cases with apraxia. This hypermetabolism could still be detected during wakefulness, although the paroxysms were less frequent. Moreover, the overall distribution of glucose consummation was "immature", disclosing, as for younger children, a higher metabolic rate in the cortical mantle than in the subcortical areas. Whether this abnormal functioning is a causal mechanism per se or a direct consequence of the tangential transmission of the paroxystic activity remains to be determined. The more plausible hypothesis seems to be that the abnormal neuronal firing interferes with maturational phenomena such as selective synaptic stabilization, particularly in those "associative" areas connected with high level functioning and for the Landau-Kleffner syndrome, the receptive areas for language.

7. TREATMENT

A wide range of anticonvulsant drugs acts very well for the seizures but not for the language problems. Corticosteroids may suppress the epileptic activity and are often used with success in the early stages of the disease, but there is usual recurrence with limited reproducibility of the effect when they are withdrawn. Speech therapy is widely used[2] and was already described in the princeps description.[12] Visual input and written language are usually used to communicate; these children easily learn the written names of objects, and this can serve as intermediate support for relearning auditory correspondence. Although sign language may help severely auditory-deprived children who sometimes develop their own alternative communication system in the early stages of the disorder, it has no effect, of course, on auditory comprehension. Classic aphasic speech therapy can only be used with children having residual oral language, and must start from a profile that emphasizes the strengths and deficits of the defective expressive and receptive speech. For children with severe communication limitations, neurosurgery has been tried, at first with a radical left temporal lobectomy, which proved unable to produce any long-lasting language improvement. A new procedure developed by Morrell[14] appears to be much more successful, providing it is used with children who have not shown recovery in the two years following the onset and whose temporal focus can be proven to be unilateral. It consists in subpial intracortical resections: tiny vertical sections are made at the cortical surface. The theoretical basis of the procedure relies on the vertical disposition of the normal brain subunits and, in opposition, on the tangential propagation of the hypersynchronous epileptic discharges: the subpial section of the neural networks with tangential contacts impedes the transmission of the deleterious currents while leaving the physiological cerebral transmission intact. The results have proven highly effective with 79% of children regaining age-appropriate language level. The success of the method is probably due to the secondary limitation of the propagation of the tangential epileptic currents at the origin of disruption of normal synaptic maturation.

REFERENCES

1. Bishop D (1985): Age of onset and outcome in acquired aphasia with convulsive disorders. Developmental Medicine and Child Neurology 27:705–712.
2. De Wyngaert E and Gommers K (1993): Language rehabilitation in the Landau-Kleffner syndrome: considerations and approaches. Aphasiology 7:475–480.
3. Deonna T, Beamanoir A, Gaillard F and Assal G (1977): Acquired aphasia in childhood with seizure disorders: a heterogeneous syndrome. Neuropediatrics 8:263–273.
4. Deonna T, Peter C and Ziegler A (1989): Adult follow-up of the acquired aphasia-epilepsia syndrome in childhood: report of 7 cases. Neuropediatrics 20:132–138.
5. Dugas M, Grenet P, Masson M, Jaquet J-P and Jaquet G (1976): Aphasie de l'enfant avec épilepsie. Évolution régressive sous traitement anti-épileptique. Revue Neurologique 132:189–193.
6. Dugas M, Masson M, Henzey M-F and Regnier N (1982): Aphasie acquise de l'enfant avec épilepsie: douze observations personnelles. Revue Neurologique 138:755–780.
7. Dulac O, Billard C and Arthuis M (1983): Aspects électro-cliniques et évolutifs dans le syndrome aphasie-épilepsie. Archives Françaises Pédiatriques 27:299–308.
8. Gerard C-L, Dugas M, Valdois S, Franc S and Lecendreux M (1993): Landau-Kleffner syndrome diagnosed after 9 years of age: another Landau-Kleffner syndrome? Aphasiology 7:463–474.
9. Holmes G and McKeever M Saunders (1981): Epileptiform activity in aphasia in childhood: an epiphenomenon? Epilepsia 22:63.
10. Kellerman K (1978): Recurrent aphasia with subclinical bioelectrical status epilepticus during sleep. European Journal of Pediatrics 128:207–212.
11. Korkman M (1996): Paper presented at the European INS meeting.
12. Landau W and Kleffner •• (1957): Syndrome of acquired aphasia with convulsive disorders in children. Neurology 7:523–530.

13. Maquet P, Hirsh E, Metz-Lutz M-N, Motte J, Dive D, Marscaux C and Franck G (1995): Regional cerebral glucose metabolism in children with deterioration of one or more cognitive functions and continuous spikes and waves discharges during sleep. Brain 118:1497–1520.
14. Morrell F, Whisler W, Smith M *et al.* (1995): Landau-Kleffner syndrome: treatment with subpial intracortical transection. Brain 118:1529–1546.
15. Otuama L, Urion D, Janicek M, Treves S, Bjornson B and Moriarty J (1992): Regional cerebral perfusion in Landau-Kleffner syndrome and related childhood aphasia. Journal of Nuclear Medicine 33:1758–1765.
16. Paeteau R, Kajola M, Korkman M, Hämälïnen M, Granstrom M and Hari R (1991): Landau-Kleffner syndrome: epileptic activity in the auditory cortex. Neuroreport 2:201–204.
17. Patry G, Lyagoubi S and Tassinari C-A (1971): Subclinical "electrical status epilepticus" induced by sleep in children. A clinical and electroencephalographic study of six cases. Archives of Neurology 24:241–252.
18. Rapin I, Mattis S, Rowan A-J and Golden G-S (1977): Verbal auditory agnosia in children. Developmental Medicine and Child Neurology 19:192–207.
19. Rodriguez I and Niedermeyer E (1982): The aphasia-epilepsy syndrome in children: electroencephalographic aspects. Clininal Electroencephalography 23:365.
20. Tassinari C, Terzano G, Capocchi G, Dalla Bernardina B, Valadier C, Vigevano F *et al.* (1985): Epileptic seizures during sleep in children. In Penry JK (ed): "Epileptic Syndromes in Infancy, Childhood and Adolescence". London: John Libbey.
21. Van De Sandt-Koenderman, Smit I, Van Dongen H and Van Hesch J (1984): A case of acquired aphasia and convulsive disorders: some linguistics aspects of recovery and breakdown. Brain and Language 21:174–183.
22. Van Dongen H, Meulstee J, Blauw K and Van Harskamp F (1989): The Landau-Kleffner syndrome, a case study with a fourteen year follow-up. European Neurology 29:109–114.
23. Van Hout A (1992): Acquired aphasias in children. In Boller F, Grafman J (eds): "Handbook of Neuropsychology", Vol 7. Amsterdam: Elsevier, pp 139–161.

22

COGNITIVE PROFILES OF CSWS SYNDROME

Eliane Roulet Perez

CHUV, Neuropediatric Unit
Rue du Bugnon 46, 1011 Lausanne
Switzerland

1. FROM AN EEG PATTERN TO AN EPILEPTIC SYNDROME

1.1. The EEG Pattern of CSWS

Continuous spike-waves during sleep (CSWS), also described under the term of "electrical status epilepticus during sleep" (ESES),[5,14] refers to a striking electroencephalographical (EEG) abnormality consisting of bilateral discharges of spike-waves (1.5–5 Hz) that almost completely replace the physiological non-REM slow-wave sleep activity (85% of the tracing is a recognized but still debated criterion). This EEG pattern can be associated with different seizure types (i.e., atypical absences, atonic episodes, nocturnal partial and generalized seizures) of variable severity. Although the underlying mechanism is still uncertain, CSWS are now considered to be a result of "secondary bilateral synchrony," i.e., rapid secondary generalization from one or several cortical epileptic foci. Such foci can be found in the waking state or during periods of REM sleep when the amount of discharges tends to decrease, and they are often located in the frontal or central regions of the cerebral cortex.

1.2. The Syndrome of CSWS

As a syndrome, epilepsy with CSWS is defined as a combination of this particular EEG pattern with epilepsy and acquired neuropsychological disorders. The age of diagnosis lies usually between 2 and 10 years, but the precise onset of CSWS is often difficult to determine since a sleep EEG is not always recorded after a first seizure. CSWS can persist for months and years, but usually remit progressively during puberty. The etiology of the epilepsy is either a static focal (sometimes multifocal) brain lesion (sequelae of meningitis, prenatal-perinatal ischemic damage, cortical dysplasia) or unknown (cryptogenic/idiopathic).

Neuropsychology of Childhood Epilepsy, edited by Jambaqué et al.
Kluwer Academic / Plenum Publishers, New York, 2001.

A still unsolved question is whether the syndrome of CSWS and acquired epileptic aphasia (AEA; or Landau Kleffner syndrome), which also consists of an acquired neuropsychological disorder accompanied by an abnormal sleep EEG with nearly continuous generalized or bitemporal epileptic discharges, are really two different syndromes or should be merged into one global entity. Discussion of this problem is beyond the scope of this chapter.[11] However, it should be kept in mind that these syndromes are classified on the basis of EEG findings in epilepsy with CSWS and of a clinical symptom in AEA, which is quite confusing. This separation would be valid only if CSWS were never found in AEA or aphasia would never be a manifestation of epilepsy with CSWS, which is definitely not the case. Since AEA is discussed in detail in the previous chapter, we shall focus here on data concerning children with CSWS who do not exhibit isolated or predominant language deterioration.

2. NEUROPSYCHOLOGICAL FINDINGS

As stated before, acquired neuropsychological impairments are a crucial finding. In fact, they are the hallmark of the CSWS syndrome. It should be pointed out, however, that there is no unique cognitive and behavioral profile associated with CSWS. Rather, there are different possible clinical pictures which will be discussed below. In most cases, deterioration is reported, but the spectrum of the severity of the disorder is quite variable. Dementia, aphasia, apraxia and psychotic behavior have all been reported, although they are rarely described in detail. Before onset of CSWS, there may be a history of normal or mildly delayed development.

2.1. Acquired Frontal Syndrome with CSWS

In a longitudinal study of four boys with CSWS,[10] we attempted to better understand the nature and evolution of the behavioral and cognitive disorder in correlation with the epilepsy. In these children, a severe neuropsychological regression appeared between 3.5 and 8 years, which was associated with a frontal focus (on the right side in three, on the left in one). Previous development was unremarkable and no brain lesion was found. Clinical seizures were of mild to moderate severity. Deterioration occurred between one and two years after the first seizures and was quite rapid in two cases (after a few weeks) but insidious in the two others (after more than 6 months).

2.1.1. Behavioral Changes. Behavioral changes were the most disturbing symptoms that first alarmed the parents. Initially, attention deficit (distractibility and inability to focus attention), hyperactivity and impulsiveness were noticed. Aggressiveness, mood swings and disinhibition appeared later. Normal play was replaced by repetitive activities and mouthing of objects and clothing. The sense of danger was lost. In our cases, we did not observe complex ritualistic behaviors, fascinations or aberrant perceptions. Behavioral disorders were often so severe in the worst phase of the disease that a formal neuropsychological examination was impossible and assessment had to be limited to observation and administration of a few subtests of various batteries (McCarthy or Wechsler scales).

2.1.2. Subtle Cognitive Changes. In one of our patients, unexpected learning difficulties in mathematics and French with well-preserved graphomotor skills were noticed by the teacher at the age of 7 before massive behavioral deterioration occurred and CSWS were diagnosed. In another child, who was tested at age 3, an insidious cognitive stagnation preceded the more marked behavioral and intellectual changes three months before the onset of CSWS. These subtle signs of cognitive impairment have to be kept in mind, especially when a neuropsychologist has to evaluate a child in whom CSWS are found after a first seizure and who still appears normal. Such changes may be erroneously attributed to adverse effects of antiepileptic medication or psychological consequences of the epilepsy. However, in this particular context, they can be the first signs of a more dramatic deterioration and thus an indication for more aggressive medical treatment.

2.1.3. Cognitive Profile. The most striking features of our children's cognitive dysfunction were severe impairments of verbal and non-verbal reasoning and temporal disorientation (Fig. 1). Language was normal in structure but not in content. Spontaneous

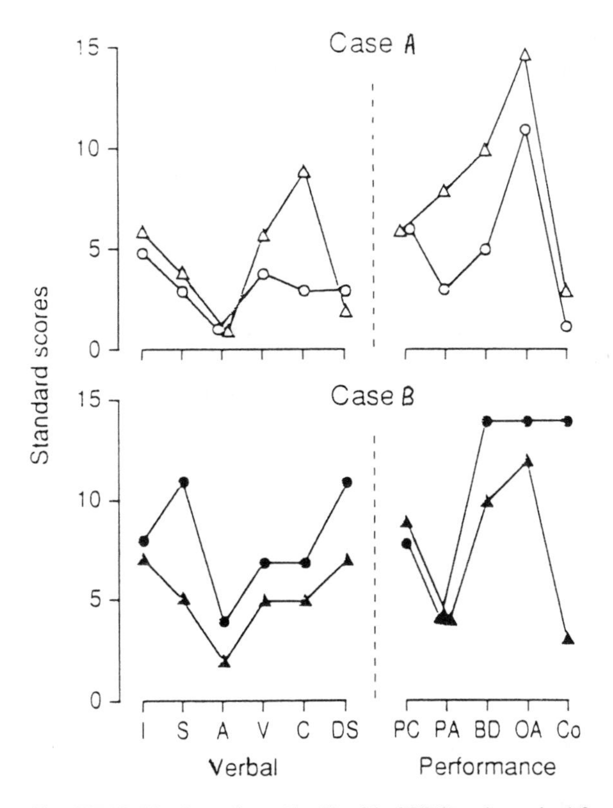

Figure 1. Cognitive profiles (WISC-R) of two boys (A, B) with CSWS and acquired frontal syndrome. Both children had very similar heterogeneous scores with poorer results in verbal subtests (lower curves) and both showed improvements with treatment (upper curves). Note the preserved (high) scores in OA and BD, which were also found in the other two children of this study. Case A was tested at 8 years 4 months (lower curve; FS IQ:58, VIQ:58, PIQ:66) and at 9 years 5 months (upper curve; FS IQ:76, VIQ:70, PIQ:88). Case B was tested at 7 years 8 months (lower curve; FS IQ:76, VIQ:67, PIQ:83) and at 9 years 7 months (upper curve; FS IQ:94, VIQ:86, PIQ:105). I: Information; S: Similarities; A: Arithmetic; V: Vocabulary; C: Comprehension; DS: Digit Span; PC: Picture Completion; PA: Picture Arrangement; BD: Block Design; OA: Object Assembly; CO: Coding (normal values: 10 ± 3).

discourse was characterized by echolalia, repetitive questioning, irrelevant comments, inappropriate associations of ideas and inability to maintain a topic. Prosody and facial expression were normal. Naming was preserved but word fluency (semantic categories) was aberrant and perseverative. Abstract verbal concepts such as analogies (Similarities subtest of the WISC-R) and causality (answers to why questions, Comprehension subtest of the WISC-R) were not understood. One child who had learned to read before he started to deteriorate maintained this skill, but his comprehension was poor. The children were able to recite numbers but were unable to use them to perform easy operations or solve simple mathematical problems. Piagetian tasks like number conservation (i.e., recognition that a quantity of items remains unchanged whatever their spatial arrangement) was not acquired in the three older children. Similarly, classification tasks (sorting of tokens according to size, color and shape, subtest of the McCarthy Scales) were failed. Time concept was blurred and temporal sequencing (narration of a known story, Picture Arrangement subtest of the WISC-R) was severely deficient. By contrast, visual recognition and visuoconstructive skills were preserved. Indeed, solving puzzles (i.e., subtest of the McCarthy, Object Assembly subtest of the WISC-R) constituted the best performance of these children (Fig. 1): the children proceeded quickly using a trial and error approach without elaborated strategies. Rote memory (songs, poems, commercials) was intact but the patients were unable to learn new material in an organized way. Perseveration was observed on many tasks (i.e., Construction subtest of the McCarthy, Coding subtest of the WISC-R, copying of letters).

In our cases, behavior improved markedly between three weeks and six months after decrease of the epileptic activity (CSWS) following treatment. Cognitive functions also improved, albeit more slowly (within 2 to 15 months after effective treatment) and remained mildly to severely impaired during follow-up.

On the basis of this neuropsychological picture and its evolution, we concluded that our children had an acquired frontal syndrome.[3] Since problem-solving abilities, abstract thinking and time concept were most disturbed, we can postulate a dysfunction of the dorsolateral type. Orbito-frontal structures were probably also involved, at least during the worst periods which were characterized by mood swings, disinhibition and perseveration.

By analogy to AEA where a bitemporal epileptic dysfunction seems the most likely cause of language impairment, we considered that frontal lobe dysfunction was due to sustained focal epileptic discharges with local propagation and secondary bilateral synchrony creating the CSWS pattern.

New evidence in favor of this hypothesis was provided by the case study of a 25-year-old man with a documented history of severe expressive language, cognitive and behavioral regression during early childhood who underwent epilepsy surgery.[12] Complex partial seizures appeared at the age of 3.5 years, and his epilepsy had clinical and EEG features of CSWS and AEA which we initially considered to be of idiopathic origin. The neuropsychological impairments improved markedly but not completely during later childhood. He had mild sequelae of an expressive aphasia, borderline intelligence and schyzotypical personality traits. Seizures relapsed unexpectedly at the age of 19 years and became refractory. Presurgical exploration revealed a lateral frontal epileptic focus with intermittent spread to the patient's Broca area (mapped with subdural electrodes) and orbito-frontal regions. Histopathology revealed a focal cortical dysplasia of mild degree which was not visible on high resolution MRI. This man's prolonged but reversible early regression and residual neuropsychological disorder were certainly the result of an active left latero-frontal epileptic focus which interfered with developing language and behavioral functions.

2.2. The Spectrum of Neuropsychological Deficits

2.2.1. Location of Epileptic Focus. If one surmises that the site of the epileptic focus determines the nature of the neuropsychological deficits and that CSWS are only markers of a very active focal process, one should expect variable clinical profiles. This is indeed the case and illustrated in Table 1, which summarizes some clinical pictures of children with CSWS reported in the literature. It may be noted that AEA and acquired frontal syndrome are most frequently encountered,[13] but other clusters of deficits have been found in single cases, pointing to different dysfunctional areas. This table is not exhaustive and other observations certainly exist. Sometimes multiple cortical areas are involved, resulting in a complex picture combining various symptoms. This is best illustrated by a child reported by Maquet *et al.*[6] who became mute, apraxic and agnosic at the age of 8 years. These symptoms were accompanied by episodes of apathy alternating with hyperactive and compulsive behavior. The EEG showed CSWS with right-sided predominance. The MRI was normal. A positron emission tomography (PET) scan performed during sleep when the patient had CSWS but no clinical seizure, revealed extended areas of glucose hypermetabolism on the right hemisphere, involving frontal, temporal, perisylvian and parietal cortex. A similar result was obtained four months later in the same condition, which indicates the persistence of the epileptic process, the physiopathology of

Table 1. CSWS: spectrum of neuropsychological deficits in relation to site of main epileptic focus

Main acquired deficit	Localization of main focus (EEG)	Comments
Deficit in verbal and non verbal reasoning, time concept, severe behavioral disorder	Frontal (left and/or right)	Uni-bilateral frontal (see text)
Aphasia (different types)	Perisylvian (bilateral)	In VAA (most frequent), focus in superior temporal gyrus = auditory cortex[7]
Drooling, speech arrests, swallowing difficulties	Rolandic	1 case, epileptic "opercular syndrome"[2]
Mild dysphasic signs, dysfluency, slow learning	Left parietal (main) Left frontal (occasional)	1 case, arachnoid cyst in the left sylvia fissure. More subtle dysfunction[4]
Dyslexia, dycalculia, global dyspraxia	Left temporo-occipital	1 case, acquired "Gerstmann syndrome"[1]
Navigational and visuospatial disorder, left hemineglect (1 of 2)	Right hemisphere	2 cases, right hemispheric dysfunction, language considered normal[15]
Visual Agnosia	Occipital bilateral, parietal right, temporal left	1 case, visuoverbal disconnection, visual perception, language and praxis normal[8]

VAA = verbal auditory agnosia.

which is still hypothetical. On the other side of the clinical spectrum, children with CSWS and foci in the central (sensori-motor) regions, like those found in some very active cases of benign partial epilepsy with rolandic spikes, do usually not present major neuropsychological and behavioral disorders.[13] More subtle changes are possible but these cases are unfortunately rarely studied prospectively.

2.2.2. Age at Onset and Duration. In addition to location of the focus, other variables, such as age at onset and duration of the active epileptic process, are important in determining the nature and severity of the concomitant neuropsychological disorder. It would be unreasonable to expect that late onset and short duration of epilepsy with CSWS would have the same consequences than early onset and prolonged exposure. Indeed, all possible variations of impairments can be found:[4] fluctuations within a normal range, stagnation of learning (raw scores of tests do not increase according to age), regression (drop in raw scores), aberrant development or non-emergence of a skill. These variable evolutions can concern different skills in one patient, depending on the maturational stage of the functions involved at disease onset. For example, a 7-year-old child may show mild dysphasic difficulties (slight regression), non-emergence of age-expected reading skills and a drop in attentional capacities.[4]

2.3. CSWS and Developmental Delay

2.3.1. Difficulties of Diagnosis. When CSWS occurs in the context of a developmental delay with or without an identified brain lesion, recognition of the underlying problem is rendered more difficult for various reasons:

1) Diagnosis of CSWS may be missed: In retarded children with behavioral problems, a sleep EEG is often difficult to obtain and not requested routinely when seizures appear.
2) Developmental delay may mask deterioration: A developmental disorder is often not precisely evaluated during the first two years and once its severity is recognized it becomes difficult to discern which of the symptoms are attributable to the underlying encephalopathy and which are the result of the CSWS. This is even more difficult when the delay has no identified etiology. Important cues are the onset of strange behaviors (disinhibition, withdrawal, perseveration) or learning arrest. These symptoms are unexpected in children with static brain lesions. Serial examinations are often necessary to document stagnation and to clarify the situation.
3) Misinterpretation of behavioral changes is possible: Aggravation of behavioral problems in a delayed child is often attributed to environmental factors or antiepileptic drugs.

2.3.2. Role in Autistic Spectrum Disorders. Since the specific role of epilepsy in neuropsychological disorders has increasingly been recognized and because some behavioral abnormalities found in CSWS resemble features seen in autistic spectrum disorders, the question was raised whether CSWS could be one of the etiologies of autistic regression.[9] Although unrecognized epilepsies can occasionally be at the origin of an autistic regression, the syndrome of CSWS has to our knowledge not been clearly documented in this specific situation. Such cases may exist but they must be differentiated from more global cognitive deterioration. It is also possible that the infantile epilepsies responsible for autistic regression are of a different type, affecting other neuronal networks than those involved in epilepsy with CSWS.

3. TREATMENT

If one accepts that there is a close relationship between the epilepsy and the neuropsychological disorders, medical treatment must be introduced with the purpose of decreasing or arresting the intense epileptic activity seen in the EEG and not only the clinical seizures. Often several drugs have to be tried, and in practice it is sometimes difficult to justify aggressive treatments, for example steroids, on the basis of an EEG or subtle behavioral or cognitive dysfunction.

Many data, however, indicate that early recognition and treatment can be effective and may prevent a dismal outcome. In this context, serial neuropsychological evaluations showing stagnation/regression or improvement in correlation with treatment and EEG findings will provide the most important information to guide the clinician.

In the future, our understanding of CSWS and AEA will probably change, but the puzzling clinical features found in these syndromes have generated a more global concept of "prolonged cognitive impairment of epileptic origin", which will certainly remain and be extended to other epilepsies.

REFERENCES

1. Badinand H, Bastuji H, De Bellecise *et al.* (1995): Case reports. In Beaumanoir A, Bureau M, Deonna T, Mira L, Tassinari CA (eds): "Continuous Spike an Waves during Slow Sleep, Electrical Status Epilepticus during Slow Sleep". Mariani Foundation Neurology Series: 3, London: John Libbey, pp 123–133.
2. Colamaria V, Sgrò V, Carabello R *et al.* (1991): Status epilepticus in benign rolandic epilepsy manifesting as anterior operculum syndrome. Epilepsia 32:329–334.
3. Damasio AR and Anderson SW (1993): The frontal lobes. In Heilman KM, Valenstein E (eds): "Clinical Neuropsychology". New York: Oxford University Press, pp 409–460.
4. Deonna T, Davidoff V, Maeder-Ingvar M *et al.* (1997): The spectrum of acquired cognitive disturbances in children with partial epilepsy and continuous spike waves during sleep. European Journal of Pediatric Neurology 1:19–29.
5. Kobayashi K, Nishibayashi N, Ohtsuka Y *et al.* (1994): Epilepsy with electrical status epilepticus during slow sleep and secondary bilateral synchrony. Epilepsia 35:1097–1103.
6. Maquet P, Hirsch E, Metz-Lutz MN *et al.* (1995): Regional cerebral glucose metabolisme in children with deterioration of one or more cognitive functions and continuous spike-and-wave discharges during sleep. Brain 118:1497–1520.
7. Morrell F, Whisler WW, Smith MC *et al.* (1995): Landau-Kleffner syndrome: treatment with subpial intracortical transection. Brain 118:1529–1546.
8. Pavao Martins I, Lobo Antunes N and Levy Gomes A (1993): Acquired visual agnosia in a child: a neuropsychological study. Approche Neuropsychologique des Apprentissages chez l'Enfant 5:70–75.
9. Rapin I (1995): Autistic regression and disintegrative disorder: How important is the role of epilepsy. Seminars in Pediatric Neurolology 2:278–285.
10. Roulet Perez E, Davidoff V, Despland PA and Deonna T (1993): Mental and behavioural deterioration of children with epilepsy and CSWS: acquired epileptic frontal syndrome. Developmental Medicine and Child Neurology 35:661–674.
11. Roulet Perez E (1995): Syndromes of acquired epileptic aphasia and epilepsy with continuous spike waves during sleep: models for prolonged cognitive impairment of epileptic origin. Seminars in Pediatric Neurolology 2:267–277.

12. Roulet Perez E, Seeck M. Mayer, E, Despland PA, De Tribolet N and Deonna T (1998): Childhood epilepsy with neuropsychological regression and continuous spike-waves during sleep: epilepsy surgery in a young adult. European Journal of Paediatric Neurolology 2:303–311.
13. Rousselle C and Revol M (1995): Relations between cognitive functions and continuous spike and waves during slow sleep. In Beaumanoir A, Bureau M, Deonna T, Mira L, Tassinari CA (eds): "Continuous Spike an Waves during Slow Sleep, Electrical Status Epilepticus during Slow Sleep". Mariani Foundation Neurology Series: 3, London: John Libbey, pp 123–133.
14. Tassinari CA, Bureau M, Dravet C, dalla Bernardina B and Roger J (1992): Epilepsy with continuous spike and waves during slow sleep, otherwise described as ESES (epilepsy with electrical status epilepticus during slow sleep. In Roger J, Bureau M, Dravet C, Dreifuss FE, Perret A, Wolf P (eds): "Epileptic Syndromes in Infancy, Childhood and Adolescence". London: John Libbey, pp 245–256.
15. Zaiwalla Z and Stores G (1995): Case reports. In Beaumanoir A, Bureau M, Deonna T, Mira C, Tassinari CA (eds): "Continuous Spike an Waves during Slow Sleep, Electrical Status Epilepticus during Slow Sleep". Mariani Foundation Neurology Series: 3, London: John Libbey, pp 196–199.

23

PRESURGICAL NEUROPSYCHOLOGICAL ASSESSMENT

Mary Lou Smith

Department of Psychology
University of Toronto at Mississauga
3359 Mississauga Road North
Mississauga, Ontario L5L 1C6
Canada and Department of Psychology
The Hospital for Sick Children
555 University Avenue, Toronto
Ontario M5G 1X8, Canada

INTRODUCTION

Epilepsy is most commonly treated with antiepileptic medications, but individuals with intractable seizures with a focal onset may be referred for epilepsy surgery. Although the first documented cases of surgery for focal epilepsy were done over 100 years ago, this method of treatment has only seriously been considered for use with children since 1975, when Davidson and Falconer demonstrated that surgery can also result in good seizure control in the pediatric age group. Since that time, children with intractable epilepsy have been referred for epilepsy surgery in growing numbers. There is increasing recognition that epilepsy surgery can be of significant benefit to children with medically intractable seizures by eliminating or significantly decreasing seizures in 50% to 90% of selected cases.[4,5]

Many children and adolescents who are candidates for epilepsy surgery have underlying neural pathology, suffer from frequent seizures that are unpredictable and intractable in nature, and endure potential adverse side-effects from multiple medications. Seizure onset commonly occurs at a time that is essential to the development of basic cognitive, behavioral, and social skills that are crucial for long-term educational, vocational, and interpersonal adaptation.[8,16] Thus, children with epilepsy are at risk for problems in multiple areas of psychosocial function, including behavioral adjustment, social competence, academic achievement, family life, and neuropsychological status.

The purpose of a neuropsychological assessment is to evaluate the cognitive, behavioral, and emotional effects of brain dysfunction; in the case of epilepsy, one might also

Neuropsychology of Childhood Epilepsy, edited by Jambaqué et al.
Kluwer Academic / Plenum Publishers, New York, 2001.

wish to document the impact of the seizures themselves and possible effects of medications on the child's functioning. A typical neuropsychological assessment consists of several components. It starts with a detailed history-taking that covers the child's developmental, medical, and educational history. This information may be drawn from several sources, such as interviews and questionnaires completed by parents, teachers, siblings, the child him- or herself; school records; and medical records. The assessment also involves the administration of tests (to be discussed in further detail later in this chapter), and the observation of the behavior of child, during the test-taking phase and perhaps in other situations.

The neuropsychological assessment is now considered an essential part of the presurgical evaluation of patients considered for epilepsy surgery. Recognition of the contribution of the neuropsychological assessment to the evaluation of the patient was first established with adult candidates for surgery, but the philosophy has largely been adopted in pediatric programs as well.[10,11,14,15] This chapter will cover issues pertaining to the goals of the presurgical assessment, special considerations for children, the domains that are typically covered in the assessment, the intracarotid amobarbital procedure in children and functional mapping.

1. GOALS OF THE PRESURGICAL NEUROPSYCHOLOGICAL ASSESSMENT

1.1. Providing Information About the Epileptogenic Focus

Early in the history of neuropsychological input to epilepsy surgery programs, an important purpose of the assessment was to provide information about the lateralization and localization of the site of epileptogenic dysfunction. This is accomplished by delineating the pattern of strengths and weaknesses within the individual's profile of cognitive, sensory and motor functioning. These patterns are interpreted with reference to knowledge of brain-behavior relations, particularly that of the specialization of different brain regions for specific functions, and in the case of children, with respect to the child's age or developmental level. The emphasis on lateralization and localization has diminished somewhat as more sophisticated technologies were developed for detecting structural, functional, and electrographic abnormalities. When the neuropsychological data are integrated with other data on the child, such as findings from EEG and structural imaging, however, it is important to determine whether there is consistency between these various sources of information. Discordant information may be diagnostic, as, for example, suggesting that there may be atypical organization within the brain of functions, such as atypical speech representation. Lack of agreement between neuropsychological results and other diagnostic data could also suggest that the epilepsy has created interference beyond the area of the seizure focus; this in itself can be used as an argument for the potential benefit of surgery.

1.2. Risks of Surgery

The neuropsychological assessment serves a number of other purposes.[10,11,14,15] It may assist in the prediction of whether the child is at risk for significant cognitive deficits if surgery were to be conducted, such as the deterioration in, or loss of, language abilities or memory. These risks need to be evaluated in deciding whether the child is a good surgical candidate, what approaches to surgery might be appropriate, and/or the extent of surgical excision that can be undertaken.

1.3. Evaluating the Outcome of Surgery

The presurgical assessment also provides a baseline against which the neuropsychological outcome of the surgery can be evaluated. With presurgical assessment results, one has a standardized and objective way of determining change in the child, as a result of surgery. To date, the literature on outcome after epilepsy surgery has focused largely on changes in seizure frequency and neurological variables. Parents and children are frequently also concerned about the effects of the epilepsy and its medical treatment on other aspects of the child's functioning, including cognitive performance, school performance, peer relationships, and self-esteem, and their wish is that improvement will also be experienced in these areas. At the Hospital for Sick Children in Toronto, we typically wait a year before conducting a post-surgical assessment, unless there are critical educational or treatment decisions requiring neuropsychological input before that time. This length of time allows ample time for the physical recovery, for outcome with respect to changes in seizure type or frequency to be apparent, and for the potential range of changes in a variety of contexts to be apparent to parents. For example, it may take some time to determine whether the child's rate of learning in the classroom has changed.

1.4. Understanding the Child

Additionally, the assessment results are valuable in assisting parents and professionals reach a better understanding of the child and reasons for the child's particular pattern of emotional, social, and behavioral functioning.[10] The difficulties the child is experiencing can be placed within the context of epilepsy for both the parents and the children themselves. Placing the cognitive and behavioral functioning of the child in the context of brain function and dysfunction can make a tremendous difference in understanding and acceptance of limitations. Parents are frequently relieved to discover that their child is not unique, that there are others with similar problems, and that there are reasons for their child to be this particular way. In addition, the pediatric neuropsychologist is frequently called in to consult to teachers and other professionals working with the child. Teachers may not understand well the particular academic problems of children with epilepsy, which can be uniquely different from the problems encountered in other types of learning disabilities. The neuropsychological status of the child may also be informative to counselors or behavioral therapists in determining the best treatment strategies.

2. SPECIAL CONSIDERATIONS IN PEDIATRIC NEUROPSYCHOLOGY

The field of neuropsychology has a longer history with respect to applications and research with adults than with children. This is true for the assessment of neurological and neurosurgical populations in general, and is particularly true for application to pediatric epilepsy. In the infancy stages of our field, it was tempting to take principles derived from adults and apply them to children. This approach was erroneous in that theories derived from the study of adults assume a mature nervous system at the time that the brain suffers some dysfunction. With children, one has to consider factors such as the effects of dysfunction within a developing nervous system, ongoing maturational changes, behavioral and structural plasticity, and the impact of environmental and social factors on the development of the child.[2,6]

Because pediatric epilepsy surgery is relatively new, little has been published about the neuropsychological features of children who are candidates for surgery, either in the preoperative or the postoperative phase. The developmental factors mentioned in the preceding paragraph suggest that the brain-behavior relationships derived from the study of adults may not apply to children. There is a large and well-established literature on the lateralizing and localizing signs seen in adults with focal epilepsy, but these may or may not be present in children with epilepsy. Recent publications on pediatric epilepsy surgery have indicated that the samples have been constituted largely of patients with lesions in extratemporal areas, whereas with adults, temporal-lobe lesions are more commonly operated on in surgery programs.[18] This suggests that there may be fundamental differences between patients seen in adult surgical centers and those that present to pediatric programs.

For example, the typical patterns of memory performance seen in adults before and after temporal lobectomy may not hold for children undergoing epilepsy surgery. It has sometimes, but not always (see Jambaqué, this volume), been reported that, preoperatively, no differences were found between children with left and those with right temporal-lobe lesions on tests of verbal memory and visual memory, whereas differences on these types of tasks have distinguished between adults with left and right temporal-lobe lesions, respectively.[1,17] In addition, although hippocampal sclerosis is commonly associated with poor memory in adults, one study has reported that children with hippocampal sclerosis show normal memory function (19, but see also Brizzolara *et al.*, this volume, for opposite evidence). It may be that the patterns of specialization within the brain for memory differ after early, intractable seizures or that we require different types of tasks to identify the components of learning and memory that may be compromised after early brain damage. Postoperative outcome is addressed further in this book in the chapter by Helmstaedter and Lendt.

For reasons other than the theoretical complexity of dealing with the developing nervous system and the small body of published findings on preoperative assessments with pediatric candidates for epilepsy surgery, the neuropsychologist is faced with other challenges in working in pediatric programs. A wide range of assessment skills and experience are necessary to prepare for the assessment of children from infancy to adolescence, with a wide range of etiologies that may result in patterns of atypical development or generally normal development with only specific deficits. The type of assessment and the choice of procedures will depend on the age of the child and the developmental level of the child.[10,15]

3. DOMAINS COVERED IN THE NEUROPSYCHOLOGICAL ASSESSMENT

In a 1993 survey[11] of clinical practice by neuropsychologists in epilepsy surgery centers around the world, responses were received from 82 neuropsychologists, of whom

54 indicated that they evaluated some children. Only 7 of the centers responding dealt primarily with children and in many of the other programs, children younger than adolescents were not seen; since that time, the number of centers performing epilepsy surgery with children has expanded. When the results for neuropsychologists who worked with children of any age were tabulated, it was revealed that there was general agreement that the assessment had to be broad-based, and to encompass multiple domains of functioning. These domains include: intelligence, language, memory, attention, problem-solving/executive function, visuo-spatial and perceptual analysis and reasoning, academic skills, motor and sensory function, behavior, personality, emotional status and adaptive functioning.

The survey indicated that in the choice of test instruments, there was consensus only in that all neuropsychologists included some measure of overall intellectual functioning. Otherwise, when looking at other specific tests included in the standard presurgical assessment, considerable variability was evident. This variability in the choice of test instruments presumably reflects a number of factors. One is the range of ages of children seen across centers; thus neuropsychologists who deal with adolescents only will have a very different test battery than neuropsychologists who see preschoolers. A second factor relates to language, as the survey was international in scope; many tests are not available in more than one language. In some epilepsy surgery centers, certain of the domains may be assessed by other professionals, such as psychiatrists, occupational therapists, physiotherapists, and speech-language pathologists. For examples of specific tests that can be used to evaluate abilities in the various domains, the reader may refer to the chapters by Jones-Gotman *et al.*[11] and Oxbury.[15]

4. SPECIAL PROCEDURES

4.1. Intracarotid Amobarbital Testing

As with adult surgical centers, the neuropsychologist in pediatric programs is frequently involved in the Intracarotid Amobarbital Procedure (IAP). The IAP is a technique which involves the injection of a drug, sodium amytal, into one hemisphere of the brain; the injection is usually delivered by a catheter through the internal carotid artery.[12] The effect of the drug is to anesthetize the injected hemisphere briefly, and to be able to test the abilities of the other hemisphere in isolation. The IAP is used most often to determine cerebral dominance for language or to evaluate the memory function of each hemisphere.

Across epilepsy surgery centers, there is generally consistency in the procedures involved in testing for language—counting, naming pictures or objects, comprehension of simple questions or commands, and reading if the child is able. The language testing included in the protocol must be individualized to the child, who may require considerable pre-test preparation. Chronological age, mental age, and delay in language development all factor into the choice of procedures and test items.

A review of the small number of studies conducted on language lateralization with the IAP revealed somewhat disparate estimates (ranging from 66% to 88%) of left hemisphere language representation in children selected for epilepsy surgery. The discrepancy in the estimates may be due to differences in testing procedures or criteria for classifying unilateral versus bilateral speech representation or to actual differences in the patient populations studied at different surgical centers. For example, it would appear that there was a higher percentage of younger children in the patient series that had the lower estimates of left language dominance[7,12,20] than in those with the higher estimates.[21,22]

With children, the assessment of memory during the intracarotid amobarbital test is not always as successful as the assessment of speech.[12] The difficulty in arriving at a valid and reliable indicator of memory status may be due to the fast-paced and stressful conditions, to the inability of young children to comprehend the requirements of the task, or to failure of children to attend sufficiently well to the test items to ensure proper registration of the materials.[7,20] In general, age and IQ are determinants of whether or not the test may yield any useful information.[21,22]

4.2. Functional Mapping

In recent years there has been an increase in epilepsy surgery centers that employ techniques for seizure recording, such as subdural or implanted electrodes, that also allow for the mapping of cognitive, sensory and motor functions. The neuropsychologist can have a critical role in the testing of eloquent functions in the context of cortical stimulation. This is almost an uncharted territory, as little has been published on functional mapping in children. Functional MRI procedures are also becoming increasingly utilized in the evaluation of surgical patients, and although the majority of work has been done with adults, there is evidence that these procedures can be used reliably to lateralize language in children[9,13] (also see chapter in this book by Hertz-Pannier and Chiron). The preoperative neuropsychological assessment can contribute to identifying which children can partake in the functional mapping procedures and the types of cognitive tasks they would be capable of completing.

5. FUTURE DIRECTIONS

The future challenges for the field of presurgical neuropsychological assessment are vast and exciting. First, we have to establish a knowledge base of the neuropsychological aspects of the childhood epilepsies. There are few data available on the neuropsychological aspects of the pediatric epilepsies that are amenable to surgical resection. More information is required about the specific neuropsychological deficits associated with epileptic lesions in the frontal, temporal, parietal, and occipital areas. We need to learn about the relation of hippocampal or medial temporal regions to the cognitive functioning of children with epilepsy. There are few group data available to use to draw conclusions about the sensitivity and specificity of the presurgical evaluation. This knowledge base will guide us in developing tasks that can fit with the requirements for the presurgical neuropsychological assessment—tests that are sensitive to the specific effects of epilepsy, and tests that can be used with children of a variety of ages and developmental levels.

We are at a point in our field where there are still numerous research questions that can be posed based on the data generated in the presurgical assessment. For example, there are questions about the duration of the window for plasticity for the structural and functional reorganization of the brain, the amount of behavioral plasticity that can be exerted for different cognitive functions, the risk-benefit ratio of plasticity, or what is known as the crowding effect; what functions are compromised and to what extent, as a result of sparing or reorganization of other functions. We require more information about the cognitive and behavioral consequences of focal lesions incurred in childhood, both as manifest during development and in terms of what the final outcome in adulthood will be. The neuropsychological correlates of seizure history variables have been investigated with adults but not as thoroughly with children: age of onset, duration, etiology, and seizure frequency. Of particular importance is the question of how these variables are related to neuropsychological outcome after surgery.

The ultimate goal of epilepsy surgery is to improve the quality of life for children and their families. Making the decision about whether to proceed with surgery is a difficult process. Parents seek information on what changes to expect as a result of surgery—what positive changes might occur, whether new deficits might be expected, and what problems might persist after surgery. The presurgical neuropsychological assessment provides a baseline for evaluating the outcome of surgery. This in turn will provide a knowledge base for counseling families on what changes to expect as a result of surgery, which will aid them in their decision-making and help prepare them for changes to come.

REFERENCES

1. Adams CBT, Beardsworth ED, Oxbury SM, Oxbury JM and Fenwick PBC (1990): Temporal lobectomy in 44 children: Outcome and neuropsychological follow-up. Journal of Epilepsy 3(Suppl):157–168.
2. Baron IS, Fennell EB and Voeller KKS (1995): Pediatric neuropsychology in the medical setting. New York: Oxford University Press.
3. Davidson S and Falconer MA (1975): Outcome of surgery in 40 children with temporal-lobe epilepsy. The Lancet 5:1260–1263.
4. Duchowny MS (1989): Surgery for intractable epilepsy: Issues and outcome. Pediatrics 84:886–894.
5. Fish DR, Smith SJ, Quesney LF, Andermann F and Rasmussen T (1993): Surgical treatment of children with medically intractable frontal or temporal lobe epilepsy: Results and highlights of 40 years' experience. Epilepsia 34:244–247.
6. Fletcher JM and Taylor HG (1984): Neuropsychological approaches to children: Towards a developmental neuropsychology. Journal of Clinical Neuropsychology 6:39–56.
7. Hempel A, Fangman M, Risse G, Mercer M and Frost MD (1993): Utility of the intracarotid amobarbital procedure in children. Epilepsia 34 (Suppl 6):88.
8. Hermann BP, Black RB and Chabria S (1981): Behavioral problems and social competence in children with epilepsy. Epilepsia 22:703–710.
9. Hertz-Pannier L, Gaillard WD, Mott SH *et al.* (1997): Non-invasive assessment of language dominance in children and adolescents with functional MRI: A preliminary study. Neurology 48:1003–1012.
10. Holmes Bernstein J, Prather PA and Rey-Casserly C (1995): Neuropsychological assessment in preoperative and postoperative evaluation. Neurosurgery Clinics of North America 6:443–454.
11. Jones-Gotman M, Smith ML and Zatorre RJ (1993): Neuropsychological testing for localizing and lateralizing the epileptogenic region. In Engel J Jr (ed): "Surgical Treatment of the Epilepsies". New York: Raven Press, pp 245–261.
12. Jones-Gotman M, Smith ML and Wieser H-G (1997): Intra-arterial amobarbital procedures. In Engel J Jr, Pedley TA (eds): "Epilepsy: A Comprehensive Textbook". Philadelphia, Lippincott-Raven, pp 1767–1775.
13. McAndrews MP, Smith ML, Logan WJ, Crawley A and Mikulis DJ (1997): Language lateralization in children with epilepsy as revealed by functional MRI. Epilepsia 38 (Suppl 8):143–144.
14. Oxbury S (1997): Assessment for surgery. In Cull C, Goldstein LH (eds): "The Clinical Psychologist's Handbook of Epilepsy". London: Routledge, pp 54–76.
15. Oxbury S (1997): Neuropsychological evaluation—children. In Engel J Jr., Pedley TA (eds): Epilepsy: A Comprehensive Textbook, Philadelphia, Lippincott-Raven, pp 989–999.
16. Seidenberg M and Berent S (1992): Childhood epilepsy and the role of psychology. American Psychologist 47:1130–1133.
17. Smith ML. Are material-specific memory disorders associated with temporal-lobe seizure foci in children? Journal of the International Neuropsychological Society 4:24.
18. Snead OC, Chen LS, Mitchell WG *et al.* (1996): Usefulness of [18F] flurodeoxyglucose positron emission tomography in pediatric epilepsy surgery. Pediatric Neurology 14:98–107.

19. Stanford LD, Chelune GJ and Wyllie E (1998): Neuropsychological functioning of children with hippocampal sclerosis. Epilepsia 39 (Suppl 6):249.
20. Szabo CA and Wyllie E (1993): Intracarotid amobarbital testing for language and memory dominance in children. Epilepsy Research 15:239–246.
21. Westerveld M, Zawacki T, Sass KJ, Spencer S, Novelly RA and Spencer D (1994): Intracarotid Amytal procedure evaluation of hemispheric speech and memory function in children and adolescents. Journal of Epilepsy 7:295–302.
22. Williams J and Rausch R (1992): Factors in children that predict performance on the intracarotid amobarbital procedure. Epilepsia 33:1036–1041.

24

NEUROPSYCHOLOGICAL OUTCOME OF TEMPORAL AND EXTRATEMPORAL LOBE RESECTIONS IN CHILDREN

Christoph Helmstaedter and Michael Lendt

University Clinic of Epileptology Bonn
Sigmund Freud Strasse 25, D-53105 Bonn
Germany

1. NEUROPSYCHOLOGY IN FOCAL RESECTIVE EPILEPSY SURGERY

During the recent years much effort has been done to evaluate the cognitive features of focal epilepsies and the consequences of epilepsy surgery on the patients' cognitive abilities. For a long time, however, epilepsy surgery was performed almost exclusively in adults. Now there is increasing evidence that an early evaluation of pharmacoresistance and an early decision for surgery might prevent hindrance or regression of cognitive development due to persisting epilepsy. Furthermore, an early surgery can be suggested to diminish the impact of epilepsy on the patients' psychosocial situation and his/her quality of life in general. Presumably an early surgery can also prevent the disease to become more severe. However, the question of the natural course, at least of focal epilepsies, remains speculative until longitudinal research is performed on this issue. While there are several studies showing that epilepsy surgery in children controls epilepsy as successfully as in adults, not much is known about the cognitive consequences of focal epilepsies and their surgical treatment in children. The impact of epilepsy and epilepsy surgery on brain functions must be assumed to differ depending on the developmental level of the child at the time of epilepsy onset and the time of surgery. Depending on the age at which brain damage occurs, the development of the respective brain areas as well as the acquisition of performances relying on these areas can be impaired. However, at the same time, the young brain is known to be plastic, and processes of plasticity interfere with the impairment and the hindrance of development. Besides age of onset and extent of the structural/functional impairment, the conditions when plasticity may become effective or not are largely unknown.[2,7,24,26]

Neuropsychology of Childhood Epilepsy, edited by Jambaqué et al.
Kluwer Academic / Plenum Publishers, New York, 2001.

2. FINDINGS IN ADULT PATIENTS

As already mentioned, most research in focal resective epilepsy surgery has been performed with adults. With consideration of the suggested advantages of an early surgery, its seems useful to take the findings in adults as a reference when discussing the results in children. In adults, localization related cognitive deficits are frequently observed before surgery. Temporal lobe epilepsies for example are characteristically associated with impaired declarative memory functions, frontal lobe epilepsies with impairment in motor control, planning and decision making, set maintenance, divided attention, and working memory.[18] Of the temporal lobe epilepsies, those with hippocampal sclerosis represent a group on their own, in that they differ from TLE with other lesions with respect to clinical (onset of epilepsy, febrile seizures etc.) and neuropsychological features (predominant encoding/retrieval problems, largely preserved attention and working memory, unspecifically reduced general intelligence).[16] Posterior epilepsies are rare and their neuropsychological correlates have not yet been systematically evaluated. Typical parietal or parieto-occipital symptoms or syndromes (receptive aphasia, acalculia etc.) that have been described after acute cerebrovascular lesions or head trauma are very rare. More often cryptic symptoms in terms of unspecific language problems, signs of legasthenia or neurological problems resulting from lesions of primary sensory areas are observed. Furthermore, parietal epilepsies can show mixed characteristics with additional temporal or frontal dysfunction.

Besides localization-related impairment, material-specific cognitive impairment can be observed according to the lateralization of epilepsy. The findings are more conclusive with regard to left hemisphere epilepsies and they are more clear-cut in temporal than in extratemporal lobe epilepsies.

The preoperative impairment can well be related to pathology. This has been particularly demonstrated for temporal lobe epilepsy. Memory deficits correlate to mesial pathology as defined by MRI volumetry, cell counts or sclerosis grading in mesial temporal structures. Furthermore different impairment is observed depending on whether hippocampal sclerosis or other pathology is present.[12,16]

Focal epilepsies directly or indirectly involving motor or language areas are frequently associated with atypical handedness and/or language dominance. This is particularly observed in patients with early onset left hemisphere lesions and may be attributable to processes of functional plasticity. However, as pointed out recently, interhemispheric plasticity can be assumed also in right hemisphere epilepsies. Furthermore, and this may be of importance in children, we must assume that the development of an atypical language dominance pattern takes time and may last for years after the lesion/epilepsy onset.[15] Figure 1 a/b demonstrates the relation between the age at onset of left hemisphere epilepsies and the occurrence of atypical language dominance in a series of 120 patients with left hemisphere epilepsies; this figure also illustrates the relationship between atypical dominance in epilepsies and the location and extent of the lesions/epilepsies within the left hemisphere.

Drugs may have positive or negative effects on cognitive performance depending on substance, dose, and combinations of drugs. Furthermore one must consider interactions of the drugs with predamaged brain tissues and seizure activity. In most cases the effect of antiepileptic drugs on cognitive functions is negligible. With the exception of intoxication and individual incompatibilities, negative side effects seem more likely in those epilepsies which are more severe and more difficult to control.[29]

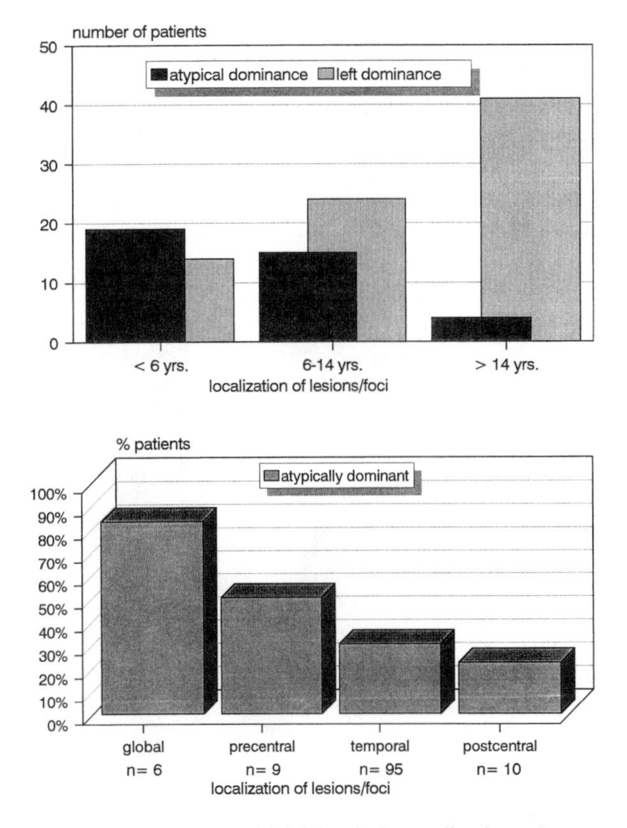

Figure 1. Language dominance in 120 patients with left hemisphere epilepsies. a. Language dominance and age at the onset of epilepsy; b. atypical dominance and the localization of left hemisphere epilepsies.

Surgery can cause additional impairment. Preexisting pathology, type and extent of the resection, and the success in postoperative seizure control are determinants of the postoperative cognitive outcome. As with the preoperative impairment, impairment due to surgery corresponds to the site and the side of the resection, particularly in left sided resections. Improvement of cognitive functions is most likely observed in functions located adjacent or contralateral to the site of surgery and which were secondarily affected by epilepsy before surgery. One can summarize the findings by concluding that losses are due to the resection of still functioning tissues and that improvement is due to a release effect linked to seizure control and to compensatory capacities of ipsi- and contralateral structures.[12]

3. FINDINGS IN CHILDREN AND ADOLESCENTS

When looking at resective epilepsy surgery in children, the current literature is still dominated by reports of the need of epilepsy surgery in children followed in number by reports that successful seizure outcome can be achieved also in the younger patients. It should be noted that most reports about epilepsy surgery in children are concerned with callosotomies and hemispherectomies. These surgical procedures, however, have their own chapters in this book and will not be considered here. Up to now, there have been only very few studies with sample sizes of at best 37 patients which, more or less systematically, evaluated the neuropsychological aspects of focal resections in children.[1,3,10,19,20,22] The results of these studies are provided in Table 1. As can be seen in this table, all studies refer to temporal lobe resections and not a single one deals with resective surgery in frontal or posterior regions. There are two reasons for this concentration on temporal lobe resections which incidentally also holds for research conducted in adults. The first reason is that the temporal lobe structures are particularly vulnerable to epilepsy. They consequently represent the most frequent type of focal epilepsy. The second reason is that the temporal-mesial structures are structurally and functionally better defined than other brain regions. They are thus more easily evaluated than other brain structures which can be affected by epilepsy.

3.1. Temporal Lobe Epilepsy

Differences in the incidence of focal epilepsies of different locations is also reflected by our own sample of 70 children who have had focal resective epilepsy surgery with a preoperative and a 12-month postoperative neuropsychological evaluation. Sixty-eight percent of these children had temporal lobe epilepsies (TLE), 20% had frontal lobe epilepsies (FLE) and 12%, posterior epilepsies (PE). Data obtained with these children as well as with adult patients will be presented in the following sections in addition to results reported in the literature (see Table 1). The adult patients were selected out of our surgical series according to their surgical site (temporal, frontal or posterior), and their early onset of epilepsy. Of these patients, 83% patients had TLE, 12% FLE, and 5% PE. (For more detailed information see Table 2).

In order to provide data which can be compared over different tests and age groups, performances were all categorized in a rating system according to normative test data. This system comprised 5 categories ranging from "highly significantly below average" to "highly significantly greater than average". Retest-data (retest interval >6 months) from nonoperated children and adults served as normative reference data for the classification of the postoperative performance changes as "gains" or "losses". Keeping in mind the methodological difficulties of standardized testing in children of different ages and the difficulties of comparing results in children with those obtained in adult patients one must consider that these data only represent trends and that detailed analyses are still required.

Looking at the preoperative findings in children with TLE, one can conclude that temporal epilepsies are rarely associated with overall reduced cognitive performance as indicated by IQ. This is also reflected by our own data which indicate an IQ below 86 in about 20% of the children with TLE. Irrespective of the hemispheric lateralization of epilepsy, verbal IQ seems more likely affected than performance IQ. Correspondingly, a high percentage (56%) of children displayed significant problems in language functions in our own sample. In contrast, visuoconstruction was affected in only 19%. Thus, the incidence of language impairment is nearly twice as high as in adult patients with early onset TLE (56% vs. 30%). Furthermore, and this is very different from adults, language is more frequently affected than memory. As demonstrated in Fig. 2a, impaired language can only in part be related to a left hemisphere lateralization of epilepsy. It is more frequent in L-TLE but can be also observed in R-TLE. An additional look at children with frontal and posterior epilepsies suggests that impaired language functions are a frequent feature of epileptic children in general (Table 3a). This observation can best be explained

Table 1. Overview of neuropsychological studies conducted on the effects of focal resection in children

study	patients/surgery	follow-up-interval	evaluated performance	preoperative	postoperative
Meyer *et al.* (1986)	37 TLE, age: 7–18 yrs, mean: 15.8 yrs surgery: ATR	after 6 months up to 10 years; m = 4.5 yrs.	WISC-IQ, WMS	– average performance in IQ and memory with a total IQ of 91.6 – as a trend verbal IQ lower than performance IQ	– improvement in IQ with shorter duration of epilepsy – no change in memory
Adams *et al.* (1990)	20 RTLE 14 LTLE age: 2–15 yrs. surgery: ATR	screening after 3–6 weeks complete after 6 months	WISC-IQ verbal memory figural memory naming comprehension	– average total IQ 90 with verbal IQ poorer than performance IQ, – impaired verbal/figural memory – no differences between RTLE and LTLE	– no changes in IQ and figural memory – decrease in verbal memory and naming in LTLE – improved comprehension in RTLE
Beardsworth & Zaidel (1994)	13 LTLE, 14.0 yrs. 16 RTLE, 12.9 yrs surgery: ATR, SAH	6 months	memory for faces	poorer memory for faces in young and adult patients with RTLE	– improvement in memory RTLE
Lewis *et al.* (1996)	11 LTLE 12 RTLE, mean age 14.5 surgery: ATR	after 1 to 8 yrs.	WISC-IQ, WMS	– reduced IQ of 82.8 – RTLE poorer figural memory than LTLE	slightly increased IQ, no change in memory
Gilliam *et al.* (1997)	33 TLE <12 yrs. surgery: cortical		WISC-IQ	no data	increase in IQ in 6 (29%) of 21 patients
Lendt *et al.* (submitted)	10 LTLE 10 RTLE age: 10–16 yrs. surgery: ATR	after 3 months and 12 months	attention verbal memory nonverbal memory	– no difference in memory between TLE-patients and controls, – no difference between LTLE and RTLE in material specific memory, impaired language in LTLE	increase in language and attention functions; no change in memory; neg. effects of continuing seizures on performance one year after surgery

abbreviations: L-TLE = left temporal lobe epilepsy, R-TLE = right temporal lobe epilepsy.
ATR = anterior 2/3 temporal lobectomy, SAH = selective amygdalo hippocampectomy.

Table 2. Patients' characteristics

	TLE		FLE		PE	
	children	adults	children	adults	children	adults
n	48	220	14	33	8	11
age (yrs. M/SD)	13/2	29/8	11/3	25/6	13/2	26/5
gender (male/female)	22/26	103/117	9/5		2/6	
onset of epilepsy (yrs. M/SD)	7/4	7/4	6/4	7/4	5/4	8/3
pathology:						
– no finding	5%	1%	8%	—	—	10%
– AHS	25%	48%	—	—	—	—
– tumor	47%	19%	21%	33%	37%	27%
– development/others	23%	32%	71%	77%	63%	63%
side of surgery (left/right)	28/20	109/111	7/7	19/14	3/5	3/8
one year follow up: seizure free*	69%	63%	43%	45%	50%	54%

M = mean.
SD = standard deviation.
AHS = hippocampal sclerosis.
seizure free* = not one seizure nor any aura since surgery.

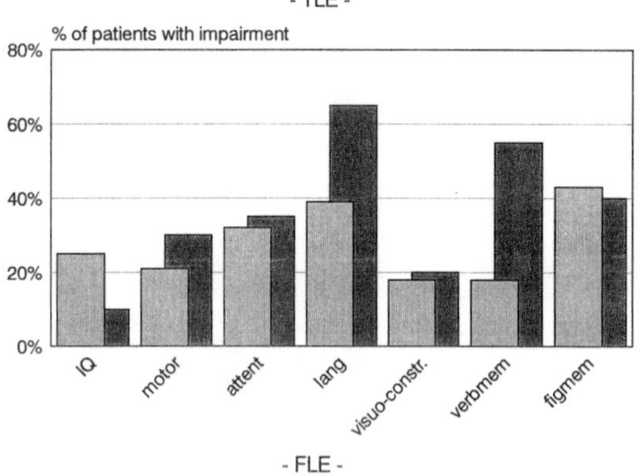

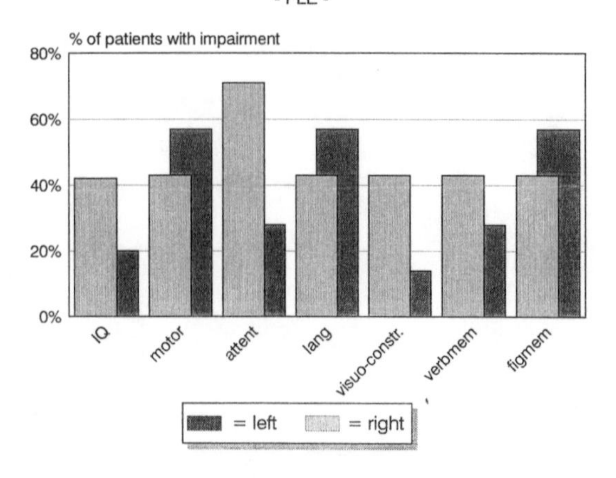

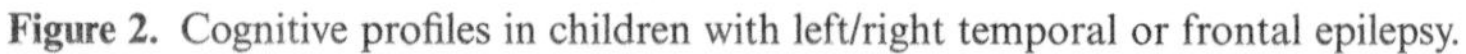

Figure 2. Cognitive profiles in children with left/right temporal or frontal epilepsy.

Table 3. Preoperative and postoperative (one year follow-up) cognitive profiles in a. Children and b. Adults

a) cognitive profiles in children and postoperative change in performance at the one year follow-up

	TLE n = 48		FLE n = 14		PE n = 8	
evaluated function	preop. impaired	postop. change	preop. impaired	postop. change	preop. impaired	postop. change
I.Q.	19%		29%		63%	
motor functions	33%	improved*	50%	—	75%	—
attention	33%	improved*	50%	improved*	75%	—
language	56%	improved*	50%	—	75%	—
visuoconstruction	19%	—	29%	—	50%	—
verbal memory	33%	—	36%	—	63%	—
figural memory	42%	—	50%	—	50%	—

*p < 0.01 (Wilcoxon text for related samples).

b) cognitive profiles and one year postoperative change in a matched sample of adult patients

	TLE n = 220		FLE n = 33		PE n = 11	
evaluated function: change	preop. impaired	postop. change	preop. impaired	postop. change	preop. impaired	postop. change
I.Q.	10%	no data	9%	no data	10%	no data
motor functions	35%	—	22%	decreased*	43%	—
attention	19%	improved*	38%	—	10%	—
language	30%	improved*	44%	—	44%	—
visuoconstruction	10%	improved*	12%	—	10%	—
verbal memory	51%	decreased*	50%	—	40%	—
figural memory	47%	—	53	—	50%	—

*significant changes with p at least <0.05 (Wilcoxon tests for related samples).

by the fact that the onset of epilepsy coincides with language development in most of these children.

In contrast to language, the findings on memory largely parallel those reported for adults. Right temporal lobe epilepsies are associated with poorer figural memory and left temporal lobe epilepsies, with poorer verbal memory.[6,9,17] In our own sample, verbal memory impairment was lateralized to the left temporal lobe but figural memory was impaired in both children with R-TLE and L-TLE. Thus, R-TLE patients show the expected impairment in figural memory, and the problem with the discriminant validity of figural memory impairment results from patients with L-TLE. We know from adult patients that atypical cerebral dominance can explain figural memory impairment in L-TLE. According to our recent results regarding the influence of sex and atypical dominance on material specific memory in a large sample of adult patients with L-TLE, only the group of left hemisphere language dominant men showed the expected pattern of impaired verbal memory and unimpaired figural memory.[14] Atypically dominant men as well as women of either language dominance showed an impairment in figural memory which was not expected with a left lateralization of the temporal focus. Suppression of right hemisphere functions due to actual language transfer as well as a disposition for a more bilateral language representation could explain these findings. Interestingly, figural memory impairment is the dominant memory problem in children with TLE. Verbal memory impairment is less frequently observed. In our patients, its incidence is also less frequent when compared to adult patients with an early onset of epilepsy (33% vs. 52%). As already mentioned, it is language rather than verbal memory which is impaired. We did not determine language dominance in each of the children but one can assume that the observed pattern of impairment in children also reflects the result of ongoing processes of functional restructuring.

In summary, the preoperative findings suggest that TLE has indeed a different cognitive manifestation in childhood than in adulthood. One conclusion from these data might be that the impairment with TLE becomes more specific with brain maturation and progressive cognitive development. An interesting parallel is given with seizure semiology, which, like neuropsychology, is more often unspecific and misleading in children than in adults.[4] However, without longitudinal observations, it is hard to bridge the gap between children with mesio-temporal lobe epilepsies and adults with a similar type of epilepsy. Like others, we also find that hippocampal sclerosis (AHS) is less frequently observed in children than in adult patients with a comparable onset of TLE in early childhood (25% vs. 48%).[6] Some argue that AHS develops with the progress of the disease and that it is a consequence of seizures rather than the cause of seizures.[5,11,21] This could also account for the observation of an increasing focalization of the cognitive impairment. On the other hand, and this possibility cannot be excluded here, the different occurrence of AHS in children and adults can be misleading due to the cross-sectional view. One can argue for example, that there are as many children as adult patients with AHS but that children with AHS have "silent" or well controlled epilepsies which will become pharmacoresitant in adulthood. The question of whether children and adults with early onset epilepsies represent different cohorts is also raised when considering extratemporal epilepsies with early onset. These children are so impaired cognitively that one can hardly expect them to improve in the following years to resemble adults with epilepsies in the same localization.

Keeping the differentiation of temporal (memory) and extratemporal functions, extratemporal functions significantly improve after surgery (Table 3a, Fig. 3). In our own sample, significant gains were observed in 30% to 40% of the patients. Postoperative gains were related to a poorer preoperative performance and to successful seizure control, which suggests that these functions were indirectly affected by epilepsy before surgery. The observation of improved extratemporal functions is fully in line with reports of a postoperative improvement in IQ (see Table 1).

Comparable improvement has been reported after temporal lobe surgery in adult patients with respect to "frontal functions". The pattern of improved functions in our adult sample is very similar to that observed in children. We can conclude from these findings that in children with TLE, the basic functions of attention, language and motor speed/coordination are secondarily affected or hindered in their development before surgery. Similar patterns of improvement in children and adults, the short follow up interval of one year after surgery, and the positive effect of a successful seizure control on performance provide some evidence that these changes reflect a release of extratemporal functions rather than plasticity or normal/accelerated development.

With regard to temporal resections, children as a group do not show the significant losses in verbal memory which are characteristically seen in adult patients. Memory functions in children remain largely unchanged or they even improve after surgery (Table 1). Adams' report[1] is the only one indicating losses in verbal memory after left temporal resections. In our own sample we find, as Adams reported,[1] that if there are negative cognitive consequences after temporal lobe surgery they are observed in memory. Twenty to 30% of the children show losses in memory when retest-reliability and practice effects in

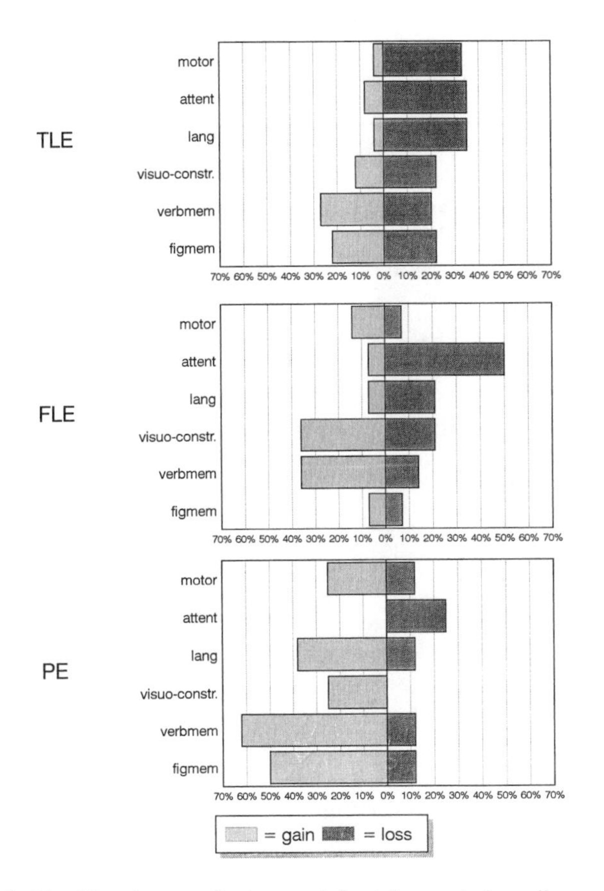

Figure 3. Cognitive changes after temporal, frontal or posterior epilepsy surgery.

nonoperated children are taken into consideration (Fig. 3a). The changes in verbal memory follow the same pattern as in adult patients, i.e., deterioration is more likely after left temporal resections.

A strong negative relationship between age at surgery and loss in verbal memory can be observed in left-temporal resected patients. As we reported elsewhere, this age effect can best be explained within the context of plasticity or compensation capacities.[12] When patients are operated during the "plasticity period", they show a much better memory outcome than patients operated at an older age. The best results are obtained when surgery is performed before puberty. However, the age effect does not account for all aspects of verbal memory. It seems restricted to learning and short-term or working memory rather than to the consolidation/retrieval processes involved in long term memory. Consolidation/retrieval is affected by surgery regardless of age. Because there is some evidence that learning and memory are differentially associated with mesio-temporal and temporo-cortical structures, one might conclude from these data that in children, compensation is more likely for cortical lesionectomy than for surgery that includes the mesial structures. However, these children were between 6 and 14 years old when they underwent surgery and, keeping in mind that earlier lesions can hinder further development of superordinate functions, it is not clear from these data whether younger children can compensate as well from a surgery that includes the mesial structures.

Besides changes in cognitive functions, we have also observed changes in behavior after temporal lobe surgery.[8] Behavioral disorders in terms of attentional deficit disorder and/or hyperactivity are frequent and a positive change was observed in the subgroup of 37 children reported in Table 1.[22] Twelve children out of 37 showed behavioral disturbances before surgery and significant improvement in behavior was obtained in 7 children. In 6 of these children, improvement was observed directly after surgery. Whereas there was evidence that the behavioral disorder was associated with neoplastic lesions and very active spike foci preoperatively, no particular predictor for the postoperative change could be discerned.

3.2. Extratemporal Epilepsies

The literature provides nothing with regard to the cognitive capabilities of children with extratemporal focal epilepsies, nor are there any studies regarding the postoperative cognitive development after focal resections. As mentioned in the adult section, even in older patients our knowledge is based on only a few studies.[13]

Therefore we evaluated children with frontal or posterior (parietal/parieto-occipital) epilepsies out of our own sample in order to investigate the effect of extratemporal epilepsy and epilepsy surgery on cognition. As in TLE, adult patients who were largely matched to children regarding onset of epilepsy, pathology, and seizure outcome served as controls.

As can be seen in Table 3a, the cognitive impairment in children with frontal (FLE) and posterior (PE) epilepsies appears more generalized than in children with TLE. Group comparisons (Wilcoxon Rank Sum Test) show that the discriminative feature of FLE is an attention deficit, and that the discriminative feature in posterior epilepsies is the poorer performance in visuo-construction and the poorer general IQ. When compared to adult patients with extratemporal epilepsies, children also seem more frequently and more severely impaired. This finding raises the question as to how the performances in the adults had been within childhood. On the one hand, it is possible that these children will reach adult performance levels in the ensuing years. On the other hand, as already discussed with regard to children and adult patients with TLE, the cross sectional view may wrongly suggest a developmental pattern, while the patient groups may actually have different type or severity of epilepsies.

The comparison between children with TLE, FLE and PE suggests a hierarchy in impairment depending on the localization of epilepsy. Early impairment in posterior receptive functions results in the most severe and generalized deficits, followed by injury of frontal areas (expressive functions) and temporal structures which are associated with memory. Specific consequences of injury to these structures for the subsequent development of the underlying functions are the most reasonable explanation for this observation.[27] In contrast to adult patients, the side of the lesion seems to have secondary relevance. However, as we had found in adult patients with FLE, children with right FLE tend to show poorer performances than children with left FLE.

Postoperatively, the group of frontal resected children shows significant improvement in attention (Fig. 3b). Although the result is not significant in group statistics, nevertheless at least 30% of the children show significant losses in verbal memory and visuo-construction. Postoperative drops in memory may be explained by negative effects of frontal surgery on working memory, and drops in visuo-construction may be due to problems in planning and executive behavior. Regression analysis indicates that losses after frontal surgery are associated with a later onset of epilepsy or an older age at the time of surgery, and gains are associated with earlier epilepsy or younger age at surgery. Furthermore, in this sample, gender seems to be associated with improvement/deterioration in visual and visuo-constructive functions.

Side of surgery only affects the changes observed in language functions. Right-sided surgery has a positive effect on language whereas left-sided surgery rather seems to produce a negative effect on this function. Similar results are seen in adult patients.[13] However, contrary to adults, children significantly improve with regard to attention and they do not show losses in motor functions as adult patients do (Table 3b). When surgery spares the structures relevant for language and motor control, frontal lobe surgery, like temporal lobe surgery, causes no serious harm on cognition in children. However, when comparing temporal and frontal surgery, a much broader range of improvement is seen after temporal lobe surgery. This is surprising considering the widespread neuronal association of the frontal lobe to almost all other parts of the brain and the superordinate control function of the frontal lobe. Considering additionally the fast irradiation of epileptic activity and seizures in frontal lobe epilepsy, one would expect more significant release effects after surgery.[24] Although our analysis of adult patients with frontal surgery does indicate a positive effect of a successful seizure control on memory for example, one can assume that a great part of the preoperative impairment of "extra frontal" functions in childhood FLE reflects the direct developmental involvement of the frontal lobe in these functions.

Coming to the small group of children with posterior epilepsies, we find that they not only represent the group with the most extended and most serious deficits before surgery but also the group with the worst outcome when we look at each individual child. Although group statistics do not reach statistical significance, the individual evaluation indicates postoperative losses in memory functions and partly also in language and motor functions (Fig. 3). Behavioral disorders before and after extratemporal resections in children have not yet been evaluated and research on this issue is badly needed.

CONCLUSION

Focal epilepsies of left or right localization seem to be associated with more diffuse impairment in children than in adults. The pattern of impairment reflects the impact of epilepsy on the development of cognitive functions in general. A hierarchy of impairment is suggested which follows the importance of the respective structures for subsequent cognitive development. Accordingly, posterior epilepsies are associated with most severe impairment followed by frontal and temporal epilepsies. Different from adult patients, the pattern of functional deficits has less localizational or lateralizational value. When compared to adult patients, one could assume that the impairment becomes more specific with further development. Particularly with extratemporal epilepsies, one can hardly bridge the gap between the impairment observed in children and that observed in adults, and it is important to note that one must be careful with conclusions regarding the development of epilepsies and cognitive impairment on the basis of cross sectional data.

The question of seizure outcome was not the primary core of this contribution but one can state that focal surgery in children is as successful in controlling seizures as focal surgery in adult patients. Seizure control and the preoperative ability status are major determinants of the postoperative cognitive outcome. This is very much the same as in adult patients.

The postoperative changes indicate a significant release of functions which were secondarily affected by epilepsy before surgery. Improvement is particularly seen after temporal lobe resections and in part also after frontal resections. Losses are more likely after extratemporal resections and impairment due to surgery seems more related to the function subserved by the removed area than to preoperative impairment attributable to epilepsy.

Regarding the familiar question regarding the age at which surgery should take place, the seizure outcome data clearly favor an early surgery. This will prevent the children from stigmatization and all other negative consequences epilepsy has in the long run. From a neuropsychological point of view, one can in part conclude to greater compensation capacities in childhood but the evidence is not yet sufficient to warrant an early surgery on the sole basis on neuropsychological findings.

REFERENCES

1. Adams CBT, Beardsworth ED, Oxbury SM, Oxbury JM and Fenwick PBC (1990): Temporal lobectomy in 44 Children: Outcome and neuropsychological Follow-Up. Journal of Epilepsy 3(Suppl 1):157–168.
2. Aicardi J (1997): Paediatric epilepsy surgery: how the view has changed. In Tuxhorn I, Holthausen H, Boenigk H (eds): "Paediatric Epilepsy Syndromes and their Surgical Treatment". London: John Libbey, pp 3–7.
3. Beardsworth ED and Zaidel DW (1994): Memory for faces in epileptic children before and after brain surgery. Journal of Clinical and Experimental Neuropsvchology 16:589–596.
4. Brockhaus A and Elger CE (1995): Complex partial seizures of temporal lobe origin in children of different age groups. Epilepsia 36:1173–1181.
5. Cendes F, Andermann F and Gloor P (1997): Atrophy of mesial structures in patients with temporal lobe epilepsy: cause or consequence of repeated seizures? Annals of Neurology 41:41–45.
6. Cohen M (1992): Auditory/Verbal and Visual/Spatial memory in children with complex partial epilepsy of temporal lobe origin. Brain and Cognition 20:315–326.
7. Duchowny MS (1994): Pediatric epilepsy surgery: Special considerations. In Wyler AR, Hermann BP (eds): "The Surgical Management of Epilepsy". Boston: Butterworth-Heinemann, pp 171–188.
8. Elger CE, Brockhaus A, Lendt M, Kowalik A and Steidele St (1997): Behaviour and cognition in children with temporal lobe epilepsy. In Tuxhorn I, Holthausen H, Boenigk H (eds): "Paediatric Epilepsy Syndromes and their Surgical Treatment". London: John Libbey, pp 311–325.
9. Fedio P and Mirsky AF (1969): Selective intellectual deficits in children with temporal lobe or centrecephalic epilepsy. Neuropsychologia 7:287–300.
10. Gilliam F, Wyllie E, Kashden J, Faught E, Kotagal P, Bebin M, Wise M, Comair Y, Morawetz R and Kuzniecky R (1997): Epilepsy surgery outcome: comprehensive assessment in children. Neurology 48:1368–1374.
11. Harvey AS, Berkovic SF, Wrennall JA and Hopkins IJ (1997): Temporal lobe epilepsy in childhood: clinical, EEG, and neuroimaging findings and syndrome classification in a cohort with new-onset seizures. Neurology 49:960–968.
12. Helmstaedter C and Elger CE (1998): Functional plasticity after left anterior temporal lobectomy: substitution and compensation of verbal memory impairment. Epilepsia 39:399–408.
13. Helmstaedter C, Gleissner U, Zentner J and Elger CE (1998): Neuropsychological consequences of epilepsy surgery in frontal lobe epilepsy. Neuropsychologia 36:333–341.
14. Helmstaedter C, Kurthen M and Elger CE (1999): Gender and language dominance as determinants of different patterns of material-specific memory performance in left temporal lobe epilepsy. Laterality 4:51–63.
15. Helmstaedter C, Kurthen M, Linke DB and Elger CE (1997): Natural atypical language dominance and language-shifts from the right to the left hemisphere in right hemispheric pathology. Naturwissenschaften 84:1–3.
16. Hermann BP, Seidenberg M, Schoenfeld J and Davies K (1997): Neuropsychological characteristics of the syndrome of mesial temporal lobe epilepsy. Archives of Neurology 54:369–376.

17. Jambaque I, Dellatolas G, Dulac O, Ponsot G and Signoret JL (1993): Verbal and visual memory impairment in children with epilepsy. Neuropsychologia. 31:1321–1337.
18. Jones-Gotman M, Smith ML and Zatorre R (1993): Neuropsychological testing for localizing and lateralizing the epileptogenic region. In Engels J (ed): "Surgical Treatment of the Epilepsies". New York: Raven Press, pp 245–262.
19. Lendt M, Helmstaedter C and Elger CE (1999): Pre- and postoperative profiles in children and adolescents with temporal lobe epilepsy. Epilepsia 40:1543–1550.
20. Lewis DV, Thompson Jr, RJ, Santos CC, Oakes WJ, Radtke RA, Friedman AH, Lee N and Swartzwelder HS (1996): Outcome of temporal lobectomy in adolescents. Journal of Epilepsy 9:198–205.
21. Mathern GW, Babb TL, Mischel PS, Vinters HV, Pretorius JK, Leite JP and Peacock WJ (1996): Childhood generalized and mesial temporal epilepsies demonstrate different amounts and patterns of hippocampal neuron loss and mossy fibre synaptic reorganization. Brain 119:965–987.
22. Meyer FB, Marsh WR, Laws ER and Sharbrough FW (1986): Temporal lobectomy in children with epilepsy. Journal of Neurosurgery 64:371–376.
23. O'Leary DS, Lovell MR, Sackellares JC, Berent S, Giordani B, Seidenberg M and Boli TJ (1983): Effects of age of onset of partial and generalized seizures on neuropsychological performance in children. Journal of Nervous and Mental Disease 171:626–629.
24. Oxbury S (1997): Neuropsychological Evaluation- Children. In Engel J Jr, Pedley TA (eds): "Epilepsy: a Comprehensive Textbook". Philadelphia: Lippincott-Raven Publishers, pp 998–999.
25. Quesney LF (1986): Seizures of frontal lobe origin. In Pedley TA, Meidrum BS (eds): "Recent Advances in Epilepsy", Vol 3. Edingbourgh: Churchill-Livingston, pp 101–110.
26. Resnick TJ, Duchowny M and Jayakar P (1994): Early surgery for epilepsy: redefining candidacy. Journal of Child Neurology 9:36–41.
27. Spreen O, Tupper D, Risser A, Tuokko H and Edgell D (1984): Human Developmental Neuropsychology. New York: Oxford University Press.
28. Stuss TD and Benson T (1986): The Frontal Lobes. New York: Raven Press.
29. Trimble MR (1996): Anticonvulsants and psychopathology. In Sackellares JCh and Berent S (eds): "Psychological Disturbances in Epilepsy". Boston: Butterworth Heinemann, pp 233–344.

25

COGNITIVE, SENSORY AND MOTOR ADJUSTMENT TO HEMISPHERECTOMY

Sophie Bayard and Maryse Lassonde*

Groupe de Recherche en Neuropsychologie Expérimentale
Département de Psychologie
Université de Montréal
C.P. 6128, Succ. Centre-Ville
Montréal, Qué., H3C 3J7, Canada

INTRODUCTION

Surgical treatment for epilepsy is considered in the presence of severe convulsive disorders that do not favorably respond to anti-epileptic medication. The identification of an epileptogenic focus unilaterally localized and the presence of a degenerative or metabolic disorder constitute the principal factors for considering hemispherectomy. The surgery is usually performed when severe unilateral motor deficits without finger movements contralateral to the diseased hemisphere are already present. This criterion makes patients suffering from Rasmussen's syndrome, Sturge-Weber disease and hemimegalencephaly, a rare congenital hypertrophy of one cerebral hemisphere, the most frequent candidates for this type of surgery.

Briefly, Rasmussen's syndrome is a progressive atrophy of a cerebral hemisphere with an onset between the age of 14 months and 14 years. At the onset, it is generally accompanied with generalized tonic-clonic or partial seizures. Status epilepticus is also frequently observed and later in childhood, simple partial motor seizures are noted.[1] Sturge-Weber disease, on the other hand, is characterized by the installation of a calcification of the hemisphere gyri. Diverse veinous malformations such as ipsilateral facial angiomatosis and ocular defects are frequently encountered. Sturge-Weber disease is related to an embryological defect occurring in the first trimester of gestation.[25] Apart from its application in the treatment of epilepsy, hemispherectomy can also be used to treat focal dysplasia, vascular accidents, cranial trauma or tumors.[1,44] In adults, hemispherectomy is rarely considered prior to a two-year period in which the effectiveness of

* To whom correspondence should be addressed

Neuropsychology of Childhood Epilepsy, edited by Jambaqué et al.
Kluwer Academic / Plenum Publishers, New York, 2001.

pharmacological therapy has been assessed and proved to be unsuccessful.[7] However, in young children who might convulse up to a hundred times a day, an early intervention may represent a way to diminish, if not avoid, the widespread damages of epilepsy on the development of cognitive functions and social adaptation. This appears essential when the diagnosis of a progressive pathology is definitely recognized as irreversible, such as in Rasmussen's syndrome.

1. ANATOMICAL AND FUNCTIONAL HEMISPHERECTOMY

In 1950, Krynauw was the first to report promising results in a series of 12 hemispherectomized patients with infantile hemiplegia.[14] Ten patients became seizure-free and stopped taking any anti-convulsive medication. Improvements in intellectual performance as well as considerable reduction of behavioral problems were observed in most patients. Following this seminal work, a series of world-wide studies confirmed the effectiveness of hemispherectomy in controlling convulsive disorders secondarily associated to cerebral tumors, vascular accidents, cranial traumas, encephalic disorders and other conditions.

However, this enthusiasm rapidly decreased when Oppenheimer and Griffith[19] identified the appearance of a superficial cerebral hemosiderosis syndrome, a serious, sometimes fatal, post-operative complication of anatomical hemispherectomy. In fact, after a post-surgical period varying from 1,5 to 36 years[13] in which epilepsy was controlled and cognitive abilities improved, mental and neurological deterioration (tremor, ataxia, incontinence, somnolence, chronic intracranial hypertension associated with hydrocephalus) appeared. The lack of adequate support of the remaining structures increased their fragility to weak shocks, causing repeated sub-dural bleedings in the hemispherectomized cavity and inflammation of the meninges in 1/4 to 1/3 of the operated patients.

In 1974, Rasmussen suggested a new surgical therapy: functional hemispherectomy. This surgery consists in a functional deafferentation of the affected hemisphere from the rest of the brain. A temporal lobectomy is first realized followed by a section of the callosal fibers in the anterior (frontal) and posterior (parieto-occipital) areas, performed by subpial aspiration through the ventricle, until the temporal excision is reached. The frontal lobes are then disconnected from the basal ganglia and the contralateral hemisphere, by an excision of white matter. The insular cortex is also removed if electroencephalographic and radiological exams reveal the presence of abnormalities. Finally, most veins and arteries are spared. Long-term follow-up studies of patients treated by functional hemispherectomy (e.g., 23, 33) do not reveal any complications related to superficial cerebral hemosiderosis. To date, the surgical mortality rate is around six percent (e.g., 6), and complete or near seizure control without any medication varies between 60 to 85 percent.[44]

This chapter will review about twenty studies that were carried out from the early seventies up to date and will examine how hemispherectomy performed between infancy and late adolescence influences cognitive, sensory and motor functions.

2. INTELLECTUAL FUNCTIONS

Aaron Smith was probably the first to have realized detailed neuropsychological assessments on large groups of hemispherectomized patients.[28] In the seventies, he reported the performance on the Wechsler intellectual scale of 44 patients who underwent left (n = 27) or right (n = 17) hemispherectomy because of infantile epileptogenic lesions.[28] He observed that this radical surgery did not significantly affect intellectual functions. In fact, the verbal IQ (VIQ) as well as the performance IQ (PIQ) and global IQ (GIQ) remained relatively unchanged after surgery (pre-hemispherectomy VIQ: 76.1, PIQ: 68, GIQ: 69; and post-hemispherectomy VIQ: 81.8, PIQ: 75.3, GIQ: 76.6).

Table 1 summarizes the results that were reported in several studies that provided individual IQ scores following hemispherectomy. In the perspective of this book, only the results pertaining to patients who had developed epileptic seizures in infancy and in adolescence were analyzed. From these reviewed studies, we first examined a sample of 24 patients for whom pre- and post-surgical IQ performances were available.[8,29,33,36,8,38,40] Overall, intellectual functions remained unchanged after surgery (pre-operative GIQ mean: 72.08, post-operative GIQ mean: 64.38) (Fig. 1). Some patients, however, demonstrated a significant drop (–20 to –58) of their IQ score after surgery. Considering the available information, these patients seem to have presented high intellectual functioning prior to the surgery or had bilateral epileptogenic foci. It is noteworthy that in patients who were operated before 1974, the significant mental deterioration could also reflect post-operative complications related to anatomical hemispherectomy.

We then explored the relationship between post-operative GIQ with age at onset of seizures. From the group of patients included in the studies reported in Table 1[8,11,15,18,22,27,30,31,33,38,45,46] the GIQ of 75 left (n = 38) and right (n = 37) hemispherectomized patients was correlated with age at which seizures appeared. GIQ did not significantly differ between the two groups (LH: 70.87; RH: 72.03) and this also applied to age at onset of seizures (LH: 3.41 yrs; RH: 4.08 yrs). Overall, no correlations could be found between age at onset of seizures (birth to 12 yrs) and post-surgical intellectual functioning generally assessed by the Wechsler scales (r = –0.17).

The same pattern was observed when correlations were computed between IQ and duration of seizures or age at surgery in a sample of 36 LH and 36 RH patients whose scores are also presented in Table 1.[8,11,15,18,2,27,29,30,31,33,38,45,46] Surprisingly, postsurgically, duration of the convulsive disorder (LH mean: 5.52 yrs; RH mean: 5.49 yrs) did not correlate with IQ scores (r = 0.04). Similarly, no correlation's were found between age at surgery (LH mean: 8.74 yrs; RH mean: 8.82 yrs) and IQ (r = –0.35).

However, the heterogeneity of the underlying neurological diseases reported in Table 1 may have masked some of these relationships. In fact, considering Ogunmekan's report[18] on ten hemispherectomized patients with Sturge-Weber disease, a direct link emerged between GIQ and duration of the convulsive disorder and/or age at surgery. Their sample was relatively homogenous in terms of the clinical type of seizures (partial motor) and age of onset (eight weeks to eight months). In this study, the two patients who demonstrated intellectual deficiency after surgery (GIQ = 52 and 40) were operated later (nine and six years) than the eight others who were operated two weeks to nine months after seizure onset (mean GIQ: 93.63).

In order to somewhat control for the heterogeneity of the studies reported in Table 1, we grouped together a sample of 30 patients with Rasmussen's syndrome (mean GIQ: 68.47; mean age at onset of seizures: 4.88 yrs; mean age at surgery: 8.08 yrs). In contrast to Ogunmekan's study (1989) with Sturge-Weber patients, no significant correlations were found between age at onset of seizures (range: 1 to 12.2 yrs, r = –0.17), age at which surgery was performed (range: 2.6 to 17.07 yrs, r = –0.15) or duration of the epileptic condition (range: 0.06 to 11.07 yrs, r = –0.04) and GIQ in this subgroup of patients with Rasmussen's syndrome.

Table 1. Overview of studies carried out on hemispherectomy between 1972 and 1997

	Onset	Surgery	Side	Pre-evaluation	Age at testing	Post-evaluation
GOTT *ET AL.* (1972)						
Malignacy	8 y	10 y	L	6 y (Lorge-Thorndike) GIQ = 110	12 y	WISC (GIQ = 56, PIQ = 55, VIQ = 63)
Encephalic disorder	6 y	7 y	R	GIQ = 124	16 y	WISC (GIQ = 67, PIQ = 60, VIQ = 80)
SMITH *ET AL.* (1975)						
Cyanotic difficulties	3 y	6 y	L		21 y	GIQ = 116 PIQ = 102 VIQ = 126
DENNIS *ET AL.* (1975)						
Intractable epilepsy	6, 9 y	9 y	R		17, 5 y	WAIS (PIQ = 78, VIQ = 86)
Intractable epilepsy	4 y	20 y	R		28, 11 y	WAIS (PIQ = 80, VIQ = 95)
Intractable epilepsy	1, 1 y	6, 11 y	R		24, 2 y	WAIS (PIQ = 69, VIQ = 83)
Intractable epilepsy	4 m	5 m	R		9, 8 y	WISC (PIQ = 92, VIQ = 96)
Intractable epilepsy	1 y	10, 6 y	L		26, 4 y	WAIS (PIQ = 85, VIQ = 85)
Intractable epilepsy	10 m	14, 6 y	L		20, 11 y	WAIS (PIQ = 89, VIQ = 91)
Intractable epilepsy	8 y	13, 11 y	L		21, 8 y	WAIS (PIQ = 82, VIQ = 87)
Intractable epilepsy	7 m	17, 1 y	L		26, 2 y	WAIS (PIQ = 72, VIQ = 78)
Intractable epilepsy	1 m	5 m	L		8, 2 y	WAIS (PIQ = 87, VIQ = 94)
ZAIDEL *ET AL.* (1977)						
Glioma	8 y	10 y	R	Kuhlman Anderson (GIQ = 100)	13 y	WISC (GIQ = 56, PIQ = 55, VIQ = 63)
Intractable epilepsy	6, 7	7, 9 y	R	Stanford Binet (GIQ = 125)	16, 6 y	WISC (GIQ = 67, PIQ = 60, VIQ = 80)
VERITY *ET AL.* (1982)						
Congenital hemiplegia	4 y	8 y	L		24 y	GIQ = 70, PIQ = 63, VIQ = 79
Congenital hemiplegia	2 y	20, 10 y	R	WAIS (GIQ = 56)	32 y	GIQ = 60, PIQ = 68, VIQ = 69
Congenital hemiplegia	14 d	7, 4 y	R	3-year level	19 y	WAIS (GIQ = 71, PIQ = 72, VIQ = 73)
Congenital hemiplegia	1 y	8, 11 y	R	WISC (PIQ < 50, VIQ = 60)	13 y	WISC (VIQ = 56)
Congenital hemiplegia	3 m	10 y	R	developmentaly retarded (10 m)		severly retarded
Intractable epilepsy	7 m	3, 2 y	L	Standford Binet (GIQ = 73)	9 y	WISC (GIQ = 69, PIQ = 78, VIQ = 69)
Intractable epilepsy	3 y	7,7 y	L	Standford Binet (GIQ = 58)	8 y	WISC (PIQ = 38, VIQ = 44)
PTITO *ET AL.* (1987)						
Subdural hematoma	birth	8 y	L		19 y	GIQ = 59, PIQ = 78, VIQ = 57
Infantile hemiplegia	birth	12 y	L		21 y	GIQ = 62, PIQ = 63, VIQ = 72

Encephalitis	2, 5 y	9 y	L		19 y	GIQ = 51, PIQ = 49, VIQ = 65
Encephalopathy	birth	14 y	R		18 y	GIQ = 65, PIQ = 83, VIQ = 63
TINUPER *ET AL.* (1988)						
Chronic encephalitis	3 y	4 y	L		14 y	
Birth injury	3 d	15 y	L	GIQ = 59		GIQ = 74
Chronic encephalitis	5 y	4 y	L	GIQ = 65	14 y	GIQ = 65
Sturge-Weber	1 y	3 y	L		8 y	
Head trauma	15 m	4 y	R		8 y	
Chronic encephalitis	11 y	17 y	L	GIQ = 69	25 y	
Encephalitis	3 y	9 y	R	GIQ = 48	19 y	GIQ = 48
Head trauma	7 y	15 y	L	GIQ = 62	19 y	GIQ = 70
Sinus thrombosis	3 y	20 y	L	GIQ = 36	25 y	GIQ = 41
OGUNMEKAN *ET AL.* (1989)						
Sturge-Weber disease	6 m	10 m	L		12, 10 y	GIQ = 92
Sturge-Weber disease	4, 5 m	5 m	L		10, 5 y	GIQ = 93
Sturge-Weber disease	2, 5 m	9 y	L		11 y	GIQ = 52
Sturge-Weber disease	4 m	6 m	L		11, 6 y	GIQ = 99
Sturge-Weber disease	2 m	3 m	R		11, 3 y	GIQ = 93
Sturge-Weber disease	4, 5	5 m	R		4, 5 y	GIQ = 93
Sturge-Weber disease	8 m	11 m	L		3, 11 y	GIQ = 90
Sturge-Weber disease	2, 5 m	3 m	R		12, 3 y	GIQ = 99
Sturge-Weber disease	3 m	12 m	R		11 y	GIQ = 90
Sturge-Weber disease	8 w	6 y	L		9 y	GIQ = 40
SMITH *ET AL.* (1990)						
Perinatal insult	3 y	6 y	R	GIQ = 54 PIQ = 49 VIQ = 66	7, 3 y	GIQ = 87, PIQ = 82, VIQ = 95
VARGHA-KHADEM *ET AL.* (1991)						
Right infantile hemoplegia	15 m	16 y	L	unchanged	18, 5 y	PIQ = 63, VIQ = 69
Left infantile hemiplegia	3 y		R	PIQ = 73 VIQ = 55	26, 5 y	PIQ = 61, VIQ = 55
Rasmussen syndrome	4, 5 y	8, 75 y	L	PIQ = 60 VIQ = 60	17 y	PIQ = 70, VIQ = 69
Rasmussen syndrome	6 y	16 y	R	PIQ = 69 VIQ = 92	16, 8 y	PIQ = 49, VIQ = 68
Rasmussen syndrome	13 y	15 y	L	secondary school (normal range)	18 y	PIQ = 80, VIQ = 55
Rasmussen syndrome	10 y	15 y	R	normal school (above average)	18 y	PIQ = 75, VIQ = 80

Table 1. *Continued*

	Onset	Surgery	Side	Pre-evaluation	Age at testing	Post-evaluation
JAMBAQUÉ *ET AL.* (1992)						
Rasmussen syndrome	6 y	18 y	L	PIQ = 80 VIQ = 71	19 y	GIQ = 46
VARGHA-KHADEM *ET AL.* (1992)						
Rasmussen syndrome	12, 1 y	15, 04 y	L	GIQ = 45	16, 11 y	GIQ = 66, PIQ = 80, VIQ = 55
Rasmussen syndrome	6 y	17, 07 y	L	GIQ = 69 PIQ = 77 VIQ = 66	20, 02 y	GIQ = 68, PIQ = 77, VIQ = 64
Rasmussen syndrome	4, 03 y	8, 09 y	L	GIQ = 65 PIQ = 60 VIQ = 74	15 y	GIQ = 68, PIQ = 70, VIQ = 69
Rasmussen syndrome	2 y	6, 10 y	R	GIQ < 45 PIQ < 45 VIQ = 54	10, 05 y	GIQ = 48, PIQ = 48, VIQ = 57
Rasmussen syndrome	6, 02 y	6, 08 y[3]	R	GIQ = 105 PIQ = 101 VIQ = 108	11, 03 y	GIQ = 80, PIQ = 71, VIQ = 94
Rasmussen syndrome	5, 11 y	10, 01 y	R	GIQ = 79 PIQ = 69 VIQ = 92	16, 06 y	GIQ = 55, PIQ = 49, VIQ = 68
Rasmussen syndrome	4, 06 y	7 y	R	GIQ = 87 PIQ = 96 VIQ = 87	7, 11 y	GIQ = 85, PIQ = 73, VIQ = 98
Rasmussen syndrome	4 y	5, 08 y	R		7 y	GIQ = 50, PIQ = 48, VIQ = 60
VINING *ET AL.* (1993)						
Rasmussen syndrome	2 y	4, 5 y	L	GIQ < 50	27, 92 y	GIQ = 60
Rasmussen syndrome	6, 5 y	10, 5 y	R	GIQ < 50	32, 5 y	GIQ = 51
Rasmussen syndrome	7 y	14, 33 y	R	GIQ = 59	26, 75 y	GIQ = 67
Rasmussen syndrome	1, 5 y	4, 25 y	L	DQ = 50	10, 75 y	GIQ = 77
Rasmussen syndrome	2, 17 y	3, 83 y	R	GIQ = 81	9, 92 y	GIQ = 82
Rasmussen syndrome	4, 58 y	5, 25 y	R	DQ < 50	1, 09 y	GIQ = 52
Rasmussen syndrome	9, 5 y	12, 92 y	L		18, 5 y	GIQ = 44
Rasmussen syndrome	2, 25 y	5, 75 y	R	DQ = 50	11, 42 y	GIQ = 77
Rasmussen syndrome	5, 25 y	7, 67 y	L	GIQ = 82	12, 67 y	GIQ = 98
Rasmussen syndrome	7 y	20, 58 y	R	GIQ = 76	25, 25 y	college graduate
Rasmussen syndrome	5, 50 y	6, 92 y	L	GIQ = 70	10, 67 y	GIQ = 65
Rasmussen syndrome	12, 2 y	14, 33 y	R	GIQ = 92	18 y	GIQ = 92
ZATORRE *ET AL.* (1995)						
Chronic encephalitis	5 y	17 y	R		20 y	GIQ = 83, PIQ = 83, VIQ = 87
Hypoxia	birth	17 y	R		18 y	GIQ = 70, PIQ = 67, VIQ = 74

Prencephalic cyst	birth	20 y	L	28 y	GIQ = 88, PIQ = 88, VIQ = 90
Chronic encephalitis	10 y	20 y	R	28 y	GIQ = 74, PIQ = 64, VIQ = 83
Middle artery occlusion	birth	13 y	R	41 y	GIQ = 79, PIQ = 75, VIQ = 84
Porencephalic cyst	6 m	25 y	R	27 y	GIQ = 93, PIQ = 99, VIQ = 90
STARK *ET AL*. (1995)					
Rasmussen syndrome	4 y	18 y	L	24 y	WAIS (GIQ = 71)
Rasmussen syndrome	9 y	12 y	L	15, 11 y	WISC-R (GIQ = 44)
Rasmussen syndrome	1, 6 y	4 y	L	7, 11 y	WISC-R (GIQ = 77)
Rasmussen syndrome	5, 6 y	7, 5 y	L	9, 1 y	WISC-R (GIQ = 82)
Rasmussen syndrome	5, 6 y	5, 9 y	L	7, 3 y	WISC-R (GIQ = 65)
Rasmussen syndrome	2 y	5, 8 y	R	8, 8 y	WISC-R (GIQ = 77)
Rasmussen syndrome	4, 7 y	5 y	R	8, 6 y	WISC-R (GIQ = 52)
Rasmussen syndrome	1 y	3, 10 y	R	7, 4 y	WISC-R (GIQ = 82)
Rasmussen syndrome	7, 6 y	10 y	R	24 y	WISC-R (GIQ = 67)
ODGEN *ET AL*. (1996)					
Sturge-Weber	birth	10 m	L	27 y	GIQ = 75, PIQ = 72, VIQ = 79
Accidental blow	9 y	15 y	R	45 y	GIQ = 87, PIQ = 82, VIQ = 93
Infantile hemiplegia	6 y	8 y	L	53 y	GIQ = 71, PIQ = 64, VIQ = 78
VARGHA-KHADEM *ET AL*. (1997)					
Sturge-Weber	birth	8, 6 y	L	14, 1 y	GIQ = 59, PIQ = 53, VIQ = 52
MCCORMICK *ET AL*. (1997)					
Birth trauma	birth	8 y	R	14 y	WAIS-R (GIQ = 68, PIQ = 75, VIQ = 64)
Thrombosis	9 y	9 y	R	22 y	WAIS-R (GIQ = 68, PIQ = 62, VIQ = 75)
Chronic encephalitis		5 y	R	20 y	WAIS-R (GIQ = 83, PIQ = 83, VIQ = 87)
STARK *ET AL*. (1997)					
Rasmussen syndrome	1, 6 y	2, 6 y	L	9, 3 y	GIQ = 71, PIQ = 71, QV = 75
Rasmussen syndrome	2 y	3, 8 y	R	10, 11 y	GIQ = 81, PIQ = 70, QV = 95

Note: y = years, m = months, w = weeks, d = days.

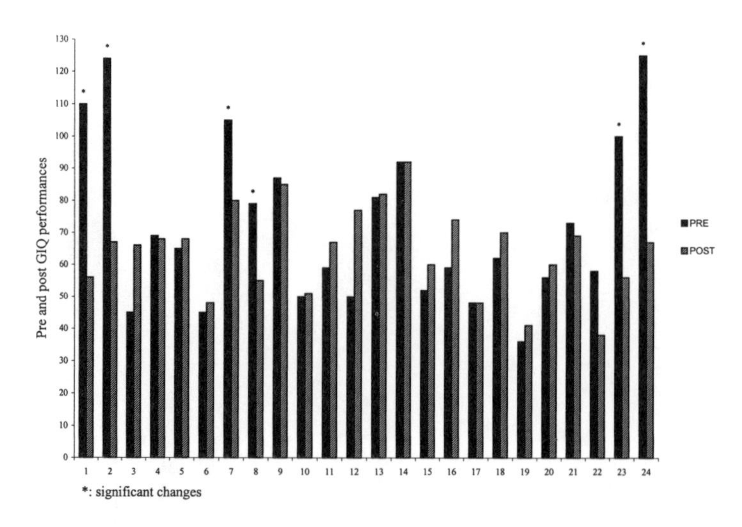

Figure 1. Pre and postoperative global IQ performances in 24 hemispherectomized patients.

However, the two groups (Sturge-Weber and Rasmussen) may not be directly comparable. In fact, seizures tend to appear earlier in Sturge-Weber disease (in the first months of life) than in Rasmussen's syndrome. Moreover, when considering the etiology of these two syndromes, it would appear that abnormalities are present at birth in the Sturge-Weber disease. The early appearance of seizures in this disorder, well before the development of certain cognitive abilities has been initiated, may thus have a more devastating effect on intellectual functions than Rasmussen syndrome, a disease that has a later and more gradual onset. These considerations warrant an early intervention especially when the diagnosis of Sturge-Weber disease is well documented[18] and the more so since recent studies[6] have shown that hemispherectomy does not constitute a higher medical risk when it is performed in the first months of life.

In contrast to the positive results obtained in patients with several epileptic syndromes (Rasmussen's syndrome, Sturge-Weber disease, infantile hemiplegia), the outcome of hemispherectomy in hemimegalencephaly, a rare congenital hypertrophy of one hemisphere, is more controversial. King *et al.*[12] were the first to report encouraging results in an infant suffering from hemimegalencephaly who underwent a right hemispherectomy at the age of 5 months. The surgery led to better psychomotor development. These authors and others[9] strongly advocated early intervention in this condition.

However, Vigevano *et al.*[39] were probably the first to consider the risk of an epileptic focus in the non-hemimegalencephalopathic hemisphere in this disease. In their series of 4 cases, two children were operated, and after surgery, they recovered from seizures. Surgery was avoided in the two others because of adequate treatment with anti-epileptic drugs. However, in the two operated cases, in spite of seizure control, severe mental and psychomotor retardation were observed. Vining *et al.*[41] and Wyllie at al.[49] also reported profound mental and psychomotor deterioration in their operated patients with hemimegalencephaly despite adequate seizure control. In fact, a recent group of studies (e.g., 10, 24) do indicate the presence of bilateral anomalies at the level of the cortex and white matter in patients presenting hemimegalencephaly. It would thus appear that the presence of structural and bilateral epileptic anomalies may account for the poor mental and psychomotor outcome sometimes observed after surgical intervention in children with hemimegalencephaly.

3. LANGUAGE FUNCTIONS

In his series of 12 hemispherectomies (10 left and 2 right) performed between the ages of 8 months to 21 years, Krynauw[14] did not report important language deficits in the ten patients who underwent left hemispherectomy. In two of them, a marked improvement of language was reported. He stated: "with regard to the language function, I suppose one might argue that hemisphere dominance had adjusted itself before removal of the affected hemisphere, and that it was the minor which was removed in all cases" (p. 267). Similarly, from the studies reviewed in Table 1, we sampled a group of 25 LH (mean age at onset of seizures: 3.27 yrs, range: 0–8 yrs; mean age at surgery: 9.69 yrs, range 5 months –17 yrs) and 28 RH patients (mean age at onset of seizures: 4.19 yrs, range: 0–10 yrs; mean age at surgery: 12.11 yrs, range: 5 months –20 years) for whom verbal IQ scores were available.[4,8,15,17,22,27,31,35,36,37,38] These results are illustrated in Fig. 2. Note that these two groups of patients were very similar on factors such as mean age at onset of seizure and mean age at surgery. No differences were obtained between RH (VIQ mean: 78.75) and LH patients (VIQ mean: 73.60) with regard to verbal IQ. Although IQ scores may only provide a crude description of linguistic functions, they nonetheless suggest that verbal intellectual abilities may be subserved by the remaining hemisphere.

In this context, numerous studies have attempted to account for cerebral reorganization in the case of early injuries by postulating the existence of equipotentiality in the acquisition of language. In 1977, however, in a review of their studies on hemispheric equipotentiality and language acquisition by right and left hemidecorticated patients, Dennis and Whitaker[4] indicated that "the two perinatal hemispheres are not equally at risk for language delay or disorder and that they are not equivalent substrates for language acquisition" (p. 103).

In fact, Piancentini and Hynd[20] note that a number of fundamental errors were present in studies of hemispherectomized patients that aimed at studying the notion of hemispheric equipotentiality for language. For example, comparisons are often made between left and right hemispherectomized patients even though they were operated at different ages. Mental deficiency is not always taken into account although this factor may influence language level in some patients. Finally, language was most often evaluated in a rudimentary fashion, thus limiting the conclusions that can be derived from these studies.

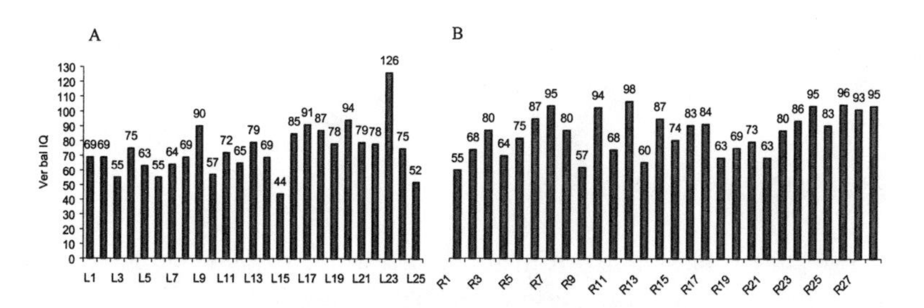

Figure 2. Postoperative verbal IQ scores obtained by A. Left (n = 25) and B. Right (n = 28) hemispherectomized patients.

Recent studies that have tried to control these factors more rigorously have highlighted the fact that the right hemisphere can, in case of surgery of the left hemisphere, support the general language functions for verbal communication on the expressive as well as on the receptive levels. For example, the five left hemispherectomized patients examined by Vining *et al.*[40] all developed fluent language after surgery despite the fact that some of them had been operated later in their childhood (mean age: 9.5 yrs). This conclusion is supported by the work of Vargha-Khadem *et al.*[37] who probably produced the most thorough study of language development following hemispherectomy. For instance, she reports the case of a patient with Sturge-Weber-Dimitri disease that started to have convulsions six days post-natally. A left hemispherectomy was performed at the age of 7 years and 7 months. Before surgery, the patient possessed the vocabulary level of a 4-year-old, communicated essentially with pointing, had an attentional deficit and a poor motivation to communicate. One month after surgery, he stopped taking any medication and he started to produce syllables and words. At the age of ten, he was capable of having an adequate conversation. However, difficulties with complex verbal material comprehension were noted. On the expressive level, he presented metaphonological, syntactic and grammatical deficits, a limited digit span (numbers and syllables), and he produced phonemic and semantic paraphasias in naming tasks. In this case, obviously, the right hemisphere supports the totality of the linguistic functions. Similarly, it would appear that speech production may generally be taken in charge by the right as well as the left hemisphere even when the surgery is performed late in childhood, sometimes even in early adolescence. However, the language deficits are much more important in patients with a resection of the left hemisphere, this being true even when the surgery was realized at a very early age.[31]

Certain linguistic abilities, in fact, cannot be completely assumed by the right hemisphere. In this context, Vargha-Kadhem *et al.*[35] have compared the performance of left and right hemispherectomized patients. They showed that even though the right hemisphere can offer sufficient support for verbal communication, certain linguistic capacities remain intimately linked with the integrity of the left hemisphere functions. The group of left hemispherectomized patients made more errors in a naming task, and syntactic and grammatical deficits were also reported. However, the deficits were less prominent in patients who had been operated early in life.

These results are in agreement with those of Stark *et al.*[31] who compared two cases of Rasmussen's syndrome where surgical intervention (LH: 2,6 yrs; RH: 3,8 yrs) was carried out soon after seizure onset (LH: 1,6 yrs; RH: 2 yrs). Before installation of the seizures, the two girls demonstrated normal cognitive and linguistic development. Postoperatively, improvement in linguistic skills was much less pronounced in the left hemispherectomized patient. Her language was characterized by word finding problems, and limited syntactic productions and comprehension. The language performances of the right hemispherectomized girl improved more rapidly and her linguistic development was higher than expected from her mental age.

4. ATTENTION AND MEMORY

Very few studies have assessed memory performance following hemispherectomy. In a group of eight patients with Rasmussen's syndrome, Vargha-Khadem and Polkey[36] pointed out that the Memory Quotient never reached the average level, fluctuating between the deficiency and low average ranges (mean: 69). Ptito *et al.*[22] also indicated visual (Benton Visual Retention Test) and verbal (Memory for Unrelated Sentences) memory problems in the four patients they studied, and this, regardless of the lateralization of the resected hemisphere. Similarly, in his large group of hemispherectomized patients (n = 44), Smith[28] found that the Benton Visual Retention Test was equally failed by left- and right-hemispherectomized patients (mean LH: 2.9, range: 0–7/10; mean RH: 2.2, range: 0–7/10). Similar conclusions can be drawn from the several studies where Digit Span subtest scores or Memory Quotients are available: both left and right hemispherectomy seem to affect memory and/or attention to the same extent (e.g., 3, 8, 29). For instance, Dennis and Kohn[3] who compared the performance of five left and four right hemispherectomized patients found that short term memory deficits were observed for both groups in the WAIS Digit Span Forward subtest, the average being 4.8 for the RH group and 5.4 for the LH group (normal average: 10). Even in the case described by Smith and Sugar[27] with superior verbal IQ (VIQ = 126), the performance on the digit span subtest remained in the lower average, being one of the lowest scores in this patient IQ subtests. These results would tend to indicate that attentional processes are almost invariably affected by the surgery and/or the pre-existing disease. These attentional limitations may, in turn, affect the patients' capacity to encode information, thus leading to apparent memory problems.

5. MOTOR FUNCTIONS

Severe epilepsy frequently induces motor deficits affecting coarse (posture, balance) as well as fine motor control (finger and hand dexterity). Generally, hemispherectomy is performed in patients who already demonstrate hemiparesis or hemiplegia in the hemibody contralateral to the affected hemisphere.

Motor deficits following hemispherectomy have been well described by Müller *et al.*[16] in cases of infantile hemiplegia syndrome. After surgery, the cranial nerve functions are generally preserved; residual motor functions seem better on the proximal than the distal level and more efficient on the legs than on the arms. This is probably due to the ipsilateral distribution of the motor projections in the axial and proximal regions of the body. The muscular spasticity, in fact, affects mostly the distal areas, and it would appear that these observations can be generalized to somato-sensory functions. Astereognosia, deficits in the guidance of exploratory movements and individual finger movements of the hand contralateral to the diseased hemisphere are often observed.

Beckung *et al.*[2] assessed motor functions (postural control, locomotion, fine motor development) in a group of 50 children operated for intractable seizures between eight months and 21 years. Postoperative improvements were already observed six months after surgery, but the most remarkable progress appeared two years later. In this large study, Beckung *et al.*[2] reports that the surgery rarely causes an amplification of the pre-existing motor deficits. On the contrary, there is a high probability of improvement in patients who become seizure-free or nearly so. Age at which surgery is performed seems to greatly condition functional motor recovery after hemispherectomy. Actually in the Beckung series, children who demonstrated unchanged or deteriorated motor functions were operated later in childhood (mean age: 13 yrs) whereas those operated at an early age (mean age: 10 yrs) showed improvements in their sensorimotor functions.

Pre-surgically, the pathological process may affect motor functions not only on the side contralateral to the affected hemisphere but also on the ipsilateral side. In a follow-up study of 32 hemispherectomized patients, Smith[26] pointed out that the postoperative

Purdue Pegboard scores with the "good" non hemiplegic hand were subnormal in 14/19 left, and 7/13 right hemispherectomized patients. Before surgery, the epileptic discharges originating from the diseased hemisphere secondary generalized to the rest of the brain and probably interfered with the fine motor functions subserved by the presumably intact hemisphere. In fact, it would appear that hemispherectomy may improve motor functions even on the side of the body ipsilateral to the hemispherectomy. Indeed, in a right hemispherectomized patient operated at 5 years of age, the performance of the hand ipsilateral to the lesion progressively increased (6, 12 and 19 months after surgery) on the Purdue pegboard test, a manual dexterity task.[26]

It is important to note that, at least in some cases,[16] age at surgery may not be the sole factor in determining the amount of recovery seen after surgery. Other factors such as the amount of brain reorganization that may have taken place prior to surgery, also need to been taken into account. Müller *et al.*[16] examined sensorimotor functions in a left-sided infantile hemiparesis patient who underwent right hemispherectomy at 18. This fifty-year-old woman demonstrated a normal performance of the axial and proximal left and right hemibody sides. This supposes that even before hemispherectomy, some of the functions that normally had to be supported by the affected hemisphere were delegated to the other one. In this context, reorganization of the motor pathways constitute an important factor which conditions residual motor functions observed after hemispherectomy as demonstrated by Benecke *et al.* (1991) using magnetic brain stimulation.

6. AUDITORY PERCEPTION

Whereas numerous reports have documented intellectual, linguistic and, to a lesser extent, mnesic functions following hemispherectomy, relatively few studies have explored certain cognitive spheres such as visuo- and auditory-perceptual analysis.

Hemispherectomized patients tend to show deficits in localizing stationary sound sources in the hemifield contralateral to the surgery but generally show a normal performance in the ipsilateral field.[21,46] However, they show a poorer performance in both hemifields when they are asked to localize moving sounds,[21] indicating that hemispherectomy may affect differently the ability to analyze moving and stationary sounds sources. It should be emphasized, however, that although hemispherectomized patients are far less accurate than non-operated subjects, auditory perception after hemispherectomy remains a relatively spared function. Furthermore, it would appear that early age of seizure onset is not a necessary condition for cerebral reorganization: a patient who was ten years old when seizures began as a result of a chronic encephalitis and who was operated at the age of 20, showed relatively good sound localization abilities.[46]

7. VISUAL PERCEPTION AND VISUO-SPATIAL FUNCTIONS

When considering visual perception, the study of hemispherectomized patients leads to the exploration of the largely documented phenomenon of "blindsight". This term refers to the residual visual capacities, without acknowledged perceptual awareness, that have been described in the hemianopic visual field contralateral to the diseased hemisphere. Indeed, certain patients remain able to accomplish in their blind field an implicit treatment of material like visual target localization, color discrimination, line orientation discrimination and simple geometric shape discrimination. Some may even be able to

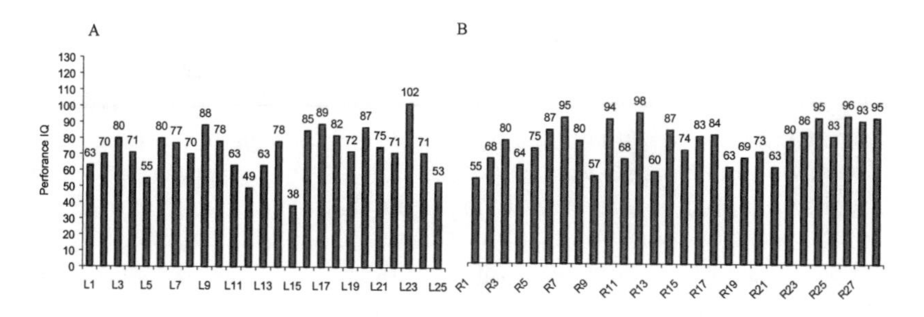

Figure 3. Postoperative performance IQ scores obtained by A. Left (n = 25) and B. Right (n = 28) hemispherectomized patients.

process more complex material such as three-dimensional shape discrimination in their hemianopic field (see Stoerig and Cowey for review).[32] However, recent studies question these findings. The "blindsight" found in some of these patients may, in fact, be extremely limited and some of their superior abilities[32,42] could find an explanation in methodological artifacts such as intraocular light scatter, uncontrolled rapid eye movements or low response criteria.[34]

Whereas post-hemispherectomy language examination generally tends to confirm the linguistic superiority of the left hemisphere, the superiority of the right hemisphere in visuo-spatial abilities is far less obvious.[36] There seems to exist a large variability in performance and this, regardless of the lateralization of the surgery: patients with either left or right hemispherectomy present visuo-spatial and visuo-perceptual deficits. Nevertheless, it would appear that the more demanding is the task in terms of visuo-spatial treatment, the more the performances of the subjects reflect a right hemisphere superiority.[36] However, these conclusions derive from studies that include a very limited number of subjects and that do not always compare equivalent groups of left and right hemispherectomized patients with respect to age at surgery and age at seizure onset. The one exception might be the study conducted by Smith[28] who administered the Hooper Visual Organization test to his large group of patients. No significant differences were observed between the left and right hemispherectomized patients (mean RH: 20.6, range = 0 to 28, mean LH: 15, range = 0 to 28/30).

A similar conclusion may be drawn from our Performance IQ (PIQ) score analysis of 25 patients with left hemispherectomy and 28 with right hemispherectomy. As shown in Fig. 3, in case of left hemispherectomy, the PIQ is not systematically superior to that found in patients with right hemispherectomy who have a comparable history in terms of age at onset of seizures or age at surgery (PIQ, LH: 72; RH: 69). It would thus appear that, as previously suggested by Smith (1977), "the right hemisphere and other residual structures after left hemispherectomy may be less committed for functional specialization" (p. 5).

CONCLUSION

In cases of unilateral hemispheric syndromes, there is now convincing evidence that hemispherectomy provides a beneficial effect for patients suffering from intractable seizures. Recent data show that the surgery may control seizures in 65 to 80% of the cases. Moreover, in many cases, antiepileptic medication is reduced or stopped after surgery.

Most of the studies reviewed above were dealing with patients showing early-onset epileptic seizures who underwent hemispherectomy before the age of 20. When assessing intellectual functions in these groups, it seems that GIQ is not related to factors such as age at onset of seizures, age at surgery, or duration of the epileptic syndrome. In addition, studies involving language demonstrate that even if the surgery is done early in life, the right hemisphere cannot take over all linguistics features normally carried out by the left hemisphere. At the motor level, positive results were observed both on the contralateral side of the lesion and the ipsilateral hemibody. The few studies that have addressed the issue of attentional and memory functions generally indicate that these processes are generally deficient after surgery and that the deficit does not seem to be material-specific. Finally, studies dealing with visuo-perceptive functions are too scarce to allow any specific conclusions.

The latter observation is a reminder of the cautions expressed by Dodrill *et al.*[5] with regard to the criteria that should be used to assess the effect of hemispherectomy on cognitive functions. In order to collect pertinent data, a neuropsychological evaluation should be performed before and after intervention. The post surgical evaluation should not be realized during the period immediately following the surgery, but maybe two years after, and a neuropsychological battery as complete as possible should be administered. The application of these criteria could perhaps allow to distinguish the cognitive deficits respectively associated with epilepsy and surgery and to browse a complete picture of the non-affected functions. It is important to point out that the majority of the studies concerned with the impact of hemispherectomy on neuropsychological functioning, including most of the work that was reported in this chapter, do not meet these criteria. For instance, some of the patients examined in these studies were assessed almost 27 years after their surgery and a number of them did not have any pre-surgical assessment.

Obviously, certain factors can limit the neuropsychological evaluation of epileptic patients, the principal one certainly being the convulsive disorder itself. It is easily conceivable that patients with continuous partial epilepsy cannot offer a satisfying level of attention during their evaluation. Some patients may even be impossible to evaluate before surgery. The medication, frequently aggressive, and its impact on the attentional capacities should also be considered. Moreover, it seems equally difficult to determine whether the progress observed after surgery is attributable to the surgery itself or to the discontinuation of anti-convulsive medications that are present in about 80% of the cases.

In conclusion, since the hemispherectomy itself does not induce an increase in cognitive functioning (as measured by IQ scales), it would probably be more appropriate to assess psychosocial features such as level of integration in the family, at school or at work. This latter type of study would first allow taking into account the real context in which these individuals interact, and would represent a more ecological observation of the positive effects of hemispherectomy on their quality of life.

REFERENCES

1. Andermann F, Freeman JM, Vigevano F and Hwang PALS (1993): Surgically remediable diffuse hemispheric syndromes. In Engel J Jr (ed): "Surgical Treatment of the Epilepsies", Second Edition. New York: Raven Press, pp 87–101.
2. Beckung E, Uvebrant P, Hedström A and Rydenhag B (1994): The effects of epilepsy surgery on the sensorimotor functions of children. Developmental Medicine and Child Neurology 36:893–901.
3. Dennis M and Kohn B (1975): Comprehension of syntax in infantile hemiplegics after cerebral hemidecortication: left-hemisphere superiority. Brain and Language 2:472–482.

4. Dennis M and Whitaker HA (1977): Hemispheric equipotentiality and language acquisition. In Segalowitz S, Frederic AG (eds): "Language Development and Neurological Theory". New York: Academic Press, pp 93–106.
5. Dodrill CB, Bruce PH, Rausch R, Chelune GJ and Oxbury S (1993): Neuropsychological testing for assessing prognosis following surgery for epilepsy. In Engel J Jr (ed): "Surgical Treatment of the Epilepsies", Second Edition. New York: Raven Press, Ldt., pp 511–518
6. Duchowny M, Jayakard P, Resnick T, Harvey AS, Alvarez L, Dean P, Gilman J, Yaylali I, Morrison G, Prats A, Birchansky S and Bruce J (1998): Epilepsy surgery in the first three years of life. Epilepsia 39:737–743.
7. Engel J Jr and Shewmon DA (1993): Who should be considered a surgical candidate? In Engel J Jr (ed): "Surgical Treatment of the Epilepsies", Second Edition. New York: Raven Press, pp 23–34.
8. Gott PS (1972): Cognitive abilities following right and left hemispherectomy. Cortex 9:266–274.
9. Humbertclaude VT, Coubes P, Robain O and Echenne B (1997): Early Hemispherectomy in a case of hemimegalencephaly. Pediatric Neurosurgery 27:268–271.
10. Jahan R, Mischel PS, Curran JG, Peacock WJ and Shields DW (1997): Bilateral neuropathologic changes in a child with hemimegalencephaly. Pediatric Neurology 17:344–349.
11. Jambaqué I, Pinard J-M, Dulac O and Ponsot G (1992): Neuropsychologie et chirurgie de l'épilepsie chez l'enfant. Approche Neuropsychologique des Apprentissages chez l'Enfant 1:9–14.
12. King M, Stephenson JBP, Ziervogel, Doyle D and Galbraith S (1985): Hemimegalencephaly—A case for hemispherectomy? Neuropediatrics 16:46–55.
13. Kalkanis SN, Blumenfeld H, Sherman JC, Krebs DE, Irizarry MC, Parker SW and Cosgrove GR (1996): Delayed complications thirty-six years after hemispherectomy: a case report. Epilepsia 37:758–762.
14. Krynauw RA (1950): Infantile hemiplegia treated by removing one cerebral hemisphere. Journal of Neurology, Neurosurgery and Psychiatry 13:243–267.
15. McCormick L, Nielsen T, Ptito M, Hassainia F, Ptito A, Villemure J-G, Vera C and Montplaisir J (1997): REM sleep dream mentation in right hemispherectomized patients. Neuropsychologia 35:695–701.
16. Müller F, Kunesch E, Binkofski F and Freund HJ (1991): Residual sensorimotor functions in a patient after right-sided hemispherectomy. Neuropsychologia 29:125–145.
17. Ogden JA (1996): Phonological dyslexia and phonoligical dysgraphia following left and right hemispherectomy. Neuropsychologia 34:905–918.
18. Ogunmekan AO, Hwang and Hoffman AJ (1989): Sturge-Weber-Dimitri disease: role of hemispherectomy in prognosis. Canadian Journal of Neurological Sciences 16:78–80.
19. Oppenheimer DR and Griffith HB (1966): Persistent intracranial bleeding as a complication of hemispherectomy. Journal of Neurology, Neurosurgery and Psychiatry 9:229–240.
20. Piacentini JC and Hynd GW (1988): Language after dominant hemispherectomy: are plasticity of function and equipotentiality viable concepts? Clinical Psychology Review 8:595–609.
21. Poirier P, Lassonde M, Villemure J-G, Geoffroy G and Lepore F (1994): Sound localisation in hemispherectomized patients. Neuropsychologia 12:541–553.
22. Ptito A, Lassonde M, Lepore F and Ptito M (1987): Visual discrimination in hemispherectomized patients. Neuropsychologia 25:869–879.
23. Rasmussen T (1983): Hemispherectomy for seizures revisited. Journal Canadien des Sciences Neurologiques 10:71–78.
24. Renowden SA and Squier M (1994): Unusual magnetic resonance and neuropathological findings in hemimegalencephalopathy: report of a case following hemispherectomy. Developmental Medicine and Neurology 36:357–361.
25. Shields D, Duchowny MS and Holmes GL (1993): Surgically remediable syndrome of infancy and early childhood. In Engel J Jr (ed): "Surgical Treatment of the Epilepsies", Second Edition. New York: Raven Press, pp 35–48.
26. Smith A (1974): Related findings in neuropsychological studies of patients with hemipherectomy and of stroke patients with chronic aphasia. Presented at the International Neuropsychological Society, Second Annual Meeting, Boston, February.
27. Smith A and Sugar O (1975): Development of above normal language and intelligence 21 years after left hemispherectomy. Neurology 25:813–818.
28. Smith A (1977): Language and nonlanguage functions after right or left hemispherectomy for cerebral lesions in infancy. Presented at the 5th Annual Meeting of the International Neuropsychological Society, February 3, Santa Fe, New Mexico.
29. Smith A (1981): Principles underlying human brain functions in neuropsychological sequelae of different neuropathological processes. In Filskov SB and Boll TJ (eds): "Handbook of Clinical Neuropsychology". New York: John Wiley & Sons, pp 175–226.

30. Stark RE, Bleilek K, Brandt J, Freeman J and Vining G (1995): Speech-language outcomes of hemispherectomy in children and adults. Brain and Language 51:406–421.
31. Stark RE and McGregor (1997): Follow-up Study of a right and a left-hemispherectomized children: implications for localization and impairment of language in children. Brain and Language 60:222–242.
32. Stoeirg P and Cowey A (1997): Blindsight in man and monkey. Brain 120:535–559.
33. Tinuper P, Anderman F, Villemure J-G, Rasmussen T and Quesney LP (1988): Functional hemispherectomy for treatment of epilepsy associated with hemiplegia: rationale, indications, results, and comparison with callosotomy. Annals of Neurology 24:27–34.
34. Tomaiuolo F, Ptito M, Marzi CA, Paus T and Ptito A (1997): Blindsight in hemispherectomized patients as reveale by spatial summation across the vertical meridian. Brain 120:795–803.
35. Vargha-Khadem F, Isaacs EB, Papaleloudi H, Polkney CE and Wilson J (1991): Development of language in six hemispherectomized patients. Brain 114:473–495.
36. Vargha-Khadem F and Polkey CE(1992): A review of cognitive outcome after hemidecortication in humans (Vol. 325). In Rose FD and Johnson DA (eds): "Recovery from Brain Damage: Reflections and Directions". Advances in Experimental Medicine and Biology.
37. Vargha-Khadem F, Carr LJ, Isaacs E, Brett E, Adams C and Mishkin M (1997): Onset of speech after left hemispherectomy in a nine-year-old boy. Brain 120:159–182.
38. Verity CM, Strauss EH, Moyes PD, Wada JA, Dunn HG and Lapointe JS (1982): Long-term follow-up after cerebral hemispherectomy: neurophysiologic, radiologic, and psychological findings. Neurology 32:629–639.
39. Vigevano F, Bertini E, Boldrini R, Bosman C, Claps D, di Capua M, di Rocco C and Rossi GF (1989): Hemimegalencephaly and intractable epilepsy: benefits of hemispherectomy. Epilepsia 30:833–843.
40. Vining EPG, Freeman JM, Brandt J, Carson BS and Uematsu S (1993): Progressive unilateral encephalopathy of childhood (Rasmussen' syndrome): a reapprasial. Epilepsia 34:639–650.
41. Vining EPG, Freeman JM, Pillas DJ, Uematsu S, Carson BS, Brandt J, Boatman D, Pulsifer MB and Zuckerberg A (1997): Why would you remove half a brain? The outcome of 58 children after hemispherectomy- The John Hopkings Experience: 1968 to 1996. Pediatrics 100:163–171.
42. Weiskrantz L (1986): Blindsight: A Case Study and Implications. New York: Oxford University Press.
43. Wyllie E, Comair YG, Kotagal P, Raja S and Ruggieri P (1996): Epilepsy surgery in infants. Epilepsia 37:625–637.
44. Wyllie E (1998): Surgical treatment of epilepsy in children. Pediatric Neurology 19:179–188.
45. Zaidel E (1977): Unilateral auditory language comprehension on the Token Test following cerebral commissurotomy and hemispherectomy. Neuropsychologia 15:1–18.
46. Zatorre RJ, Ptito A and Villemure J-G (1995): Preserved auditory spatial localization following cerebral hemispherectomy. Brain 118:879–889.

NEUROPSYCHOLOGICAL AND PSYCHO-SOCIAL CONSEQUENCES OF CORPUS CALLOSOTOMY

Hannelore C. Sauerwein*[,1], Maryse Lassonde[1], Olivier Revol[2], Francine Cyr[3], Guy Geoffroy[4], and Claude Mercier[5]

[1]Groupe de Recherche en Neuropsychologie Expérimentale
Département de Psychologie
Université de Montréal
C. P., 6128, Succ. Centre-Ville
Montréal, Qué. H3C 3J7, Canada
[2]Département de Pédopsychiatrie
Hôpital Neurologique et Neuro-chirurgical Pierre Wertheimer
69394 Lyon Cedex 03, France
[3]Département de Psychologie
Université de Montréal
C. P., 6128, Succ. Centre-Ville
Montréal, Qué. H3C 3J7, Canada
[4]Service de Neurologie
Hôpital Sainte-Justine
3175 Ch. Côte Sainte-Catherine
Montréal, Qué., H3T 1C5, Canada
[5]Service de Neurochirurgie
Hôpital Sainte-Justine
3175 Ch. Côte Sainte-Catherine
Montréal, Qué., H3T 1C5, Canada

INTRODUCTION

Corpus callosotomy has become a widely accepted surgical alternative for patients with intractable epilepsy who are unsuitable for focal cortical resection.[24,28] Although most beneficiaries of this surgery have been, and still are, adults the procedure has

* To whom correspondence should be addressed

Neuropsychology of Childhood Epilepsy, edited by Jambaqué et al.
Kluwer Academic / Plenum Publishers, New York, 2001.

gradually found its place in the treatment of severe, medically refractory epilepsy in children. Since the young brain presumably differs from the adult brain with respect to its adaptive capacities, the medical, neuropsychological and behavioral consequences of the surgery in children may not be the same as those seen in adult patients.

The main objective of callosotomy is to eliminate or reduce generalized seizures, thereby improving the patients' social adaptability and quality of life. However, there is as yet little consensus among centers with regard to the criteria for patient selection. Some think that the surgery is overused;[7] others feel that it is underutilized[22] in children. On the clinical level, there is reasonable agreement about the efficacy of corpus callosotomy in a subset of patients with unilateral lesions and major motor seizures or drop attacks.[11,22,23,30] On the other hand, several studies, including our own, have shown that children with multiple foci and diffuse brain disease may also benefit from the procedure.[13,20,22,29,30] Opinions also differ among centers regarding the kind of outcome that constitutes worthwhile improvement compared to the patient's current life situation. Unlike focal resection, callosotomy rarely renders the patient seizure-free, and outcome may not match expectations. Furthermore, evidence from patients having undergone temporal lobectomy suggests that freedom from seizures does not necessarily translate into a better quality of life.[3] For the same reason, several centers have chosen to exclude patients with severe mental retardation from the surgery,[12] even though many of these patients respond well.[8,13,20,23]

Comparisons of outcome in children are difficult for various reasons. Most studies have reported isolated cases or data from mixed samples (mostly adults). Furthermore, there are considerable differences between institutions with regard to the extent of the callosal section and the surgical procedure performed. In addition, the length of the follow-up varies among the cases reported in the literature. As Campbell[7] pointed out, the time elapsed since the surgery is a critical factor in the assessment of the outcome in view of the fact that there is considerable fluctuation of the seizures in individuals with intractable epilepsy. Moreover, it is not known whether or not the plasticity of the immature brain may work to the disadvantage of the patients by providing new pathways for the spreading of epileptic activity.

Finally, there is a paucity of studies that have addressed neuropsychological, behavioral and quality of life issues in the evaluation of the outcome of callosotomy in children. Most of the attention has focused on clinical outcome. A review by Nordgren[22] of the outcome in children aged 16 years and younger, reported by different centers, has demonstrated that the majority of the children had a greater than 80% reduction in their seizures with improvements of all seizure types. When cognitive or behavior changes were reported, these were generally considered to be an improvement with respect to the patients' preoperative level of functioning. The few exceptions (two cases in 50) that developed problems with speech and verbal memory were those with left hemisphere lesions in whom atypical speech representation was suspected. Recent studies of children and young adults operated at various centers in Australia[23] and Brazil[8] have yielded similar results.

Starting in 1978, the Department of Neurosurgery at the Hôpital Sainte-Justine in Montreal was among the first in North America to perform corpus callosotomy in a larger, predominantly pediatric patient population.[13,14,15] Over the past two decades, our team has concerned itself with the study of the long-term neuropsychological sequelae of the procedure in these children, both in terms of its effect on cognition and behavior.[20,27] and in terms of the potential compensation that may take place following hemispheric disconnection.[18,19,21] Emphasis has also been placed on the social adjustment and on the quality of life of the patients and their families. The follow-up of the first 38 patients (19 males and 19 females) to be reported here ranges from 3 to 17 years with an average length of 11 years. Sixteen of these patients (9 males and 7 females) and their families also participated in a study aimed at assessing quality of life following callosotomy.

1. METHOD

1.1. Subjects

The sample consisted of 22 children, aged 3.5 to 12 years (median age: 7 yrs), 8 adolescents, aged 13 to 16 years (median age: 15.5 yrs) and 8 young adults, aged 17 to 24 years (median age: 19.5 yrs) at the time of the surgery. Sixteen of the patients, aged 5 to 20 years (median age: 11 yrs) at the time of the surgery, participated in the quality of life study. Summary information about the patients is presented in Table 1. All patients had severely disabling, often life-threatening seizures of diverse etiologies that had resisted non-invasive therapy for at least three years. In many cases, adequate seizure control had never been achieved for any length of time. The frequent and severe seizures had a devastating effect on the children's cognitive and social development and interfered with their daily life activities and their education. Other selection criteria were generalized or secondarily generalized seizure type(s) and absence of a single focus in a resectable area. The majority of the patients (87%) had tonic-clonic seizures, mostly in combination with other seizure types. Forty-two percent had falling seizures which caused frequent injuries

Table 1. Summary information about the patients

Age at seizure onset (Mdn.: 15.5 mo.)		*Duration prior to surgery (Mdn.: 9 yrs)*	
Before age 1:	15 (39%)	2.5 to 6 yrs:	14 (37%)
Between 1 and 2 yrs:	7 (18%)	7 to 10 yrs:	9 (24%)
Between 2 and 5 yrs:	14 (37%)	11 to 15 yrs:	7 (18%)
After age 5:	2 (5%)	16 to 19 yrs:	8 (21%)
Age at callosotomy (Mdn.: 11.5 yrs)		*Length of post-op follow-up (Mdn.: 11 yrs)*	
3 to 6 yrs:	11 (29%)	3 to 6 yrs:	10 (26%)
7 to 12 yrs:	11 (29%)	7 to 10 yrs:	5 (13%)
13 to 16 yrs:	8 (21%)	11 to 14 yrs:	14 (37%)
17 to 24 yrs:	8 (21%)	16 to 19 yrs:	9 (24%)
Distribution of seizure type		*Seizure profile*	
Tonic-clonic (TC):	33 (87%)	One type (TC):	7 (18%)
Atonic:	18 (47%)	Two types:	18 (47%)
Absences:	16 (42%)	Three or more:	13 (34%)
Myoclonic:	10 (26%)	Frequent status:	5 (13%)
Partial complex:	5 (13%)	Focal origin:	24 (63%)
Tonic:	1 (3%)	Lennox-Gastaut:	7 (18%)
Pre-op mental status		*Social adjustment level*	
Severe deficiency:	17 (45%)	Chronic care facility:	2 (5%)
Moderate deficiency:	11 (29%)	Center/group home:	6 (16%)
Mild deficiency:	6 (16%)	At home, supervised:	30 (79%)
Borderline:	1 (3%)	At special school:	10 (26%)
Dull range:	2 (5%)	Normal schooling:	0
Low average:	1 (3%)	Working:	0

and required constant supervision. Five patients presented frequently with status epilepticus. In 24 cases (63%) the seizures originated from a single focus. Seven patients had diagnostic features compatible with Lennox-Gastaut syndrome.

Low IQ was not an exclusion criterion for corpus callosotomy in the present study. In fact, similar to other studies in pediatric populations[8,23] most of our patients (90%) were retarded. Half of them had an IQ below 30. Severity of mental retardation was linked to age at seizure onset: children with early onset (before the age of 2) were more severely retarded than those with later onset. Most of the children in this group had congenital pathology (e.g., tuberous sclerosis, hydrocephalus, cortical dysplasia, etc.). Longitudinal studies of the IQ profiles of those children that had developed normally prior to seizure onset revealed that prolonged exposure to intractable seizures was associated with stagnation of their mental development, thus increasing the lag between their chronological and their mental age over the years. At the time of the callosotomy, six patients lived in a rehabilitation center or a group home and two patients were permanently hospitalized. The other patients lived at home under constant supervision. Those able to attend school, were in special schools for the mentally handicapped.

The surgeries were performed between October 1978 and December 1990. In all but one of the cases, the corpus callosum was divided in a single operation following the procedure outlined by Geoffroy *et al.*[13] In three patients, the splenium was spared. One patient had received an anterior callosotomy earlier which had failed to produce satisfactory seizure control. Three other patients had undergone anterior temporal lobectomy several years prior to the callosotomy.

1.2. Material and Procedure

The patients were assessed pre- and postoperatively by a multi-disciplinary team composed of a neurologist, a neurosurgeon, a clinical psychologist, clinical and experimental neuropsychologists and a child psychiatrist. Clinical support staff, parents, teachers, social workers and resource personnel were at hand to supply additional information about the patients and their families. Initial assessment took place during the month leading up to the surgery and at various intervals after the callosotomy. During the first three weeks following surgery, the patients were observed on a daily basis, starting shortly after awakening from the anesthesia, in order to monitor the rate of recovery of motor and language functions and to assess the extent of functional disconnection. Once they were discharged from hospital, they were seen at intervals of six to eight months for the first two years and on a yearly basis thereafter up to five years. After five years, the follow-up was handled on an individual basis, depending on the availability of the patients and their caregivers. The results of these studies have been published previously.[13,14,18,19,20,27]

Between August 1995 and September 1996, all patients were contacted for reassessment. Detailed reports of their neurological status, including EEG records and CT scans, pertaining to the past five years, were obtained from the hospital files and from their attending neurologists. For patients living in other parts of the province or the country who were unable to come to our center extensive interviews with parents, psychologists, neuropsychologists or psycho-educators were conducted by telephone. Furthermore, the records of several patients living in specialized centers who had periodic in-house assessments were made available to us. These data were used in the current analysis of the outcome. On the other hand, the quality of life data reported in this study are only those obtained from patients and parents who were interviewed at our center.

Measures were obtained in three major areas: neuropsychological, behavioral and psycho-social. The results were evaluated in relation to the neurological outcome of the surgery.

Neuropsychological evaluation ranged from observation and administration of developmental scales in the severely retarded patients to the administration of a battery of standardized tests in the better cases (for description of the tests see 31). The following functions were evaluated by selecting the tests appropriate to the child's mental age and capacity: 1) intellectual development (Griffiths, Leiter, McCarthy, Stanford-Binet, WISC-R, WAIS-R); 2) attention and memory (Digit span, Benton's Visual Retention Test, Rey's Complex Figure and Auditory Verbal Learning Test, subtests of the McCarthy and Stanford-Binet intelligence scales, Tactual Performance Test, Trail Making Test); 3) expressive language (verbal of different intelligence scales, Token Test, observation of discourse and gestural forms of communication); 4) receptive language (Boehm's Basic Concepts for school and for pre-school children, Peabody Picture Vocabulary Test); 5) perception and perceptuo-motor skills (Hooper Visual Organization Test, Benton, Rey's Complex Figure, drawings, selected performance subtests of the intelligence scales) and 6) motor functions (Purdue Pegboard, Finger Tapping Test and motor subtests of the Griffiths and McCarthy scales). When possible, the patients were also submitted to a number of simple tactile transfer tasks, such as intra- and intermanual comparisons and naming of familiar objects and geometric shapes, cross localization of touch (see 19), to determine the degree of interhemispheric disconnection in the long term.

Behavioral adjustment and social competence were assessed by means of the Child Behavior Check List by Achenbach and Edelbrock[2] completed by parents, care-givers, teachers or resource personnel involved in the patients rehabilitation, as well as by observation of the patient in different social settings (hospital, home, school, institution, etc.).

Postoperative quality of life was measured with the aid of various inventories, including three self-constructed scales designed to assess the degree of satisfaction of the patient and his family with the outcome of the surgery. The first, a health-related quality of life (HQoL) measure was a 5-point analogous scale on which parents rated the well-being of the patient from very poor to excellent. The other two scales termed Patients' or Parents' Perception of the Child's Surgery (PATPCS and PARPCS, respectively) consisted of five items concerning the impact of the surgery on the emotional, educational and social adjustment of the patient to be rated on a 5-point scale. At the end of the scale, patients and parents were invited to rate the child's overall QoL before and after the surgery on a percentage basis. The other two scales: Quality of Life of Principal Caregiver (QoLPC: 17) and Parenting Stress Index (PSI: 1), evaluated the personal investment in terms of time, energy and emotional well-being of the individual(s) directly involved in the care of the patient. Since quality of life was assessed only after the callosotomy in the earlier cases, the results of these two scales were compared to data obtained from a closely matched sample of children and adolescents with intractable seizures who had never undergone brain surgery (unoperated control group). The questonnaires were administered to patients (when possible) and parents in a semi-structured clinical interview during which the participants also had the opportunity to relate their personal experiences.

1.3. Criteria for Neurological and Neuropsychological Outcome

Clinical outcome of the callosotomy was based on a modified version of the classification proposed by Engel *et al.* (10: see Table 2). For neuropsychological and

Table 2. Criteria for neurological and neuropsycholocial outcome of callosotomy

	NEUROLOGICAL OUTCOME	NEUROPSYCHOLOGICAL OUTCOME
Class I:	Free of disabling seizures, abolition or significant reduction (2–3/mo.) of partial seizure	Important improvement of behavioral adjustment, gains in cognitive and psycho-social functioning
Class II:	Rare disabling seizures, reduction of seizures of at least 50%	Improvement of attention, behavioral adjustment and psycho-social functions, greater autonomy
Class III:	Abolition of status epilepticus, general reduction of seizures, better medical management	Improved state of vigilance, improvement of sensorimotor functions, more responsive
Class IV:	No deterioration, little improvement in the long term	No consistant changes in the long term
Class V:	More seizures or addition of new seizure types	Mental deterioration and/or development of new behavior problems

behavioral outcome in children, no generally accepted guidelines exist so far in the literature. In our previous studies,[20] individual improvement had been determined by using the ratio of improved functions over the total number of functions tested which allowed us to include those children who could not be subjected to a complete battery of tests before the surgery.

In the present study, the interest focused on the more global benefits of the surgery with respect to the patients' cognitive and psycho-social adaptation in the long term. For this reason, a classification system was devised according to which class I outcome included only those patients who showed persistent and important long-term cognitive and social benefits, such as gains of 5 or more points in IQ and improvement in social adjustment, defined as the level of independence relative to their age. Class II outcome was assigned to patients who, without any appreciable long-term gains in cognitive functions, showed important improvements in attention, behavior and psycho-social adaptation to the effect that they would be able to be accepted in a group home and/or to find work in a supervised environment. Class III outcome denoted patients who showed modest but worthwhile improvements in their general behavior and adaptive capacities. Class IV outcome applied to children or young adults who showed few or no gains in the long term. Also included in this group were patients who had shown considerable fluctuation in their functioning over the years and who therefore could not be assigned with confidence to any of the first three categories. Class V was reserved for patients whose postoperative losses exceeded gains.

Ad hoc criteria for quality of life were not established since this variable seems to depend to a great extent on the expectations of the patients and/or their families and their level of satisfaction regarding the outcome of the surgery.

2. RESULTS

2.1. Neuropsychological and Social Outcome

The outcome of the surgery is presented in Table 3. Thirty patients (79%) showed an overall improvement in neuropsychological and psycho-social functioning. An equal number of cases derived long-term biological benefits from the procedure. Clinically,

Table 3. Neuropsychological outcome in relation to neurological outcome, pre-operative mental state, age at seizure onset and age at callosotomy

Neuropsychological outcome		*Neurological outcome*	
Class I:	3 (8%)	Class I:	15 (39%)
Class II:	9 (24%)	Class II:	7 (18%)
Class III:	18 (47%)	Class III:	8 (21%)
Class IV:	7 (18%)	Class IV:	7 (18%)
Class V:	1 (3%)	Class V:	1 (3%)
Total improved:	30 (79%)	Total improved:	30 (79%)

*Outcome in relation to mental state**

Pre-op IQ level	*Class I*	*Class II*	*Class III*	*Class IV*	*Class V*
Severe deficiency:	— (5)	— (1)	10 (5)	6 (6)	1 —
Moderate deficiency:	1 (5)	5 (4)	5 (1)	— (1)	— —
Mild deficiency:	— (3)	2 (2)	3 —	1 —	— (1)
Borderline:	1 (1)	— —	— —	— —	— —
Dull range:	1 (1)	1 —	— (1)	— —	— —
Low average:	— —	1 —	— (1)	— —	— —

*Outcome in relation to age at seizure onset**

Age at onset	*Class I*	*Class II*	*Class III*	*Class IV*	*Class V*
before age 1:	— (3)	2 (1)	7 (4)	6 (6)	— (1)
between 1 and 2 yrs:	1 (4)	1 (2)	3 (1)	1 —	1 —
between 2 and 5 yrs:	2 (7)	6 (3)	6 (3)	— (1)	— —
after age 5:	— (1)	— (1)	2 —	— —	— —

*Outcome in relation to age at callosotomy**

Age at surgery	*Class I*	*Class II*	*Class III*	*Class IV*	*Class V*
3 to 6 yrs:	2 (6)	— (1)	7 (2)	2 (2)	— —
7 to 12 yrs:	1 (5)	3 (2)	5 (2)	2 (2)	— —
13 to 16 yrs:	— (1)	1 (2)	5 (2)	2 (2)	— (1)
17 to 24 yrs:	— (3)	5 (2)	1 (2)	1 (1)	1 —

*Neurological outcome in brackets.

there was a significant reduction of generalized seizures. Drop attacks were abolished in all but two of the cases. Absences and partial complex seizures were also reduced in many patients. Three patients (8%) with class I outcome have been seizure-free since the surgery; two of them no longer require medication. Two others have only had one or two focal attacks in 10 years, even though one of them has developed a new independent focus in the opposite hemisphere. Several patients with class II neurological outcome have further improved over the years, an effect that may be in part accounted for by a new class of antiepileptic drugs, which several of the patients in the present study were taking at the time of the reassessment. Another patient with initial class II outcome had a temporary relapse in conjunction with a newly diagnosed orbital tumor. Following successful removal of the neoplasm, her seizures became once more well controlled. Seven patients (18%) did not improve in the long term, even though some of them have shown periodic improvements lasting up to two years at a time. These included mostly patients with multi-focal lesions. In one of these patients Rhett syndrome was diagnosed. Only one patient who experienced post-surgical complications (infection) showed an increase in seizures. This patient has developed a new anterior focus in addition to her temporal lesion.

Although neuropsychological outcome paralleled clinical improvement, the pattern was not identical. Only three of the patients (8%) had class I outcome. Two of these patients, now young adults, live independently, one is married and the mother of an infant. The third patient, a child of 9 years, is attending a regular school. Patients with class II outcome have shown few gains in the cognitive domain but they have made important progress in their psycho-social adjustment. Most of them live in group homes and work in sheltered workshops. One of these young adults has his own apartment and works for a small private company. Two others have found supervised employment in the private sector. The remaining patients attend a specialized center or live with their families with less need for supervision. Patients with class III outcome remain dependent on their caregivers. However, their greater alertness and improved motor control have facilitated their integration into specialized institutions. Sixteen (89%) attend centers or schools for the handicapped. Two live in group homes. Seven (18%) of the patients have failed to show any consistent long-term improvements. These are mostly patients in whom no satisfactory seizure control has been achieved in the long term. Three of them remain in a chronic care facility. The only deterioration was seen in a 15-year-old girl who developed postoperative mutism. This left-handed patient with left hemisphere atrophy is probably one of the cases of crossed dominance where compensation for impaired language functions in the lesioned hemisphere seems to have operated through the corpus callosum. Preoperative amytal exploration of speech lateralization was not possible in this severely retarded patient who had very little expressive speech (mostly echolalia) preoperatively. Ironically, this patient has class I neurological outcome which has considerably improved her overall quality of life, as perceived by her caregivers.

To determine which of the functions improved or deteriorated after surgery, the performances on each of the measures were examined separately. The profile of the 38 patients is presented in Fig. 1. The most important gains were seen in the children's behavioral adjustment. The patients were more alert and more autonomous. Behavior problems such as aggressiveness and hyperactivity abated. Next to behavior, motor performance was the domain in which greatest improvements were noted. Many children, who had been hypotonic before the surgery and who were wheelchair-bound to protect them from injury, became ambulatory. Gains in the cognitive domain were more modest. Half of the children showed slight improvements in memory. These were essentially attributable to an increased attention span that allowed the children to concentrate more adequately on the materiel to be learned. Expressive and receptive language also improved. Some of the younger children, who had only used gestural language, started to express themselves verbally. Intellectual capacities did not change to any extent with

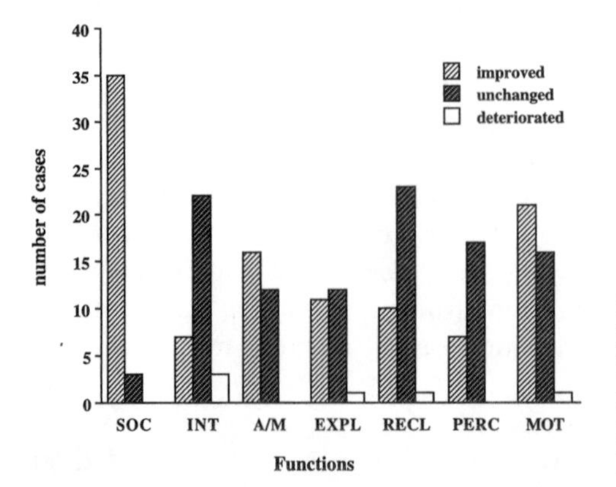

Figure 1. Postoperative outcome of neuropsychological functions. SOC: Social Adjustment; INT: Intellectual Abilities; A/M: Attention and Memory; EXPL: Expressive Language; RECL: Receptive Language; PERC: Perceptual Abilities; MOT: Motor Functions.

the exception of five patients who showed an increase in these functions after the operation. Three of these children have further progressed over time. One child has gained 20 IQ points in the course the first two years. This patient was one of several who had developed normally before seizure onset. A deterioration of IQ was observed in three patients, including the patient who has been rendered mute by the operation. The other two patients showed decrements of 10 to 14 points in their performance IQ. Both of these patients had a postoperative edema in the right hemisphere which affected their speed of performance. Both also had a documented right hemisphere lesion, and it is possible that callosally-mediated compensation was interrupted by the surgery similar to the adult patients described by Campbell *et al.*[6]

Analysis of the outcome with respect to IQ, age at seizure onset and age at the time of the callosotomy (Table 3) revealed a relationship between the patients' IQ and neuropsychological outcome: a larger proportion of children with higher preoperative IQ had class I and class II outcome. There was no such link between IQ-level and seizure outcome. As far as age at seizure onset was concerned, our data showed that most patients with early onset (before age 1) had class III and class IV neuropsychological outcome, even though their clinical outcome was more favorable in many cases. Age at callosotomy had an effect on outcome to the extent that only children operated before age 12 had class I outcome. However, a large proportion of patients operated as young adults had class II neuropsychological outcome, suggesting that age at callosotomy is not a powerful predictor of the patients' social adjustment following callosotomy.

Age at surgery, however, had an important effect on the course of recovery from postoperative sequelae. While all patients had initially displayed variable degrees of kinetic hemiapraxia of the non-dominant hand, the most common transient sequelae of hemispheric retraction during surgery, the younger patients had recovered more rapidly and more completely than the older patients.[18,19,20] Furthermore, they had shown very few, if any, of the typical disconnection deficits in simple interhemispheric transfer tasks, whereas all patients aged 13 years and older had exhibited the disconnection syndrome that is usually seen in adult patients.[4] These symptoms were still demonstrable at this latest testing session.

2.2. Quality of Life Following Callosotomy

Parents and patients reported a 40% benefit in life comfort following callosotomy on the PTPCS and PRPCS scale. Although the benefit was closely related to seizure outcome, inspection of individual results revealed that there was no systematic relationship between the success of the surgery and the improvement of the life situation of the patient and his family as perceived by the parents. In fact, class III improvement was associated with higher ratings in subjective quality of life than class I or II outcome in several cases. Most parents of patients with modest clinical and neuropsychological outcomes reported significant benefits. These included more shared activities such as traveling or eating at a restaurant, none of which had been possible before the surgery because of the patient's serious condition or behavior problems. In contrast, postoperative QoL was considered poor by parents who had hoped that their children would show greater progress in their cognitive development once better seizure control was achieved. Nevertheless, compared to non-operated controls, fewer operated patients were reported as having a poor health-related quality of life (HQoL: 15% call. vs. 30% contr.). Furthermore, on the QoLPC scale, a smaller percentage of parents or caregivers of the operated group reported that their own quality of life was affected (18% call. vs. 60% contr.). On the other hand, there was no difference between the two groups with respect to the stress and the emotional impact of the child's condition on the family (PSI: mean call. = 45.47; mean contr. = 46.0) which suggests that caring for a polyhandicapped child remains stressful for the family, even when a better medical and behavior management has been achieved.

3. DISCUSSION

The results of the present study reaffirm our earlier contention that callosotomy is a viable option for children and adolescents with severe intractable epilepsy.[13,14,20] Seventy-nine percent of the patients benefited from the surgery both clinically and behaviorally. These findings are congruent with those of other studies in similarly large series. In a review by Spencer,[2] 87% of the patients (children and adults) showed worthwhile clinical improvement following complete callosotomy. Sass *et al.*[25] reported clinical and neuropsychological benefits in all four patients (one adolescent and three young adults) having undergone total single-stage callosotomy as well as in 64% of patients with partial section. Satisfactory seizure control was also observed in 65% of children and adults operated at five Australian centers.[23] In the studies reviewed by Nordgren[22] 74% patients with complete callosal section had favorable seizure outcome. Similarly, 73.5% of the children reported by the Brazilian team[8] benefited of the surgery; 81% of the latter showed marked improvement in behavior. More importantly, no deterioration in mental state was observed by these authors.

In the present study, preoperative intellectual potential was the most reliable predictor of the degree of postoperative cognitive and behavioral gains. Children with mild or moderate mental deficiency had better neuropsychological outcomes than the more severely retarded ones. Twelve patients (31%) progressed from complete dependency to independent living (3 cases), life in group homes with work in special facilities (8 cases) or normal school attendance (1 case). Nevertheless, we maintain that low IQ should not be an exclusion criterion in view of the social and behavioral benefits that were associated with improved seizure control in the retarded patients. We agree with Cendes *et al.*[8] that behavior improvement is one of the major benefits of callosotomy in children.

Earlier seizure onset was related to more modest neuropsychological and behavioral outcome than later onset. These results are most likely attributable to the severity of the underlying neurological pathology which imposes greater limitations on the child's cognitive and social development. These patients were also more severely retarded. Longer exposure to uncontrollable seizures was found to lead to stagnation or deterioration of cognitive functioning which was irreversible in most cases. Only a few patients (10%) in the present study continued to progress after the surgery. Early intervention may help prevent further deterioration and, in doing so, may prove to be more cost-effective in the long run.

Early callosotomy was associated with milder postoperative sequelae and fewer disconnection deficits. All patients aged 13 years and older displayed the typical disconnection syndrome in the tactile modality which was still present after many years, similar to the callosotomized adults studied by Bogen and Vogel[5] and Goldstein *et al.*[16] The patients are usually unaware of the deficits since unrestricted exposure to the perceptual scene assures bilateral representation of sensory experience.

Postoperative mutism occurred in one of our young adults. This is one of the cases of crossed dominance where left hemisphere damage resulted in incomplete transfer of language functions. In these cases, surgical disconnection of the hemispheres abolishes interhemispheric compensation. Several authors[10,15,26] have suggested to exclude patients in whom atypical language lateralization is suspected and who cannot be submitted to amytal testing. No other permanent disability was created by the surgery.

Patients who failed to show clinical and neuropsychological benefits (18%) were mainly those with multiple foci and diffuse neurological pathology. The observation that some of these patients showed important improvements over long periods of time before relapsing may reflect the development of new epileptogenic foci or structural lesions. This was the case in one of our patients who developed a tumor. In two other patients, a new independent focus was identified. One of these patients presented more seizures postoperatively, an outcome that was not predictable from the preoperative EEG pattern or seizure profile.

However, the general picture was one of improvement. This was also reflected by the patients' and parents' perception of the outcome of the surgery. Quality of life reportedly improved for many of the patients, including those with more modest behavioral changes. The link between age and quality of life improvement reported by Claverie and Rougier,[9] suggesting that younger patients derive greater benefits, was not observed in our small sample. The authors attributed this effect to parental attitudes which tended to be more positive when the child was still young and had not yet been entrenched for many years in the medicosocial system. A negative correlation between outcome and QoL benefits was obtained for several of our cases who showed notable clinical improvement but fewer cognitive gains than expected by their parents. Similar results have occasionally been reported for patients with focal cortical resection.[3] Observations like this stress the necessity for preoperative establishment of realistic goals and postoperative counseling to help the patients and their families to adjust their expectations.

REFERENCES

1. Abidin RR (1990): Parenting Stress Index—Short Form. Test Manual. Charlottesville, Va.: Pediatric Psychology Press.
2. Achenbach TM and Edelbrock C (1993): Manual for the Child Behavior Check List, Burlington, VT: Department of Psychiatry.
3. Bladin PF (1992): Psychosocial difficulties and outcome after temporal lobectomy. Epilepsia 33:898–907.
4. Bogen JE (1985): The callosal syndrome. In Heilman KM and Valenstein E (eds): "Clinical Neuropsychology". New York: University Press, pp 308–359.
5. Bogen JE and Vogel PJ (1975): Neurological status in the long term following complete cerebral commissurotomy. In Michel F and Schott B (eds): "Les syndromes de disconnection calleuse chez l'homme". Lyon: Hôp Neurol, pp 227–5116.
6. Campbell AL jr., Bogen JE and Smith A (1981): Disorganization and reorganization of cognitive and sensorimotor functions in cerebral commissurotomy: compensatory roles of the forebrain commissures and cerebral hemispheres in man. Brain 104:493–551.
7. Campbell R (1991): Corpus callosotomy is an overutilized procedure in children. Journal of Epilepsy 4:67–71.
8. Cendes F, Ragazzo PC, da Costa V and Martins LF (1993): Corpus callosotomy in treatment of medically resistant epilepsy: Preliminary results in a pediatric population. Epilepsia 34:910–917.
9. Claverie B and Rougier A (1995): Life comfort and psychosocial adjustment linked to age at the time of anterior callosotomy. Journal of Epilepsy 8:321–331.
10. Engel J jr., Van Ness PC, Rasmusen TB and Ojeman LM (1993): Outcome with respect to epileptic seizures. In Engel J jr. (ed): "Surgical Treatment of the Epilepsies", Second Edition. New York: Raven Press, pp 609–621.
11. Gates JR, Rosenfield WE, Maxwell RE and Lyons REW (1987): Response of multiple seizure types to corpus callosum section. Epilepsia 28:2834.

12. Gates JR, Wada JA, Reeves AG, Lassonde M, Papo I *et al.* (1993): Reevaluation of corpus callosotomy. In Engel J jr. (ed): "Surgical Treatment of the Epilepsies", Second Edition. New York: Raven Press, pp 637–648.
13. Geoffroy G, Lassonde M, Deslisle F and Décarie M (1983): Corpus callosotomy for intractable epilepsy in children. Neurology 33:891–897.
14. Geoffroy G, Lassonde M, Sauerwein H and Décarie M (1986): Effectiveness of corpus callosotomy for control of intractable epilepsy in children. In Lepore F, Ptito M and Jasper HH (eds): "Two Hemispheres, One brain". New York: Alan Liss, pp 361–68.
15. Geoffroy G, Sauerwein H, Lassonde M and Décarie M (1990): Inclusion and exclusion criteria for corpus callosotomy in children. Journal of Epilepsy 3 (Suppl):205–11.
16. Goldstein MN, Joynt RJ and Hartley RB (1975): The long-term effects of callosal sectioning: report of a second case. Archives of Neurology 32:52–23.
17. Hooyman N, Gonyea J and Montgomery R (1985): The impact of in-home termination on family caregivers. Gerontologist 25:141–145.
18. Lassonde M, Sauerwein H, Chicoine AJ and Geoffroy G (1991): Absence of disconnexion syndrome in callosal agenesis and early callosotomy: brain reorganization or lack of structural specificity during ontology? Neuropsychologia 29:481–95.
19. Lassonde M, Sauerwein H, Geoffroy G and Décarie M (1986): Effects of early and late transection of the corpus callosum in children. Brain 109:953–67.
20. Lassonde M, Sauerwein H, Geoffroy G and Décarie M (1990): Long-term neuropsychological effects of corpus callosotomy in children. Journal of Epilepsy 3 (Suppl):279–286.
21. Lassonde M, Sauerwein H, McCabe N, Laurencelle L and Geoffroy G (1988): Extent and limits of cerebral adjustment to early section or congenital absence of the corpus callosum. Behavioural Brain Research 30:165–81.
22. Nordgren RE (1991): Corpus callosotomy is an underutilized procedure in children. Journal of Epilepsy 4:73–80.
23. Reutens DC, Bye AM, Hopkins IJ, Danks A, Somerville E *et al.* (1993): Corpus callosotomy for intractable epilepsy: Seizure outcome and prognostic factors. Epilepsia 34:904–909.
24. Robert DW, Rayport M. Maxwell RE, Olivier A and Marino jr. R (1993): Corpus callosotomy. In J Engel jr. (ed): "Surgical Treatment of the Epilepsies". New York; Raven Press, pp 519–526.
25. Sass K, Novelly R, Spencer D and Spencer SS (1990): Post callosotomy language impairments in patients with crossed cerebral dominance. Journal of Neurosurgery, 72:85–90.
26. Sass K, Spencer DD, Spencer SS, Novelly RA, Williamson PD and Mattson RH (1988): Corpus callosotomy for epilepsy. II. Neurologic and neuropsychological outcome. Neurology 38:24–28.
27. Sauerwein HC and Lassonde M (1994): Cognitive and sensori-motor functioning in the absence of the corpus callosum: neuropsychological studies in callosal agenesis and callosotomized patients. Behavioural Brain Research 64:229–240.
28. Spencer SS (1988): Corpus callosum section and other disconnection procedures for medically intractable epilepsy. Epilepsia 29 (Suppl):589–595.
29. Spencer DD and Spencer SS (1989): Corpus callosotomy in the treatment of medically intractable secondarily generalized seizures of children. Cleveland Clinic Journal of Medicine 56 (Suppl 1):S69–S78.
30. Spencer SS, Spencer DD, Williamson PD, Sass KJ, Novelly RA and Mattson RH (1984): Corpus callosotomy for epilepsy. I. Seizure effects. Neurology 38:19–24.
31. Spreen O and Strauss E (1991): A Compendium of Neuropsychological Tests. New York: Oxford University Press.
32. Williamson PD (1985): Corpus callosum section for intractable generalized epilepsy: Criteria for patient selection. In Reeves AG (ed): "Epilepsy and the Corpus Callosum". New York: Plenum Press, pp 243–257.
33. Wilson DW, Reeves A and Gazzaniga M (1978): Division of the corpus callosum for uncontrollable epilepsy. Neurology 28:649–553.

27

COGNITIVE SIDE-EFFECTS OF ANTIEPILEPTIC DRUGS

Albert P. Aldenkamp

Faculty of Social and Behavioural Sciences
University of Amsterdam and Department of Behavioral Science and Psychological Services
Epilepsy Centre Kempenhaeghe
P.O. Box: 61, NL-5590 A.B. Heeze
The Netherlands

INTRODUCTION: THE COGNITIVE SIDE-EFFECTS OF ANTIEPILEPTIC DRUGS

The possibility that cognitive impairment may develop as a consequence or aftermath of epilepsy was raised as early as 1885 when Gowers described "epileptic dementia" as one of the pathological sequelae of seizures. Nonetheless the topic was not coupled to antiepileptic drug (AED) treatment until the 1970s.

Antiepileptic drug treatment may be associated with a variety of side-effects. Some effects appear immediately after the start of drug exposure, such as nystagmus, but are relatively benign because they show habituation,[27] or are reversible when they are dose dependent. Others may be of insidious onset, emerging only after extended periods of treatment (i.e., "chronic side-effects"). A multitude of such chronic side-effects have been documented,[39] but most effects concern the central nervous system. This chapter reviews some our knowledge about a specific subgroup of such CNS-related chronic side-effects, i.e., the cognitive side-effects of AED-treatment.

Such effects are considered as much more moderate than, for instance, some of the idiosyncratic reactions to drugs. Nonetheless, a number of studies have claimed that drug-induced cognitive impairment may have a much greater impact on daily life functioning than had hitherto been suspected,[44,45,46] for example through the impact on critical functions, e.g., learning in children,[4] driving capacities in adults (often requiring milliseconds precision), or on vulnerable functions such as memory in the elderly. Moreover, as cognitive side-effects represent the long-term outcome of AEDs, the effects may increase

Neuropsychology of Childhood Epilepsy, edited by Jambaqué et al.
Kluwer Academic / Plenum Publishers, New York, 2001.

with prolonged therapy, which may contribute to the impact on daily life functioning in refractory epilepsies.[15]

The interest in the cognitive side-effects of AED-treatment is of recent origin and the first studies date from the early 1970's,[21,26] probably stimulated by the widening range of possibilities for drug treatment during that period. Although some general reviews have been published (references in the next paragraph), we considered it appropriate to attempt a first systematic review of the literature on "cognitive side-effects of antiepileptic drugs" in lines of evidence-based-medicine, i.e., reviewing the empirical data that were published in peer-reviewed journals.

1. METHODOLOGY OF THE LITERATURE REVIEW

Potentially relevant studies were identified through computerized and manual searches of the English-language literature published from January 1970 through December 1994. A computerized search of the DIMDI database was conducted. In addition, the bibliographies of several reviews on the same topic were examined.[19,22,36,41,44,45,47,48,49] Criteria for selection of the papers were: a) English language reports of original research, published in peer-reviewed journals in the period 1970–1994; this period was chosen because, after 1970, studies were conducted when most of the current AEDs had become available and modem cognitive tests had become widely used; b) studies that report psychometrically assessed cognitive functions (excluding e.g., clinical observations); c) only current AEDs (excluding experimental drugs that have been removed from study programs, such as zonisamide or felbamate); d) only studies on patients with epilepsy (excluding AED studies in e.g., psychiatric patients). An exception was made for studies in healthy volunteers, as such studies offer several potential advantages and might suggest hypotheses worth pursuing with regard to epilepsy.

This search yielded 1284 titles. After the initial search (published in Vermeulen and Aldenkamp, 53), each new published paper up to 1998 was included expanding the database to 1357 titles.

The evaluation of the separate articles showed, however, that the results could not be simply assembled, because of validity concerns that appear to be inherent to this area of research. An example is the complication that is associated with conclusions drawn from "negative findings" ("no significant differences"). In general, such "non-significant" results tend to be regarded as disappointing, but research on side-effects is a case where actually non-significance is appreciated, because the results might then be interpreted as demonstrating the absence of harmful cognitive effects. This follows from the tendency to test cognitive effects in equivalence studies; hence, in evaluating whether a new drug is equivalent to an established drug with a suggested favorable cognitive profile. "Non-significance" is then used as evidence for similarity of the cognitive profiles of the old and new drugs. An important point to consider here in drawing conclusions from "no-effect" results is the statistical power of the study. That is, such conclusions only make sense if the study has a reasonable a priori chance (80% or better) of detecting a cognitive side-effect, when such a side-effect is actually present. This probability, i.e., the statistical power, is heavily dependent on sample size. In order to obtain a power of 0.80, i.e., to achieve an 80% chance of detecting small, medium and large differences between two independent means (0.20, 0.50 and 0.80 standard-deviation respectively in line with the convention proposed by several authors (e.g., 14, 16), the necessary number of patients per group using the 5% significance level, is 393, 64 and 26 respectively. Most reviews claim that the cognitive effects of AEDs are generally small, requiring a large sample size (64 being the minimal size) for achieving statistical power. Nonetheless, small sample sizes (20 patients or less) are used in the majority of studies that claim "no effect" findings. In reality, these studies could only have detected cognitive effects of such magnitude that they would presumably be obvious to the clinician, even without any psychological tests. With "no-effect" claims it is therefore worth checking whether they might not simply reflect inadequate statistical power.

In weighing the evidence from the studies in our review, the approach taken was first to disregard studies that contain certain basic deficiencies that render them uninformative with regard to cognitive AED effects, because they fail to rule out too many plausible alternative explanations using the following criteria:

1) Statistical power of the study
In some studies the description of methods and results arguably fell below currently accepted standards of scientific communication. We therefore only included studies that met the criteria of describing the design, number of subjects, outcome measures, and provided sufficient statistical detail to evaluate the power of the study and the validity of the results. If present, power calculations were carried out. Studies that failed to reach acceptable levels of power (0.80 according to the conventions proposed by Cohen, 14), may lead to false negatives (i.e., unjustifiably rejecting the possibility of cognitive side-effects) and were not included in our review.
2) Period of drug exposure
The use of healthy volunteers has the advantage that factors such as seizure frequency and severity do not confound the interpretation of the results. Volunteers offer the best opportunity to study absolute cognitive effects of AEDs, i.e., effects as opposed to no treatment, which is rarely possible in epilepsy. However, these studies also suffer from validity problems, the most important being that the period of drug exposure is limited. There is evidence that in most AEDs, "early" cognitive side-effects may develop only during a short period, i.e., during the first few days or weeks of drug exposure. After this period normalization occurs, possibly due to the development of so-called positive tolerance or habituation.[27] Although little is known about how tolerance to the cognitive effects of AEDs develops, a failure to take this factor into account may lead to overestimation of cognitive side-effects of AEDs.[20] This important point has to be taken into account when considering studies with healthy volunteers who are typically given AEDs during a few weeks at most, long term studies not being feasible. Our examination of the normal-volunteer cognitive studies revealed that the majority of these studies use fairly short periods of drug exposure, often no longer than one day. It was decided to ignore all claims from studies that used periods of drug exposure of one day or less.
3) Studies using polytherapy
Although polytherapy is the most common treatment in refractory epilepsies, it introduces critical complications in identifying the exact cause of observed cognitive changes. Interactions between antiepileptic drugs became evident soon after routine measurement of serum levels came into practice. Such interactions can alter therapeutic efficacy and thus, conceivably, cognitive functioning.[12] Moreover polytherapy is typically given to patients suffering from refractory epilepsy, and the threat of a "seizure confound" is thus always

serious. "Seizure confound" refers to a major validity concern: the failure to separate seizure effects from "genuine" AED-effects. Typically, cognitive AED effects are studied in add-on studies, where a new drug is introduced into a polytherapy regime (e.g., 29). In this type of study, the seizure confound is even stronger and the effects on cognitive tests are a potpourri of positive and negative seizure effects, AED-effects and drug-drug interactions that can never be disentangled. It is therefore impossible to use this type of study to draw inferences about cognitive side-effects of AEDs. In general it is found that add-on polytherapy studies underestimate the cognitive effects of AEDs as the positive cognitive effects of seizure control often mask the cognitive side-effects of the drug.

We consequently decided to omit 1) studies that did not provide sufficient statistical detail or did not reach acceptable power; 2) acute drug studies (with a drug exposure period of 1 day or less) and 3) polytherapy studies.

Using these criteria our analysis only provided data on established drugs that have been used in clinical practice for a considerable period. No information is available on the newer AEDs. The studies were evaluated in three steps:

a) Absolute drug effects, i.e., the effects of drug treatment against no-treatment (non-drugs) in the same subjects, which represents the most valuable information.
b) Relative drug effects, i.e., the comparison of the cognitive side-effects of a drug with the effects of another drug. This comparison does not rule out the possibility that they may not be different because they both impair cognitive function to the same extent.
c) Information on dose-effects: the relationship with dose-increase or dose decrease.

2. RESULTS OF THE LITERATURE REVIEW

2.1. Absolute Drug Effects

For phenobarbitone (PHB) only one study[28] is available (after applying our criteria) allowing the evaluation of absolute effects, i.e., the differences between PHB and a non-drug condition. This study shows relatively serious memory impairment (short-term memory recall) in 19 patients with epilepsy.

For phenytoin (PHT) five studies are available[31,33,42,51,52] comparing PHT with a non-drug condition. These studies all reveal PHT-induced cognitive impairment in the areas of attention, memory and especially mental speed. The magnitude of the reported effects is moderate to large. A caveat is, however, in order as all these studies were carried out in normal volunteers, which opens the possibility that these effects represent short-term outcomes of the drug before habituation.

For carbamazepine (CBZ) there is no consistent report about absolute cognitive-effects. Two studies, one in normal-volunteers[51] and one in patients with epilepsy[3] report "no cognitive impairment" compared to a non-drug condition. This is challenged by Meador and coworkers[32,33] who report impairments of memory, attention and mental speed, the areas that may also be affected by phenytoin.

Table 1. Absolute cognitive side-effects of antiepileptic drugs

Type of AED	Type of impairment	N. subjects E = epilepsy, nv = volunteers
– *phenobarbitone*		
MacLeod *et al.* (1978)	short-term memory	19 (e)
– *phenytoin*		
Smith and Lowrey (1975)	memory/mental speed	10 (nv)
Thompson *et al.* (1980)	memory	8 (nv)
Thompson *et al.* (1981)	memory/attention/mental speed	8 (nv)
Meador *et al.* (1991)	attention/mental speed	21 (nv)
Meador *et al.* (1993)	impairment of memory	15 (nv)
– *carbamazepine*		
Thompson *et al.* (1980)	no impairment	8 (nv)
Meador *et al.* (1991)	mental speed/attention	21 (nv)
Meador *et al.* (1993)	impairment of memory	15 (nv)
Aldenkamp *et al.* (1993)	no impairment	56 (e)
– *valproate*		
Thompson and Trimble (1981)	mental speed	10 (nv)
Craig and Tallis (1994)	no impairment	12 (e)
Prevey *et al.* (1996)	mild psychomotor slowing	18 (e)
Aldenkamp *et al.* (1993)	psychomotor slowing	17 (e)

For valproate (VPA) three studies[18,37,50] allow the interpretation of absolute effects and show mild to moderate impairment of psychomotor and mental speed.

2.2. Relative Side-Effects

For phenobarbitone, comparisons with other AEDs are available from four studies,[13,24,30,54] all with epileptic patients. One of these shows more impairment for PHB than for phenytoin (PHT) or carbamazepine (CBZ) on visuomotor and memory tests[24] and two other studies show convincing and clinically highly relevant impairments of intelligence scores after long-term PHB treatment in comparison with valproate (VPA).[13,54] Only the study by Meador *et al.*[30] does not indicate differences between PHB and PHT or CBZ.

For phenytoin the results of head-to-head comparisons are somewhat more confusing. Using an ingenious long-term treatment and withdrawal design, Gallassi and coworkers[24] found more cognitive impairment than CBZ. On the other hand, no differences with CBZ, VPA and even with PHB are reported.[23,30,31,32,33]

For carbamazepine, again we have to consider conflicting results of the Italian study by Gallassi and coworkers, showing a more favorable profile compared with PHT and PHB[24] and the US-based study by Meador and coworkers[30,31,32,33] that showed no differences between CBZ, PHT and PHB.

Finally, for valproate, the comparison with other drugs shows lower performances on memory and visuomotor tests as compared to CBZ[24] and a favorable profile compared to PHB on intelligence tests.[13,54] One study does not show any difference with PHT.[23]

Table 2. Relative cognitive side-effects of antiepileptic drugs

Type of AED	Type of impairment	N. subjects E = epilepsy, nv = volunteers
– *phenobarbitone*		
Gallassi *et al.* (1992)	memory/visual motor (compared with CBZ and PHT)	29 (e)
Vining *et al.* (1987)	impairment of intelligence (compared with VPA)	21 (e)
Meador *et al.* (1990)	no difference with PHT/CBZ	15 (e)
Callandre *et al.* (1990)	impairment of intelligence (compared with VPA)	32 (e)
– *phenytoin*		
Gallassi *et al.* (1992)	intelligence and memory (compared with CBZ)	29 (e)
Meador *et al.* (1990, 1991, 1993)	no differences with CBZ/PHB	21 (nv), 15 (nv), 15 (e)
Forsythe *et al.* (1991)	no differences with VPA	20 (e)
– *carbamazepine*		
Gallassi *et al.* (1992)	more favorable profile (compared with PHT/ PHB)	29 (e)
Meador *et al.* (1990, 1991, 1993)	no differences with PHT/PHB	15 (nv), 21 (nv), 15 (e)
– *valproate*		
Gallassi *et al.* (1992)	visuomotor and memory (compared with CBZ)	29 (e)
Viking *et al.* (1987)	higher intelligence (compared with PHB)	21 (e)
Calandre *et al.* (1990)	higher intelligence (compared with PHB)	32 (e)
Forsythe *et al.* (1991)	no differences with PHT	20 (e)

2.3. Dose-Effects Relationships

The cognitive side-effects of antiepileptic drugs do not only develop at higher dose: there is indeed no significant correlation between dose and cognitive impairment for phenytoin (in the large cohort study of Stevens *et al.*, see 43) and valproate.[9,38] This is in agreement with the finding that, generally, side-effects of VPA are not related to dose or serum level.[25] This also concurs with the observation that no differences are reported between conventional and controlled-release formulations for VPA.[11] Only for CBZ are some effects of dose reported, with improvement of cognitive function at higher dose[11] and with the use of controlled-release formulations.[2] One study[8] evaluated the effects of switches between different generic formulations of CBZ and revealed no large effects. No information is available for PHB.

3. REMAINING ESTABLISHED DRUGS

No studies were available—after applying our methodological criteria—for clonazepam and clobazam, two drugs that are still commonly used in clinical practice. Clinical reports suggest the possibility of cognitive impairment for clonazepam, especially

Table 3. Dose relation; cognitive side-effects of antiepileptic drugs

Type of AED	Type of impairment	N. subjects E = epilepsy, nv = volunteers
– *phenobarbitone*		
no studies		
– *phenytoin*		
Stevens *et al.* (1974)	no dose effects	107 (nv)
– *carbamazepine*		
Aldenkamp *et al.* (1987)	more variablity at peak levels with conventional CBZ compared with a controlled-release formulation	11 (e)
Amman *et al.* (1990)	improvement at higher doses	50 (e)
Aldenkamp *et al.* (1998)	no differences between generic forms of CBZ	12 (nv)
– *valproate*		
Amman *et al.* (1987)	no dose effects	46 (e)
Read *at al.* (1998)	no dose effects	12 (nv)
Brouwer *et al.* (1992)	no difference with controlled-release formulation	12 (e)

increased irritability and attentional disorders, whereas clobazam seems to have a more favorable cognitive profile.

4. NEWER ANTIEPILEPTIC DRUGS

Considering the rather meager data that are available for valid inspection of the cognitive effects of established AEDs (most of these in clinical use for several decades), we cannot expect that valid and reconfirmed empirical data may be available for the relatively newer AEDs: vigabatrin, lamotrigine, oxcarbazepine, tiagabine, gabapentin and topiramate. Most of the available data for these drugs were obtained in add-on polytherapy designs, the gold standard type of design to test the efficacy of a new drug in early clinical studies. It is remarkable that no normal volunteer studies are available for any of these drugs. Nonetheless we may summarize the clinical experience and—mostly anecdotal information—that exists for the six drugs that have recently been employed for clinical use. Most of the information presented here is based on the exchange of data during the international workshop "Cognitive effects of the newer antiepileptic drugs", organized during the International Epilepsy Congress in Sydney, 1995.[6]

For vigabatrin, the absence of cognitive side-effects (compared to the existing first-line drugs) is claimed, but the data come from only one center: Riekinnen and coworkers in Finland[34,40] and are not based on high-powered studies. There are a few additional studies that did not, however, pass our aforementioned criteria, mostly because they were carried out in polytherapy designs.

Anecdotal clinical information suggests no cognitive impairment for lamotrigine; in some patients, even improvement of performance is reported[7] which may be in line with the claimed psychotropic effect of the drug. There is, however, no empirical evidence from controlled cognitive studies to support this claim.

The claim of absence of cognitive effects in oxcarbazepine, a compound related to carbamazepine, is based on two studies.[1,17] The latter study reports cognitive improvement (focused attention and speed) in 12 patients in comparison with a non-drug condition, but did not control for the beneficial cognitive effects of improved seizure control.

For tiagabine, a favorable cognitive profile is reported, based, however, only on anecdotal clinical information. No empirical information is available.

We are more certain of the favorable cognitive profile of gabapentin which has, however, the disadvantage that all controlled studies have been performed in a limited number of centers in the USA. The data were presented during the workshop in Sydney.[6]

For topiramate, no information is yet available from controlled cognitive studies. Two cognitive studies, one in Europe and one in the USA, started in 1997, both as first-line add-on randomized studies with sample sizes exceeding 60 patients.

For the drugs that are still in the experimental phase, we have information about rufinamide (CGP 33.101) showing improvement of cognitive functions in lower doses (improvement on reaction-time tests) and a possibility of impairment of short-term memory at higher doses. The company did not yet release these data, although the study had an impressive sample size (>200).[5] For (UCB L059) levetiracetam, a noötropic effect is claimed. We only have data from a small pilot study that precludes any valid conclusions.[35]

5. CONCLUSIONS

5.1. General Conclusions

A systematic review on published empirical studies using the outlined criteria allows reliable conclusions about established drugs only, i.e., AEDs that were introduced for clinical use before the 1970s. Even for these drugs, a disappointing small number of studies pass criteria of design, methodology and statistical analysis that are in line with common scientific conventions.[14,16] From a large database of 1357 papers, a rather small group of studies remained that potentially allow valid inferences about the cognitive effects of established AEDs. For the newer drugs no reliable conclusions are possible.

All established AEDs have reported absolute cognitive side-effects, i.e., all the investigated drugs have effects when compared to no treatment. These effects are definitely large for PHB and possibly larger for PHT than for CBZ or VPA. But even these last two drugs, that are generally considered to be drugs with a safe cognitive profile, have cognitive effects, mostly resulting in a mild general psychomotor slowing.

The respective differences between the four investigated AEDs can be considered as small, with the exception of the cognitive effects of PHB that has a less favorable cognitive profile when compared to PHT, VPA and CBZ.

Possibly the most marked findings of our review is the fact that all AEDs have cognitive effects and that the effects of PHT are more moderate than has been suggested previously.[44,45]

5.2. Cognitive Side-Effect Profile Per Antiepileptic Drug

Phenobarbitone is the only AED with specific absolute effects: when compared with no treatment, memory functions are affected. Long-term treatment may bring about impairment or delay of psychological development, leading to intellectual deterioration. No information is available on dose-effects.

Phenytoin has a much milder impact on cognitive functions than had hitherto been suspected. When compared with no treatment, PHT-induced attentional deficit and mental slowing are observed but these are not markedly different from the effects produced by CBZ and VPA. No correlation with dose is found.

Carbamazepine has similar absolute effects as reported for PHT and in most reports, no differences have been found with regard to the cognitive effects produced by VPA and PHT. There is some evidence that dose-effects may occur and that controlled-release formulations have a more favorable cognitive profile than conventional forms. No differences were found between different generic forms of CBZ.

Valproate has mild psychomotor slowing as absolute effect and has a similar cognitive profile when compared with CBZ and PHI. In contrast with CBZ, dose-relationships have not been obtained and there is no effect when switched from conventional to controlled-release formulations.

From the newer antiepileptic drugs there is a suggestion of equivalence with CBZ and VPA for oxcarbazepine, gabapentin and vigabatrin. For the other drugs in clinical practice: lamotrigine, topiramate and tiagabine, no empirical information is available.

For the remaining drugs: rufinamide and levetiracetam further research has to be encouraged.

5.3. Design for Future Studies

One important topic is the design for such future studies. The effects of individual agents can only be assessed satisfactorily in monotherapy and in subjects with well-controlled seizures. A frequently applied design is to assess untreated patients that are reassessed when they reach steady-state treatment with a given drug. This design is often not appropriate, though, as the seizures will not be sufficiently controlled at baseline. Therefore the test-scores will be negatively biased during baseline measurement and scores will improve after initiation of treatment as an effect of seizure control, which may mask a potential side-effect of the drug. A satisfactory design is the first-line add-on design: placebo (or contrast-drug) and test-drug are randomly applied as add-on to a baseline drug. The research hypothesis is often that the test drug is equivalent to the baseline-drug (and thus does not introduce new or more serious cognitive side-effects) and is also not different from the contrast-drug. Using a baseline drug, patients are included with satisfactory seizure control at baseline. Other polytherapy studies should not be employed. A powerful design is the monotherapy cross-over study in which the baseline assessments are carried out with the baseline drug after which cross-over is performed to the test drug. An additional advantage of these repeated measurement designs is that a relatively limited number of patients is required.

REFERENCES

1. Alkiae M, Kaelviaeinen R, Sivenius J, *et al.* (1992): Cognitive effects of oxcarbazepine and phenytoin monotherapy in newly diagnosed epilepsy: one year follow-up. Epilepsy Research 1:199–203.
2. Aldenkamp AP, Alpherts WCJ, Moerland MC, *et al.* (1987): Controlled release of carbamazepine: Cognitive side-effects in patients with epilepsy. Epilepsia 28:507–514.
3. Aldenkamp AP, Alpherts WCJ, Blennow G, *et al.* (1993): Withdrawal of antiepileptic medication-effects on cognitive function in children—the results of the multicentre "Holmfrid" study. Neurology 43:41–51.

4. Aldenkamp AP (1995): Cognitive side-effects of antiepileptic drugs. In Aldenkamp AP, Dreifuss FE and Renier WO (eds): "Epilepsy in Children and Adolescents". New York: CRC-Press Publishers, Boca Raton, pp 161–183.
5. Aldenkamp AP (1996): The cognitive profile of the experimental antiepileptic drug COP 33.101. Basel, Presentation at the Final Investigators Meeting, March 14.
6. Aldenkamp AP and Trimble MR (1996): Cognitive side-effects of antiepileptic drugs: Fact or Fiction? Abstract Epilepsia 37:82.
7. Aldenkamp AP, Mulder OG and Overweg J (1997): Cognitive effects of Lamotrigine as first line add-on in patients with localized related (partial) epilepsy. Journal of Epilepsy 10:117–121.
8. Aldenkamp AP, Rentmeester TH, Hulsman J, Majoie M, Doelman J, Diepman L, Schellekens A, Franken M and Olling M (1998): Pharmacokinetics and cognitive effects of carbamazepine formulations with different dissolution rates. European Journal of Clinical Pharmacology 54:85–192.
9. Amman MG, Werry JS, Paxton JW, *et al.* (1987): Effect of sodium valproate on psychomotor performance in children as function of dose, fluctuations in concentration and diagnosis. Epilepsia 28:115–124.
10. Amman MG, Werry JS, Paxton JW, *et al.* (1990): Effects of carbamapezine on psychomotor performance in children as a function of drug concentration, seizure type and time of medication. Epilepsia 3151–3160.
11. Brouwer OF, Pieters MSM, Bakker AM, *et al.* (1992): Conventional and controlled release valproate in children with epilepsy: a cross-over study comparing plasma levels and cognitive performance. Epilepsy Research 13:245–253.
12. Brown SW (1994): Clinical aspects of antiepileptic treatments. In Aldenkamp AP, Dreifuss FE, Renier WO *et al.* (eds): "Epilepsy in Children and Adolescents". New York: CRC-Press Publishers, pp 98–112.
13. Calandre EP, Dominguez-Granados R, Gomez-Rubio M, *et al.* (1990): Cognitive effects of long-term treatment with phenobarbital and valproic acid in school children. Acta Neurologica Scandinavica 81:504–506.
14. Cohen J (1977): "Statistical Power Analysis for the Behavioral Sciences". New York: Academic Press.
15. Committee on Drugs (1985): Behavioral and cognitive effects of anticonvulsant therapy. Pediatrics 76:644–647.
16. Cook O and Campbell DI (1979): "Quasi-Experimentation; Design and Analysis Issues for Field Settings". Boston: Houghton Miffin Company.
17. Curran HV and Java R (1993): Memory and psychomotor effects of oxcarbazepine in healthy human volunteers. European Journal of Clinical Pharmacology 44:529–533.
18. Craig I and Tallis R (1994): Impact of vaiproate and phenytoin on cognitive function in elderly patients: results of a single-blind randomized comparative study. Epilepsia 35:381–390.
19. Dodrill CB (1991): Behavioral effects of antiepileptic drugs. In Smith D, Treiman D and Trimble M (eds): "Advances in Neurology", Vol 55. New York: Raven Press, pp 213–224.
20. Dodrill CB (1992): Problems in the assessment of cognitive effects of antiepileptic drugs. Epilepsia 33 (Suppl 6):29–32.
21. Dodrill CB and Troupin AS (1977): Psychotropic effects of carbamazepine in epilepsy: a double-blind comparison with phenytoin. Neurology 27:1023–1028.
22. Evans RW and Gualtieri CT (1985): Carbamazepine: a neuropsychological and psychiatric profile. Clinical Neuropharmacology 8:221–241.
23. Forsythe I, Butler R, Berg I, *et al.* (1991): Cognitive impairment in new cases of epilepsy randomly assigned to carmabapezine, phenytoin and sodium valproate. Developmental Medicine and Child Neurology 8:221–241.
24. Gallassi R, Morreale A, Di Sarro R, *et al.* (1992): Cognitive effects of antiepileptic drug discontinuation. Epilepsia 33 (Suppl 6):41–44.
25. Herranz JL, Arteaga R and Armijo JA (1982): Side effects of sodium valproate on monotherapy, controlled by plasma levels. A study in 88 pediatric patients. Epilepsia 23:203–214.
26. Ideström, Schalling D, Carlquist U, *et al.* (1972): Behavioral and psychological studies: acute effects of diphenylhydantoin in relation to plasma levels. Psychological Medicine 2:111–120.
27. Kulig B and Meinardi H (1977): Effects of antiepileptic drugs on motor activity and learned behavior in the rat. In Meinardi H and Rowan AJ (eds): "Advances in Epileptology". Amsterdam: Swet and Zeitlinger, pp 98–104.
28. MacLeod CM, Dekaban AS and Hunt E (1978): Memory impairment in epileptic patients: selective effects of phenobarbital concentration. Science 202:1102–1104.
29. McKee PJW, Blacklaw J, Forrest G, *et al.* (1994): A double-blind placebo-controlled interaction study between oxcarbazepine and carbamazepine, sodium vaiproate and phenytoin in epileptic patients. British Journal of Clinical Pharmacology 37:27–32.

30. Meador KJM, Loring DW, Huh K, *et al.* (1990): Comparative cognitive effects of anticonvulsants. Neurology 40:391–394.
31. Meador KJM and Loring DW (1991): Cognitive effects of antiepileptic drugs. In Devinski O, Theodore W (ed): "Epilepsy and Behavior". New York: Wiley-Liss, pp 151–170.
32. Meador KJM, Loring DW, Allen ME, *et al.* (1991): Comparative cognitive effects of carbamazepine and phenytoin in healthy adults. Neurology 41:1537–1540.
33. Meador KJM, Loring DW, Abney OL, *et al.* (1993): Effects of carbamazepine and phenytoin on EEG and memory in healthy adults. Epilepsia 34:153–157.
34. Mervaala E, Partanen J, Nousiainen U, Sivenius J and Riekinnen P (1989): Electrophysiological effects of gamma-vinyl GABA and carbamazepine. Epilepsia 30:189–193.
35. Neijens LGJ, Alpherts WCJ and Aldenkamp AP (1995): Cognitive effects of a new pyrrolidine derivative (Levetiracetam) in patients with epilepsy. Progress in NeuroPsychopharmacology and Biological Psychiatry 19:411–419.
36. Novelly RA, Schwartz MM, Mattson RH, *et al.* (1986): Behavioral toxicity associated with antiepileptic drugs: concepts and methods of assessment. Epilepsia 27:331–340.
37. Prevey ML, Delaney RC, Cramer JA, Cattanach L, Collins JF and Mattson RH (1996): Effect of valproate on cognitive function. Comparison with carbamazepine. The Department of Veterans Affairs Epilepsy Cooperative Study 264 Group. Archives of Neurology 53:1008–1016.
38. Read CL, Stephe LJ, Stolarek LH, Paul A, Sills GJ and Brodie MJ (1998): Cognitive effects of anticonvulsant monotherapy in elderly patients: a placebo-controlled study. Seizure 7:159–162.
39. Reynolds EH (1975): Chronic antiepileptic toxicity: a review. Epilepsia 16:19–352.
40. Riekkinen PJ, Kalviasinen A, Aikia M, Partanen J, Saksa M and Sivenius J (1990): Cognitive and electrophysiological effects of vigabatrin and carbamazepine. Epilepsia 31:620.
41. Smith DB (1991): Cognitive effects of antiepileptic drugs. In Smith D, Treirnan D and Trimble M (eds): "Advances in Neurology", Vol 55. New York: Raven Press, pp 197–212.
42. Smith WL and Lowrey JB (1975): Effects of diphenylhydantoin on mental abilities in the elderly. Journal of the American Geriatric Society 23:207–211.
43. Stevens JH, Schaffer JW and Brown CC (1974): A controlled comparison of the effect of diphenyihydantoin and placebo on mood and psychomotor functioning in normal volunteers. Journal of Clinical Pharmacology 14:543–551.
44. Trimble MR (1983): Anticonvulsant drugs and psychosocial development: phenobarbitone, sodium valproate, and benzodiazepines. In Morselli PL, Pippenger CE and Penry JK (eds): "Antiepileptic Drug Therapy in Pediatrics". New York: Raven Press, pp 201–217.
45. Trimble MR (1987a): Anticonvulsant drugs: mood and cognitive function. In Trimble MA and Reynolds EH (eds): "Epilepsy, Behaviour and Cognitive Function". Chichester: John Wiley and Sons, pp 135–145.
46. Trimble MR (1987b): Anticonvulsant drugs and cognitive function: a review of the literature. Epilepsia 28 (Suppl 3):37–45.
47. Trimble MR and Thompson PJ (1981): Memory, anticonvulsant drugs and seizures. Acta Neurologica Scandinavica 64:31–41.
48. Trimble MR and Thompson PJ (1983): Anticonvulsant drugs, cognitive function and behaviour. Epilepsia 24 (Suppl 1):S55–S63.
49. Trimble MR and Cull C (1988): Children of school age: the influence of antiepileptic drugs on behavior and intellect. Epilepsia 29 (Suppl 3):S15–S19.
50. Thompson PJ and Trimble MR (1981): Sodium valproate en cognitive functioning in normal volunteers. British Journal of Clinical Pharmacology 12:819–824.
51. Thompson PJ, Huppert F and Trimble MR (1980): Anticonvulsant drugs, cognitive function and memory. Acta Neurologica Scandinavica (Suppl 80):75–80.
52. Thompson PJ, Huppert FA and Trimble MR (1981): Phenytoin and cognitive functions: effects on normal volunteers and implications for epilepsy. British Journal of Clinical Psychology 20:155–162.
53. Vermeulen J and Aldenkamp AP (1995): Cognitive side-effects of chronic antiepileptic drug treatment: a review of 25 years of research. Epilepsy Research 22:65–95.
54. Vining EP, Mellitis ED, Dorsen MM, *et al.* (1987): Psychologic and behavioral effects of antiepileptic drugs in children: a double-blind comparison between phenobarbital and valproic acid. Pediatrics 80:165–174.

28

EPILEPSY, COGNITIVE ABILITIES AND EDUCATION

Christine Bulteau*[1,2], Isabelle Jambaqué[1], and Georges Dellatolas[2]

[1]Service de Neuropédiatrie
Hôpital Saint Vincent de Paul
82 Avenue Denfert Rochereau
75674 Paris Cedex 14, France
[2]INSERM U169, Hôpital Paul Brousse
16 Avenue Paul Vaillant, 94800 Villejuif
France

INTRODUCTION

Over the past 40 years, learning difficulties have been consistently reported in children with epilepsy.[16] This appears to be a clear consequence of mental retardation in a sizable proportion of patients. Seizures may also interfere with learning when children miss school repeatedly. However, many children with normal intelligence quotient (IQ), whose seizures are brought under control, still experience learning difficulties and failure in school. Therefore, the mechanisms underlying schooling underachievement are multifold and a detailed analysis is required for each individual patient.

There is a growing proportion of children who are faced with these problems since the treatment of childhood epilepsy is constantly improving due to the combination of a more rational use of antiepileptic drugs, surgery and diet. However, many of these seizure-controlled patients are left with complex neuropsychological sequelae that may also interfere with school achievement and complicate the educational prognosis.

Based on etiology and syndrome diagnosis, it is increasingly likely to determine early in the course of the disease which type of cognitive function is at risk, should be monitored and eventually undergo rehabilitation. Therefore, detailed and eventually repeated neuropsychological assessments are required in addition to the epileptological diagnosis, in order to design specific and adapted programs of rehabilitation.

* To whom correspondence should be addressed

Neuropsychology of Childhood Epilepsy, edited by Jambaqué et al.
Kluwer Academic / Plenum Publishers, New York, 2001.

Standardized test batteries are now available to assess various cognitive functions, even in the very young child. In addition to the evaluation of cognitive defects, it is most important to identify the patients' capacities in order to help them overcome their difficulties using adapted strategies.

1. EDUCATION

Poor academic achievement is widely reported in children with epilepsy, although many of them have normal intelligence and stable intellectual abilities.[9,30,25] The frequency of school underachievement is difficult to evaluate because of the heterogeneity of school systems throughout the world and because of methodological differences between studies.[16] Given the incidence of epilepsy in children (between 4 and 8 per 1,000), it is highly probable that most teachers will have at least one student with epilepsy in their career.[6] Aldenkamp *et al.*[2] reported that 30% of children with therapy-resistant epilepsy were receiving special education as compared to 7% in a matched non-neurological control group. Moreover, only 33% reached the High school system whereas 68% did in the control group. School underachievement was reported in a proportion of 61% in the group with idiopathic epilepsy, which is usually considered as the most "benign" form of epilepsy.[35] In a French survey of 133 children with idiopathic generalized epilepsy, 11% of the cases were in the category of "impossible schooling", which was associated with persisting seizures, behavioral disorders and difficulties related to low socio-economic status.[12]

Teachers who provide information about these children typically describe them as "just slow learners", hyperactive and inattentive. They also perceive them as having behavioral problems, and refer to them as being solitary, immature, poorly motivated and with a low self-esteem.[18,25] The first convincing evidence regarding learning difficulties came from the Isle of Wight studies where 18% of the epileptic children over the age of eight years showed a delay of two or more years in reading comprehension.[29] Similar findings were noted in 20% of children with uncomplicated epilepsy who performed one year behind their chronological age in reading comprehension.[40] Several authors have evaluated achievement for a wide range of academic areas in a sizable group of children with epilepsy and normal IQ in regular schools; all reported marked deficiencies in mathematics followed by deficiencies in spelling, reading, writing and general knowledge.[2,18,32] Even after taking into consideration IQ levels, these difficulties remained in 15 to 50% of children, depending on the various academic subscales being assessed.[30,25]

2. FACTORS UNDERLYING ACADEMIC VULNERABILITY

Several factors underlying academic vulnerability have been suggested, including age at epilepsy onset, type and frequency of seizures, laterality of epileptic foci and side effects of antiepileptic drugs (AED).[18,32,37,28,34,38] O'Leary *et al.*[27] found that children whose generalized seizures developed before the age of 5 years performed significantly worse on verbal and performance IQ tests and in the trail making tests than children whose seizures had begun later. Children with partial seizures have variable prognoses depending on seizure control and laterality of the epileptic focus. When focal spike discharges are located in the left hemisphere, children may be more prone to reading difficulties, inattention and dependency, and other behavioral disturbances may be noted.[32,33]

Preliminary results in a population of 251 children with epilepsy have shown that the type of epileptic syndrome, defined according to the International Classification of Epilepsies and Epileptic Syndromes,[11] correlated with school achievement and IQ scores. Children with idiopathic generalized or localization-related epilepsy had higher IQ scores and better academic performances than children with symptomatic or cryptogenic generalized epilepsy or with undetermined epilepsy and epileptic syndromes, whether focal or generalized.

Moreover, subclinical epileptiform EEG discharges may affect cognitive performance, and this may be very important to consider in the context of learning difficulties.[21,22] Significant impairment of speed performance (for example on either a simple reaction time or a choice reaction time test) were found in 68% of patients with generalized discharges as compared to only 33% of patients with focal discharges.[28] For the past 10 years, the simultaneous video recording of EEG and test performance has unraveled specific effects of subclinical discharges on reading, mental arithmetic and motor dexterity.[31] The nature of the cognitive task seems an important factor in the modulation of the epileptiform discharges; cognitive activity and increased concentration both reduce epileptogenic activity in some patients, while stress may increase it in others. In some children, epileptiform discharges increase during rest and scholastic activity; during reading they occur relatively less frequently and with a shorter total duration over the left hemisphere than the right one.[22]

Antiepileptic medication is also an important factor that may influence cognitive functioning. Stores and Hart[32] found that epileptic children attending regular schools and treated with phenytoin for at least 2 years, had reading skills significantly lower than those taking other AEDs. From a questionnaire completed by teachers, Bennett-Levy found that they perceived children with epilepsy as having poor concentration and mental processing and as being less alert than their classmates.[4] When children who had been epileptic and were no longer taking any AED were compared with normal controls, their scores were similar on all measures except that they were significantly less alert. The impact of drug treatment on higher-order cognitive function is not clearly elucidated since AEDs may have slightly suppressing effect on motor speed.[1]

Nonseizure-related factors such as sex, associated behavioral and psychosocial problems and parental attitudes have also been implicated;[36] low self-esteem, depression and adaptation problems in the family have been reported in patients with benign childhood epilepsy.[30] Moreover, personality assessments indicate mentalisation troubles—inhibition, alexithymia—or, too much excitability which is bound to be linked to narcissistic need and poor body image.[7] School children with epilepsy could be victims of prejudicial attitudes from educational environment.[5] Emotional impulsiveness, instability and labile mood are among the most frequently reported behavioral disorders making school integration very difficult.[17,33,23,10] Low socioeconomic or cultural levels are considered by some authors to play a major role in the underachievement of children with epilepsy.[25]

3. COGNITIVE ABILITIES AND EDUCATION IN CHILDREN WITH EPILEPSY

In this context, systematic and comprehensive neuropsychological assessments of children with epilepsy become an important part of their overall management. Cognitive abilities and school achievements are rarely compared. A number of studies have shown that children with epilepsy have lower mean IQ scores than controls.[19,24] Some authors have claimed that epilepsy brings about nonspecific intellectual dysfunction affecting attention/arousal and information processing.[3,26,39] Other studies reported specific cognitive disorders that reflected the epileptic syndrome or the site and side of the epilepsy.[14,8,20] Forceville *et al.*[15] showed that adult patients with epilepsy have specific impairments in the Wechsler Coding, Digit Span and Information subtests, whereas mentally retarded patients without epilepsy have problems in the Arithmetic, Vocabulary and Information subtests. Vermeulen *et al.*[39] were the only ones who failed to show any specific pattern of learning impairment in a population of learning disabled children with and without epilepsy; however, children with epilepsy responded more slowly to simple auditory and visual reaction tasks and were also slower in a multiple decision reaction task and a visual searching test.

Speed of information processing, memory, vigilance, attention, alertness and motor fluency are cognitive domains particularly vulnerable to epileptic factors. Reading problems, for example, may be affected at several levels of the information processing system such as visual scanning, the ability to transform information from one modality to another, reasoning and verbal comprehension.

4. MANAGEMENT

A proper assessment of this complex situation may be very difficult, requiring the collaboration of a team consisting of an epileptologist, a psychologist, a neuropsychologist and a teacher. Most epileptic children with borderline IQ are in regular schools although they have learning difficulties. Neuropsychological evaluation must be part of the initial evaluation of a child with epilepsy and can provide a basis for planning cognitive rehabilitation early in life. Memory deficits, reading difficulties, slowness, attentional disorder and planning problems are frequently reported. A complete qualitative and quantitative analysis of cognitive functions is necessary to identify which factors are involved in academic vulnerability. Reading difficulties may result from visuo-perceptual deficits in occipital lobe epilepsy, from reasoning in frontal lobe epilepsy, or from slowness due to side effects of AEDs. For some children one could expect that early specialized teaching could be preferable to a marginal situation in a regular school. Conversely, it can be argued that specialized education "marginalizes" the child.

Another issue of importance is the information given to parents and teachers, which can help to monitor education and control inadequate behaviors. Information concerning the child's abilities should be given early to parents. Analysis of psychosocial difficulties and their appropriate treatment can reverse a vicious cycle of seizures, learning difficulties and psychosocial problems.

The most convincing approach has been put forth by Strang who proposes a treatment plan to facilitate the development of adaptive behavior.[34] Besides the medical history and neuropsychological assessment, a broad range of behavioral skills (communication, daily living, socialization, motor functions) assessed by a questionnaire filled by the child's teacher and his/her parents are evaluated. Three purposes are outlined: (a) to gather specific information concerning the child's educational symptomatology; (b) to explain the adaptive behavioral significance of the child's test results; and (c) to specify tailored recommendations for remedial education. Such an interdisciplinary assessment forms the basis of the treatment plan determining the educational, psychological and socially oriented interventions appropriate to each child suffering from cognitive impairment. No child must be simply considered as an "epileptic"; instead, each child should be seen as a person, whose quality of life can be substantially improved by the development of interdisciplinary interventions.

REFERENCES

1. Aldenkamp AP, Alpherts WCJ, Blennow G, Elmqvist D, Heijbel J, Nilsson HL, Sandstedt P, Tonnby B, Wahlander L and Wosse E (1993): Withdrawal of antiepileptic medication in children-effects on cognitive function: the multicenter Holmfrid Study. Neurology 43:41–50.
2. Aldenkamp AP, Alpherts WCJ, Dekker MJA and Overweg J (1990): Neuropsychological aspects of learning disabilities in epilepsy. Epilepsia 31 (Suppl 4):S9–S20.
3. Alpherts WCJ and Aldenkamp AP (1990): Computerized neuropsychological assessment of cognitive functioning in children with epilepsy. Epilepsia 31 (Suppl 4):S35–S40.
4. Bennett-Levy J and Stores G (1984): The nature of cognitive dysfunction in school children with epilepsy. Acta Neurologica Scandinavica 69 (Suppl 4):79–82.
5. Baumann RJ, Wilson JF and Wiese HJ (1995): Kentuckians' attitudes toward children with epilepsy. Epilepsia 36:1003–1008.
6. Blom S, Heisbel J and Bergfors PG (1978): Incidence of epilepsy in children. Epilepsia 19:343–350.
7. Bobet R (1996): L'enfant épileptique et sa famille: aspects psychologiques. Approche Neuropsychologique des Apprentissages chez l'Enfant (Hors série): 59–61.
8. Bornstein RA, Pakalnis A and Drake ME (1988): Verbal and nonverbal memory and learning in patients with complex localization related seizures of temporal lobe origin. Journal of Epilepsy 1:203–208.
9. Bourgeois BFD, Prensky AL, Palkes HS, Talent BK and Busch SG (1983): Intelligence in epilepsy: a prospective study in children. Annals of Neurology 14:438–444.
10. Carlton-Ford S, Miller R, Brown M, Nealeigh N and Jennings P (1995): Epilepsy and children's social and psychological adjustment. Journal of Health and Social Behavior 36:285–301.
11. Commission on classification and terminology of the International League Against Epilepsy (1989): Proposal for revised classification of epilepsies and epileptic syndromes. Epilepsia 30:389–399.
12. Desguerre I, Chiron C, Loiseau J, Dartigues JF, Dulac O and Loiseau P (1994): Epidemiology of idiopathic generalized epilepsies. In Malafosse A, Genton P, Hirsh E, Marescaux C. (eds). "Idiopathic Generalized Epilepsies": Clinical and Genetic Aspects. London: John Libbey and Co. Ltd, pp 19–25.
13. Fawrell TR, Dodrill LB and Batzel LW (1985): Neuropsychological abilities of children with epilepsy. Epilepsia 26:395–400.
14. Fedio P and Mirsky AF (1969): Selective intellectual deficits in children with temporal lobe or centrencephalic epilepsy. Neuropsychologia 7:287–300.
15. Forceville EJM, Dekker MJA, Aldenkamp AP, Alpherts WCJ and Schelvis AJ (1992): Subtest profiles of the WISC-R and WAIS in mentally retarded patients with epilepsy. Journal of Intellectual Disability Research 36:45–49.
16. Henriksen O (1990): Education and epilepsy: assessment and remediation. Epilepsia 31 (Suppl 4):S21–S25.
17. Hermann BP (1982): Neuropsychological functioning and psychopathology in child with epilepsy. Epilepsia 23:S45–S54.
18. Holdsworth L and Whitmore K (1974): A study of children with epilepsy attending ordinary schools/ their seizure patterns, progress and behaviour in school. Developmental Medicine and Child Neurology 16:746–758.
19. Huttenlocher PR and Hapke RJ (1990): A follow-up study of intractable seizures in childhood. Annals of Neurology 28:699–705.
20. Jambaqué I, Dellatolas G, Dulac O, Ponsot G and Signoret JL (1993): Verbal and visual memory impairment in children with epilepsy. Neuropsychologia 31:1321–1337.
21. Kasteleijn-Nolst Trenité DGA, Siebelink BM, Berends SGC, van Strien JW and Meinardi H (1990): Lateralized effects of subclinical epileptiform EEG discharges on scholastic performance in children. Epilepsia 31:740–746.
22. Kasteleijn-Nolst Trenité DGA, Smit AM, Velis DN, Willemse J and van Emde Boas W (1990): On-line detection of transient neuropsychological disturbances during EEG discharges in children with epilepsy. Developmental Medicine and Child Neurology 32:46–50.

23. Kim WJ (1991): Psychiatric aspects of epileptic children and adolescents. Journal of the American Academy of Child Adolescent Psychiatry 30:874–886.
24. Masur DM and Shinnar S (1992): The neuropsychology of childhood seizure disorders. In Segalowitz SJ and Rapin I (eds): "Handbook of Neuropsychology, vol 7: Child Neuropsychology". Amsterdam: Elsevier, pp 457–470.
25. Mitchell WG, Chavez JM, Lee H and Guzman B (1991). Academic underachievement in children with epilepsy. Journal of Child Neurology 6:65–72.
26. Mitchell WG, Zhou Y, Chavez JM and Guzman BL (1992): Reaction time, attention, and impulsivity in epilepsy. Pediatric Neurology 8:19–24.
27. O'Leary DS, Seidenberg M, Berent S and Boll TJ (1981): Effects of age of onset of tonic-clonic seizures on neuropsychological performance in children. Epilepsia 22:197–204.
28. Rugland AL (1990): Neuropsychological assessment of cognitive functioning in children with epilepsy. Epilepsia 31 (Suppl 4):S41–S44.
29. Rutter M, Graham P and Yule WA (1990): A neuropsychiatric study in childhood clinics. Clinics in Developmental Medicine, Nr. 35–36, London: SIMP with Heinemann.
30. Seidenberg M, Beck N, Geisser M *et al.* (1987): Academic achievement of children with epilepsy. Epilepsia 27:753–759.
31. Siebelink BM, Bakker DJ, Binnie CD *et al.* (1988): Psychological effects of subclinical epileptiform EEG discharges in children. II. General intelligence tests. Epilepsy Research 2:117–121.
32. Stores G and Hart J (1976): Reading skills of children with generalized or focal epilepsy attending ordinary school. Developmental Medicine and Child Neurology 18:705–716.
33. Stores G (1978): School-children with epilepsy at risk for learning and behavior problems Developmental Medicine and Child Neurology 20:502–508.
34. Strang JD (1990): Cognitive deficits in children: adaptive behavior and treatment techniques. Epilepsia 31 (Suppl 4):S54–S58.
35. Sturniolo MG and Galletti F (1994): Idiopathic epilepsy and school achievement. Archives of Disease in Childhood 70:424–428.
36. Taylor DC and Lochery M (1991): Behavioral consequences of epilepsy in children: Developing a psychosocial problem. Advances in Neurology 55:153–162.
37. Trimble MR (1987): Anticonvulsivant drugs and cognitive function: a review of the literature. Epilepsia 28:S37–S45.
38. Trimble MR (1990): Antiepileptic drugs, cognitive function, and behavior in children: evidence from recent studies. Epilepsia 31 (Suppl 4):S30–S34.
39. Vermeulen J, Kortstee SWAT, Alpherts WCJ and Aldenkamp AP (1994): Cognitive performance in learning disabled children with and without epilepsy. Seizure 3:13–21.
40. Yule W (1980): Educational achievement. In Kulig BM, Meinardi H, Stores G (eds) "Epilepsy and Behavior". Lisse/ Berwyn: Swets & Zeitlinger, pp 162–8.

29

QUALITY OF LIFE IN EPILEPTIC CHILDREN

Hannelore C. Sauerwein

Groupe de Recherche en Neuropsychologie Expérimentale
Département de Psychologie
Université de Montréal
C.P. 6128, Succ. Centre-Ville
Montréal, Qué. H3C 3J4
Canada

INTRODUCTION

Quality of life issues are a relatively recent concern in epilepsy research as disease management tends to shift from a treatment-centered to a patient-oriented approach. Health-related quality of life (HQOL) is a multidimensional construct that encompasses physical, emotional, mental and social aspects of well-being and functioning as viewed from the perspective of the patient, his family and care-givers. Assessment of HQOL takes into account the patient's subjective perception of the consequences of his illness and its treatment as well as his adjustment to his condition and to the reactions of his social environment.[70]

Epilepsy has a profound effect on the patient's everyday life and social adjustment. Epileptic patients tend to score lower on various scales measuring HQOL than patients with other chronic diseases. This may reside in the fact that different disorders impose dissimilar limitations on lifestyle and social well-being. For instance, a study by Austin *et al.*,[7] comparing asthma patients with epilepsy patients, showed that the impact of the illness on HQOL varied between the two populations: while the former reported more physical complaints, the latter mentioned significantly more psychosocial problems. In fact, it is now widely recognized that the impact of epilepsy extends well beyond the seizure status.[63] The impact of epilepsy also differs between age groups according to differences in life roles, demands and responsibilities. In children and adolescents, self-care, mobility and interpersonal relationships are important aspects of the patient's quality of life,[1] whereas independence and vocational prospects appear to be the most common concerns of young adults. In a study by Hayden *et al.*,[41] 64 percent of the 517 epileptic adults interviewed reported that their life is at times affected by the illness. Not surprisingly, the severer the epilepsy, the more numerous are the problems encountered by

Neuropsychology of Childhood Epilepsy, edited by Jambaqué et al.
Kluwer Academic / Plenum Publishers, New York, 2001.

the patient. In a survey by Thompson and Oxley,[79] lack of social contacts and employment opportunities were the most frequent problems reported by patients with severe seizure disorder.

Data on children are still scarce as scales designed to measure HQOL in pediatric populations are just beginning to be developed.[29] Typically, the assessment relies on parents' identification of areas of concern. In a study by Hoare and Russel,[48] parents rated the impact of the epilepsy on various aspects of their child's life. The most frequent concerns were medication effects, school performance (particularly reading and mathematics), cognitive abilities, self-care skills, injury, alertness, moodiness, teasing and friendship. Seizures and academic performance were also primary worries of the families of 443 elementary and junior high-school children in a survey conducted by Hanai.[40] Indeed, there is ample evidence in the literature that children with epilepsy are more susceptible to the development of cognitive and behavior problems, learning disabilities and social problems (see 43).

In this chapter, the impact of various factors on the quality of life of epileptic children, adolescents and young adults will be discussed. Furthermore, methods of assessment of HQOL in epilepsy as well as comprehensive care programs aimed at improving the quality of life of epileptic patients and their families, will be presented.

1. FACTORS IMPACTING ON THE QOL OF THE EPILEPTIC PATIENT

1.1. Seizure Factors

Seizures impose restrictions on cognitive and social functioning. Individuals experiencing frequent seizures feel stigmatized by their epilepsy.[9] Studies in adult patients have revealed that the degree of seizure control is the most important factor determining self-reported quality of life: low seizure frequency and seizure severity correlate with higher ratings on various HQOL scales.[21,22,59,63] This is particularly evident in surgical cases where HQOL improvement generally parallels seizure outcome, being highest in patients with the most satisfactory outcome and lowest in those with the poorest outcome.[57,58,63] Similar results have been obtained in children and adolescents who have undergone epilepsy surgery for refractory seizures.[52] Furthermore, Hoare and Russel[48] found that parents of children with early onset of epilepsy, longer duration of epilepsy and higher seizure frequency had more concerns regarding their child's quality of life. Carpay *et al.*,[18] suggested that seizure frequency may also secondarily affect the quality of life in children because of the restrictions imposed by parents and physicians to prevent seizure-related injuries. In their study, improved seizure control resulted in an important reduction of restrictions, which presumably improved the quality of the patients' life.

Apart from frequency and severity, the unpredictability of seizures is a major concern of epileptic patients. In fact, Hayden *et al.*[41] remarked that for some patients the unpredictability of the seizures seems to have a far greater impact on the quality of life than their frequency. The lack of control over one's body can lead to low self-confidence and the perception of diminished competency, which in turn may result in a variety of difficulties that risk to impede achievement and self-fulfillment.

1.2. Treatment Factors

Seizure factors and treatment factors are interrelated and cannot be easily separated. There is no doubt that antiepileptic drugs (AED) can significantly reduce seizure frequency, thus contributing to a greater sense of well-being. However, antiepileptic therapy itself may have an adverse effect on cognitive development and behavior in children (i.e., 27, 62), especially when polypharmacotherapy and/or higher doses are required. A group of newer drugs in particular has been found to cause hyperactivity and aggressive behavior in certain patients.[11] However, these deleterious effects have to be weighed against the beneficial effects on seizure frequency, seizure severity and mood.[74] Recent research indicates that the magnitude of side effects produced by various AEDs is negligible with monotherapy within the therapeutic range.[62]

1.3. Cognitive and Behavior Factors

Childhood epilepsy coincides with critical periods of development during which cognitive and social skills are acquired. Impaired cognitive functioning impedes the development of adaptive behavior.[30,76] As has been shown in previous chapters (i.e., 45, 50, 68), cognitive deficits are overrepresented in children with epilepsy. The severity of the impairments is related to the type of seizures, the medication used to treat the condition and the underlying pathology.[30,77] Most vulnerable are children with early onset of epilepsy, poor seizure control, several seizure types, polypharmacotherapy, congenital malformations of the CNS, perinatal injury, infection of the brain or its covering membranes and certain epileptic syndromes that are known to have a devastating effect on the child's cognitive and adaptive development (i.e., 31). Many of these children have mental retardation.[31,59]

The cognitive functions most frequently impaired in epileptic children are attention, memory, problem-solving abilities, language skills and mental flexibility[3,50,71,76] Antiepileptic medication[27] and epilepsy surgery[5,16,60,69] may alleviate or exacerbate the deficits. Furthermore, learning disabilities are frequent among epileptic children.[14,66,77] Reading, arithmetic and reading comprehension are most frequently affected.[4,14] In addition, children may have behavioral problems, such as irritability, hyperactivity or aggressiveness that may be associated with a particular syndrome[40,50,68] or medication.[11,27] The problems are compounded when other handicaps (sensory, motor) or psychiatric illnesses co-exist.

Cognitive and/or behavior problems compromise the quality of life of children and young adults by imposing restrictions on education, mobility and vocation. A study by Austin *et al.*[7] has shown that children with epilepsy succeed less well in school than children with asthma. Furthermore, evidence suggests that epileptic children tend to underestimate their intellectual abilities which may lead to lowering of their occupational aspirations.[83] This is born out by observations that young people with epilepsy are less likely to finish school or pursue vocational training, choosing instead to be unskilled manual workers.[33] Working with adolescents enrolled in a vocational preparation program, Lipinski[56] noted that the epileptic trainees showed less perseverance and motivation to succeed than the average young person. As for higher education, according to longitudinal studies conducted in Scandinavia, fewer epileptic patients attend university and attain academic positions.[73] Employment is among the lowest for epileptic adults, including those with good seizure control.[23,33] Unemployment is generally associated with lower self-esteem and reduced life satisfaction.

1.4. Psychological and Psychiatric Factors

Children with epilepsy have a high incidence of psychological, behavioral and psychiatric problems that risk to adversely affect their quality of life. Anxiety, depression and a variety of maladjusted behaviors have been reported.[34,35,46,53,55] In some types of epilepsy, disorganization of behavior may occur (i.e., 45, 68). Other types may be accompanied by autistic disorders,[19,28,51,61] psychotic episodes[54,75] or aggressive behavior.[44] Again, these problems are more frequent in children with refractory seizures and multiple drug regimens. In many cases, the emotional and psychiatric problems are more limiting than cognitive impairments as far as social adjustment and employability are concerned.[38]

1.5. Psychosocial Factors

Epilepsy has an impact on all domains of the patient's life: daily activities, school, work, recreation, personality development, interpersonal relationships, friendship, family life and marriage. Moreover, looking after a chronically ill child is burdensome and may put a strain on relationships within the family.[36,47] Many parents of epileptic children become overly protective, thereby limiting the child's autonomy and self-expression. Since epilepsy frequently begins in early childhood, personality development may be stunted. Lack of exposure to social situations as well as reduced demands and expectations lead to inadequate communication skills and low self-esteem in the growing child and young adult. The patient assumes a disability status which is often more limiting than the epilepsy itself. In fact, there is evidence that the handicapping effect of epilepsy is to a large extent mediated by the patients' own perception of himself and his condition:[21,22] epileptic individuals have a tendency to evaluate themselves more negatively.[21,22,65] This negative self-concept has been attributed to the concept of "stigma" which is considered to be the source of low self-esteem.[82] Stigma refers to social disapproval of an attribute that is viewed as negative or unacceptable.[26] The person who possesses this attribute may be subjected to discrimination.

Although stigmatization of epilepsy and other disabilities has somewhat diminished in today's society, epileptic people are still exposed to it at some time in their lives. In the study by Hoare and Russel[48] reviewed above, parents who completed a quality of life questionnaire considered "teasing" a major concern. In his chapter "Psychosocial Components of Childhood Epilepsy", Taylor[78] relates the case of a teenage girl who developed "severe and lasting school phobia" after she experienced a first seizure accompanied by micturition while waiting in the line-up for lunch at the school cafeteria. The author goes on to remark that: "This girl found that her entry into the epileptic role created subsequently such overwhelming expectations of her being in that role that her role as her normal self was not sustainable. The clearest stigmata of epilepsy are the seizures, but there are also the pills and regimens, the constraints and the vigilance, the days absent and unaccounted for, which suddenly or gradually create the climate for modern equivalents of ancient demonology to be exhibited, in this instance relabelling as 'physically handicapped' and separation within a special school" (pp 125–126). For young adults with epilepsy, discrimination in the work place is still the primary reason for underemployment and career restrictions.[17,20]

Other research has shown that the belief that epilepsy is stigmatized is sufficient to depress self-esteem in epileptic people.[82] In fact, Jacoby[49] has argued that perceived stigma may be more important in depressing self-esteem than actual instances of discrimination.

Stigmatization, whether perceived or real, leads to strategies of concealing the condition from other people, thus complicating the life of the patient.

Finally, the patient's perception of his quality of life also includes his expectations, e.g., the perceived discrepancy between his current and expected status, and this aspect may be relatively independent of his objective physical state.[15] Using the examples given by Cramer,[24] a patient with poor seizure control who is able to live independently and work part-time as a volunteer may be content with his life, whereas a patient whose rare seizures force him to accept a job with less responsibility and opportunity for advancement, may feel severely disadvantaged by his epilepsy. Similarly, a child with occasional absence seizures who is overprotected may perceive his/her quality of life as being poor. By the same token, patients who have been rendered seizure-free by neurosurgery sometimes report that their quality of life has not improved accordingly. In fact, studies concerned with the patients' psychosocial adjustment following successful neurosurgery[12] have shown that individuals who find themselves suddenly cured of their epilepsy may go through a period of intense distress as they are trying to cope with the raised expectations of their families and entourage as well as with the loss of certain privileges that were associated with their former patient status. Evidently, the situation is still worse for those patients whose hopes for improvement did not materialize due to surgical failure.[12]

Taken together, these examples demonstrate that the psychosocial consequences of epilepsy are complex and constitute a risk factor for maladjustment and reduced life-satisfaction.

2. ASSESSMENT OF HQOL IN EPILEPTIC CHILDREN

Assessment of HQOL is gradually becoming an integral part of the treatment program for various chronic disorders. The aims of such assessment are 1) to provide physicians and health professionals with a measurement of the patient's difficulties in order to be able to adjust the divers therapeutic interventions to current needs, thereby improving the patient's care and quality of life; 2) to identify the mechanisms underlying the interaction of disease and its psychological and psychosocial impact and 3) to guide policy-makers in the decisions that determine the allocation of resources.[59]

While the recent literature is abound with scales designed to assess HQOL in adult patients with epilepsy,[9,72] few instruments exist that evaluate this dimension in children.[48] In fact, there is no single HQOL scale that relates specifically to epileptic children. The scales employed to evaluate HQOL in epileptic children are usually instruments of a more general nature, such as the Child Attitude Towards Illness Scale[6] on which specific items related to epilepsy may be tagged on. One of the reasons for the lack of disease-specific scales for epileptic children may be that many of these children are too young or too retarded to be able to complete a self-report. Indeed, most scales for pediatric populations are designed to be completed by the parents or the principal care-giver. Achenbach and Edelbrock's[2] widely used Child Behavior Checklist and the Health Utility Index[29] may serve as examples. A methodological problem common to all scales for children with chronic illnesses, pointed out by Hoare and Russell,[48] is that quality of life measures are often defined in negative terms (i.e., absence of dysfunction) which may underestimate the impact of the epilepsy on children who previously functioned in the higher ranges of mental abilities. Ideally, the scale should be relevant for epileptic children, broad enough to accommodate cross-cultural aspects, easy to complete, available in different versions for children, parents, teachers and other professionals and include norms for pediatric populations.

3. COMPREHENSIVE CARE PROGRAMS FOR EPILEPTIC PATIENTS

The care of the epileptic patient requires an individualized, multidisciplinary approach, involving the patient and his family, professionals in the health field (i.e., neurologist, nurse, neuropsychologist, clinical psychologist, speech therapist, occupational therapist, physiotherapist, social worker), teachers, resource teachers, vocational trainers, employers and governmental agencies. As stressed by Durwen and Diehl[32] and others,[37] early intervention is important in order to avoid the development of psychological dysfunction and social problems as well as unnecessary long-term costs.

The starting point for a patient-oriented comprehensive care program is to make the patient and his family as much as possible active partners in the treatment and rehabilitation. One way in which the patient may participate in the management of his epilepsy is to adjust his lifestyle to avoid situations or activities that are likely to bring on a seizure. There is general consensus among clinicians and researchers that noncompliance with the drug regimen for various reasons, ranging from forgetting to intentional change of the prescription, is the most frequent trigger of seizures,[13,25,39] accounting for 30 to 50 per cent of seizure reoccurrence.[56] Other potential triggers are sleep deprivation, anxiety, stress, hunger (hypoglycemia), TV, video games, alcohol and recreational drugs. However, in order to recognize these triggers and learn to avoid them, the patient needs first and foremost information. Thus, any integrated program of comprehensive care has to begin with educating the patient and his family about his epilepsy.

3.1. Interventions with Children

An individualized treatment plan for children, like the one described by Strang,[76] should start with an assessment of the child's neuropsychological and adaptive functioning in order to identify his/her strength an weaknesses which in turn will determine the kind of professional and educational services he/she will require. Proper school orientation is often difficult for epileptic children.[42] Intelligent children may be classified as "learning disabled" because of memory or attentional problems and directed to specialized schools. Conversely, children with obvious cognitive deficiencies are sometimes retained in a regular class-room although they might be better served in a specialized school where they could benefit from educational programs tailored to their individual needs. In this case, the fear of "marginalizing" the epileptic child has to be weighed against the potential benefits that can be derived from individualized multidisciplinary intervention. Evidently, any problems that could impede learning and social adaptation should be treated first. These include sensory or motor problems, as well as behavior and psychological problems such as hyperactivity, defiance, aggressiveness, anxiety and depression.

3.2. Interventions with Young Adults

Health researchers working with adolescents and young adults have found that group interventions are more effective than individual interventions for these age groups.[37] One such program directed towards young people is the Modular Service Package for Epilepsy or "MOSES" that has recently been described by Ried.[67] The program, developed by medical advisors, non-medical professionals and representatives of self-help groups, teaches basic medical facts and self-care, as well as social, cognitive, emotional and behavioral aspects of epilepsy in an interactive way. The service package, which is designed for groups of 7 to 10 young patients without severe cognitive impairments, is currently used in Germany, Switzerland and Austria. More extensive modular programs for young adults are offered at various epilepsy centers in Europe, North America, Australia and Japan.[11,38,56,80] Patients enrolled in these programs receive medical care, counseling and rehabilitation. They learn among others coping strategies, self-reliance, social kills and independent living skills, such as shopping, cooking, cleaning, banking, using public transportation and more. Vocational training as an interface between school and work place, as well as on-the-job coaching, is offered by some centers where government funds are available for such programs.[11,56,80] Furthermore, patients may benefit from self-help groups and epilepsy networking. The programs evaluated so far have shown their efficacy in improving the quality of life of epileptic children and youth by raising their self-esteem and increasing their employment opportunities.[11,38,56,80]

3.3. Interventions with Parents

Parents of epileptic children require factual information about epilepsy. This may be provided by physicians, nurses, social workers, epilepsy associations and pamphlets. Fraser and Clemmons[37] have pointed out that parents often fail to understand the medical lingo and may not follow physicians' instructions out of ignorance or uncertainty. It is therefore important that they obtain information from all available sources since a clear understanding of the disorder and its treatment requirements is likely to result in greater compliance with the medication regimen which will ultimately benefit the child.

Parents also require guidance regarding the demands they can make on the epileptic child. They need to learn to avoid the pitfalls of overprotection, to adjust their expectations and to set realistic goals for the child. Finally, they have to be provided with means and strategies helping them to cope with a disabled child. These include financial support where it is needed, practical help, counseling and self-help groups. Continued follow-up of the family by a clinical social worker will serve to reassure them along the way and facilitate their adjustment to the conditions created by the changing needs of the child.

3.4. Interventions with Teachers, Trainers and Other Professional Groups

As much as the patients and their families, teachers, school nurses, school counselors and other professionals working with epileptic children should be educated about epilepsy and its cognitive, behavioral and psycho-social consequences. Teachers should be informed of the most frequent learning disabilities and behavior problems in epileptic children and of strategies that can be employed to enable the child to function more efficiently in the class-room. For example, epileptic children, especially those on heavy medication, tend to be slower or tire more easily. They should be given more frequent breaks and more time for in-class assignments. Teachers can also facilitate the integration of the epileptic child in the class-room by informing the class-mates about the disorder. This can be done through explanations, videos or discussions, possibly with a neurologist, nurse or social worker present who can answer specific questions about epilepsy. Similar procedures could be used with employers or potential employers of young adults. These interventions may positively affect attitudes and beliefs, thus favoring greater social acceptance of the epileptic patient.

Finally, patient groups, parents groups and epilepsy associations should continue to lobby for funds supporting epilepsy research and intervention programs for epileptic people. Bringing epilepsy concerns to the forefront of public awareness may also help to dispel long-standing fears and misconceptions. Demystifying epilepsy is a major step towards improving the quality of life of the epileptic person.

REFERENCES

1. Aaronson N (1988): Quality of life: what is it? How should it be measured? Oncology 2:69–74.
2. Achenbach TM and Edelbrock C (1993): Manual for the Child Behavior Check List, Burlington, VT: Department of Psychiatry.
3. Addy DP (1987): Cognitive function in children with epilepsy. Developmental Medicine and Child Neurology 29:394–397.
4. Aldenkamp APS, Alphert WCJ, Dekker MJA and Overweg J (1990): Neuropsychological aspects of learning disabilities in epilepsy. Epilepsia 31 (Suppl 4):9–20.
5. Andermann F, Freeman JM, Vigevano F and Hwang PA (1993): Surgical remedial diffuse hemispheric syndromes. In Engel J jr. (ed.): "Surgical Treatment of the Epilepsies", Second Edition. New York: Raven Press, pp 87–101.
6. Austin JK and Huberty TJ (1993): Development of the Child Attitude Toward Illness Scale. Journal of Pediatric Psychology 18(4):467–480.
7. Austin JK, Shelton Smith M, Risinger MW and McNelis AM (1994): Childhood epilepsy and asthma: Comparaison of quality of life. Epilepsia 35(3):608–615.
8. Baker GA, Camfield C, Camfield P, Cramer JA, Elger CE, Johnson AL *et al.* (1998): Commission on outcome measurement in epilepsy, 1944–1997: Final report. Epilepsia 39(2):213–231.
9. Baker GA, Jacoby A, Buck D, Stalgis C and Monet D (1997): Quality of life of people with epilepsy: a European study. Epilepsia 38(3):353–362.
10. Beran RG and Gibson RJ (1998): Aggressive behavior in intellectually challenged patients with epilepsy treated with lamotrigine. Epilepsia 39(3):280–282.
11. Beran RG, Major M and Veldze L (1987): Evaluation of the first 18 months of a specific rehabilitation programme for those with epilepsy. Clinical and Experimental Neurology 23:165–170.
12. Bladin PF (1992): Psychosocial difficulties and outcome after temporal lobectomy. Epilepsia 33:898–907.
13. Buchanan N (1993): Noncompliance with medication amongst persons attending a tertiary referral epilepsy clinic: implications, management and outcome. Seizure 2(1):79–82.
14. Bulteau C, Jambqué I and Dellatolas G (this volume): Epilepsy, cognitive abilities and education.
15. Calman KC (1984): Quality of life in cancer patients: A hypothesis. Journal of Medical Ethics 10:124–127.
16. Campbell AL jr., Bogen JE and Smith A (1981): Disorganization and reorganization of cognitive and sensorimotor functions in cerebral commissurotomy: compensatory roles of the forebrain commissures and cerebral hemispheres in man. Brain 104:493–551.
17. Caroll D (1992): Employment among young people with epilepsy. Seizure 1(2):127–1131.
18. Carpay H, Vermeulen J, Stroink H, Brouwer OF, Peter AC, van Donselaar CA *et al.* (1997): Disability due to restrictions in childhood epilepsy. Developmental Medicine and Child Neurology 39(8):521–526.
19. Carracedo A, Martin Mucrica F, Garcia Penas JJ, Ramos J, Cassinello E and Calvo MD (1995): Autistic syndrome associated with refractory temporal epilepsy. Revue Neurologique 23:1239–1241.
20. Chaplin JE, Wester A and Tomson T (1998): Factors associated with the employment problems of people with established epilepsy. Seizure 7(4):299–303.
21. Collings JA (1990): Psychosocial well-being and epilepsy: an empirical study. Epilepsia 31(4):418–426.
22. Collings JA (1995): The impact of epilepsy and self-perception. Journal of Epilepsy 8:164–171.
23. Cooper M (1995): Epilepsy and employment—employers' attitudes. Seizure 4(3):193–199.
24. Cramer JO (1994): Quality of life for people with epilepsy. Neurologic Clinics 12(1):1–13.

25. Cramer JO (1999): Compliance and epilepsy related lifestyle. Abstracts of the 10th International Bethel-Cleveland Symposium: Comprehensive Care for People with Epilepsy, Bielefeld, Germany, April 14–18, 1999: p 32.
26. Crocker J and Major B (1989): Social stigma and self-esteem: The self-protective properties of stigma. Psychological Review 96:608–630.
27. Cull CA and Trimble M (1989): Effects of anticonvulsant medications on cognitive functioning in children with epilepsy. In Hermann BP and Seidenberg M (eds.): "Childhood Epilepsies: Neuropsychological, Psychosocial and Intervention Aspects". New York: John Wiley & Son, pp 105–118.
28. Deonna T, Ziegler AL, Despl PA and Van Melle G (1986): Partial epilepsy in neurologically normal children: clinical syndromes and prognosis. Epilepsia 27:241–247.
29. Dobkin PL and Trudel JG (1995): Quality of life in pediatric patients. Canadian Health Psychologist 3(1):25–27.
30. Dodrill CB (1992): Neuropsychological aspects of epilepsy. Psychiatric Clinics of North America 2:383–394.
31. Dulac O (this volume): Classification, management and mechanisms of seizures and epilepsies.
32. Durwen HF and Diehl LW (1993): Medizinisch-psychosoziale Rehabilitation und Prävention bei Patienten with Epilepsien. Versicherungsmedizin 45(3):93–99.
33. Elwes RD, Marshall J, Beattle A and Newman PK (1991): Epilepsy and employment. A community based survey in an area of high unemployment. Journal of Neurology, Neurosurgery and Psychiatry 54(3):200–203.
34. Endermann M (1997): Learned helplessness in epilepsy: A contribution toward the prediction of depression levels in a clinical setting. Journal of Epilepsy 10:161–171.
35. Ettinger AB, Weisbrot DM, Nolan EE, Gadow KS, Vitale SA, Andriola MR *et al.* (1998): Symptoms of depression and anxiety in pediatric epilepsy patients. Epilepsia 39(6):595–599.
36. Ferrari M (1989): Epilepsy and its effect on the family. In Hermann BP and Seidenberg M (eds.): "Childhood Epilepsies: Neuropsychological, Psychosocial and Intervention Aspects". New York: John Wiley & Son, pp 159–172.
37. Fraser RT and Clemmons DC (1989): Vocational and psychosocial interventions for youth, with seizure disorders. In Hermann BP and Seidenberg M (eds.): "Childhood Epilepsies: Neuropsychological, Psychosocial and Intervention Aspects". New York: John Wiley & Son, pp 201–219.
38. Fraser RT, Clemmons D and Trejo W (1983): Program evaluation in epilepsy rehabilitation. Epilepsia 24(6):734–746.
39. Gomes M da M and Maia Filio H de S (1998): Medication-taking behavior and drug self regulation in people with epilepsy. Arquivos de Neuro-Psichiatria 56(4):714–719.
40. Hanai T (1996): Quality of life in children with epilepsy. Epilepsia 37 (Suppl 3):28–32.
41. Hayden M, Penna C and Buchanan N (1992): Epilepsy: patient perception of their condition. Seizure 1:191–197.
42. Henriksen O (1990): Education and epilepsy: Assessment and remediation. Epilepsia 31 (Suppl 4):21–25.
43. Hermann BP and Seidenberg M (1989): Childhood Epilepsies: Neuropsychological, Psychosocial and Intervention Aspects. New York: John Wiley & Son.
44. Hermann BP, Schwartz MS, Whitman S and Karnes WE (1980): Agression and epilepsy: seizure type comparisons and high-risk variables. Epilepsia 21:15–23.
45. Hernandez MT, Sauerwein HC, De Guise E, Lortie A, Jambaqué I, Dulac O *et al.* (this volume): Neuropsychology of frontal lobe epilepsy in children.
46. Hoare P (1984): The development of psychiatric disorder among school-children with epilepsy. Developmental Medicine and Child Neurology 26:3–13.
47. Hoare P (1993): The quality of life of children with chronic epilepsy and their families. Seizure 2:269–275.
48. Hoare P and Russell M (1995): The quality of life in children with chronic epilepsy and their families: preliminary findings with a new assessment measure. Developmental Medicine and Child Neurology 37(8):689–696.
49. Jacoby A (1994): Felt versus enacted stigma: a concept revisited. Social Science and Medicine 38:269–274.
50. Jambaqué I (this volume): Neuropsychology of temporal lobe epilepsy in children.
51. Jambaqué I, Mottron L, Ponsot G and Chiron C (1998): Autism and visual agnosia in a child with right occipital lobectomy. Journal of Neurology, Neurosurgery and Psychiatry 6:555–560.
52. Keene DL, Higgins MJ and Ventureyra EC (1997): Outcome and life prospects after surgical management of medically intractable epilepsy in patients under 18 years of age. Childs Nervous System 13:530–535.

53. Kim WJ (1991): Psychiatric aspects of epileptic children and adolescent. Journal of the American Academy of Child and Adolescent Psychiatry 30(6):874–886.
54. Kyllerman M, Nyden A, Praquin N, Rasmussen P, Wetterquist AK and Hedstrom A (1996): Transient psychosis in a girl with epilepsy and continuous spikes-and-waves during slow sleep. European Child and Adolescent Psychiatry 5:216–221.
55. Lindsay J, Ounsted C and Richards P (1984): Long term outcome in children with temporal lobe seizures. Developmental Medicine and Child Neurology 26:25–32.
56. Lipinski CG (1990): Berufsvorbereitung und Ausbildung bei jugendlichen Anfallkranken. Wiener Klinische Wochenschrift 102(8):213–217.
57. MacLachlan RS, Rose KJ, Derry PA, Bonnar C, Blume WT and Girvin JP (1997): Health-related quality of life and seizure control in temporal epilepsy. Annals of Neurology 41(4):482–489.
58. Malmgren K, Sullivan M, Ekstedt G, Kullberg G and Kumlien E (1997): Health-related quality of life after epilepsy surgery: a Swedish multicenter study. Epilepsia 38(7):830–838.
59. Meador KJ (1993): Research use of the new quality-of-life in epilepsy inventory. Epilepsia 34 (Suppl 4):34–38.
60. Milner B (1975): Psychological aspects of focal epilepsy and its neurosurgical management. In Purpura DP, Penry JK and Walter RD (eds.): "Advances in Neurology". New York: Raven Press, pp 299–321.
61. Neville BG, Harkness WF, Cross JH, Cass HC, Burch VC, Lees JA *et al.* (1997): Surgical treatment of severe autistic regression in childhood epilepsy. Pediatric Neurology 16:137–140.
62. Nichols ME, Meador KJ and Loring DW (1993): Neuropsychological effects of antiepileptic drugs: a current perspective. Clinical Neuropharmacology 16(6):471–481.
63. O'Donoghue MF, Duncan JS and Sander JW (1998): The subjective handicap of epilepsy. A new approach to measuring treatment outcome. Brain 121:317–343.
64. Pfäfflin M, May TW, Mayer M, Thorbecke R, Specht U, Schöndienst M and Wolf P. (1999): Evaluating comprehensive care: Experience with and first results of the Bethel study. Abstracts of the 10th International Bethel-Cleveland Symposium: Comprehensive Care for People with Epilepsy, Bielefeld, Germany, April 14–18, 1999: p 36.
65. Regan KJ, Banks GK and Beran RG (1993): Therapeutic recreation programmes for children with epilepsy. Seizure 2(3):195–200.
66. Renier WO (1990): Learning disabilities and behavioural problems in children with epilepsy. Wiener Klinische Wochenschrift 102:218–222.
67. Ried S (1999): Modular Service Package Epilepsy (MOSES): Abstract of the 10th International Bethel-Cleveland Symposium: Comprehensive Care for People with Epilepsy, Bielefeld, Germany, April 14–18, 1999: p 25.
68. Roulet Perez E and Deonna T (this volume): Cognitive profiles of Continuous Spike Waves in Sleep (CSWS) syndrome.
69. Sauerwein HC, Lassonde M, Revol O, Cyr F, Geoffroy G and Mercier C (this volume): Neuropsychological and psycho-social consequences of corpus callosotomy.
70. Schipper H, Clinch J and Powell V (1990): Definitions and conceptual issues. In Spiker B (ed.): "Quality of Life Assessment in Clinical Trials". New York: Raven Press, pp 11–24.
71. Seidenberg M (1989): Neuropsychological functionning of children with epilepsy. In Hermann BP and Seidenberg M (eds.): "Childhood Epilepsies: Neuropsychological, Psychosocial and Intervention Aspects". New York: John Wiley & Son, pp 71–82.
72. Selai CE and Trimble MR (1995): Quality of life assessment in epilepsy: the state of the art. Journal of Epilepsy 8:332–337.
73. Sillanpää M (1983): Social functioning and seizure status of young adults with onset of epilepsy in childhood. Acta Neurologica Scandinavica 96:1–81.
74. Smith D, Chadwick D, Baker G, Davis G and Dewey M (1993): Seizure severity and the quality of life. Epilepsia, 34 (Suppl 5):31–35.
75. So NK, Savard G, Andermann F, Olivier A and Quesney LF (1990): Acute post-ictal psychosis: a stereo EEG study. Epilepsia 31(2):188–193.
76. Strang JD (1990): Cognitive deficit in children: adaptative behavior and treatment techniques. Epilepsia 31 (Suppl 4): 54–58.
77. Sturniolo MG and Galletti F (1994): Idiopathic epilepsy and school achievement. Archives of Disease in Childhood 70:424–428.
78. Taylor DC (1989): Psychosocial components of childhood epilepsy. In Hermann BP and Seidenberg M (eds.): "Childhood Epilepsies: Neuropsychological, Psychosocial and Intervention Aspects". New York: John Wiley & Son, pp 119–142.

79. Thompson PJ and Oxley J (1988): Socioeconomic accompaniments of severe epilepsy. Epilepsia 29 (Suppl 1):9–18.
80. Thorbecke R (1987): Berufliche Eingliederung von Menschen with Epilepsie. Rehabilitation 26(1):20–27.
81. Van Hout B, Gagnon D, Souetre E, Ried S, Remy C, Baker G *et al.* (1997): Relationship between seizure frequency and costs and quality of life of outpatients with partial epilepsy in France, Germany, and the United Kingdom. Epilepsia 38(11):1221–1226.
82. Westbrook LE, Bauman LJ and Shinnar S (1992): Appplying stigma theory to epilepsy: a test of a conceptual model. Journal of Pediatric Psychology 17(5):633–649.
83. Zuchowicz M and Majkowicz M (1976): Zainteresowania zawodowe dzieci z padaczka i mozliwosc ich realissacji (Occupational interests of epileptic children and possibilities of their realization). Neurologia i Neurochirurgia Polska 10(2):213–217.

INDEX